About Island Press

Since 1984, the nonprofit Island Press has been stimulating, shaping, and communicating the ideas that are essential for solving environmental problems worldwide. With more than 800 titles in print and some 40 new releases each year, we are the nation's leading publisher on environmental issues. We identify innovative thinkers and emerging trends in the environmental field. We work with world-renowned experts and authors to develop cross-disciplinary solutions to environmental challenges.

Island Press designs and implements coordinated book publication campaigns in order to communicate our critical messages in print, in person, and online using the latest technologies, programs, and the media. Our goal: to reach targeted audiences—scientists, policymakers, environmental advocates, the media, and concerned citizens—who can and will take action to protect the plants and animals that enrich our world, the ecosystems we need to survive, the water we drink, and the air we breathe.

Island Press gratefully acknowledges the support of its work by the Agua Fund, Inc., Annenberg Foundation, The Christensen Fund, The Nathan Cummings Foundation, The Geraldine R. Dodge Foundation, Doris Duke Charitable Foundation, The Educational Foundation of America, Betsy and Jesse Fink Foundation, The William and Flora Hewlett Foundation, The Kendeda Fund, The Andrew W. Mellon Foundation, The Curtis and Edith Munson Foundation, Oak Foundation, The Overbrook Foundation, the David and Lucile Packard Foundation, The Summit Fund of Washington, Trust for Architectural Easements, Wallace Global Fund, The Winslow Foundation, and other generous donors.

The opinions expressed in this book are those of the author(s) and do not necessarily reflect the views of our donors.

Modeling the Environment

Second Edition

Modeling the Environment

··

Second Edition

Andrew Ford

ISLANDPRESS

Washington | Covelo | London

Library of Congress Cataloging-in-Publication Data

Ford, Andrew (Frederick Andrew)
   Modeling the environment / Andrew Ford. — 2nd ed.
      p.   cm.
   Includes bibliographical references and index.
   ISBN-13: 978-1-59726-472-3 (cloth : alk. paper)
   ISBN-10: 1-59726-472-5 (cloth : alk. paper)
   ISBN-13: 978-1-59726-473-0 (pbk. : alk. paper)
   ISBN-10: 1-59726-473-3 (pbk. : alk. paper)   1.   Environmental sciences—Simulation methods.
I. Title.

   GE45.D37F67   2010
   363.7001'1—dc22

                                        2009032257

Printed on recycled, acid-free paper

Manufactured in the United States of America
10 9 8 7 6 5 4 3 2 1

Keywords: Converters, cyclical behavior, Daisyworld, environmental systems, epidemic dynamics, exponential growth, feedback loops, homeostasis, Kaiba, mathematical models, Mono Lake, oscillations, system dynamics, stocks and flows, S-shaped growth, Stella, Vensim

For Amy

Contents

Preface

It is May of 2009 as this book goes into production. This month's newspapers are filled with stories of complicated and challenging problems. We read about rapid climate change and the dangerous changes that await us if we don't reduce greenhouse gas emissions substantially. We read about the increasing deaths from swine flu, and we are warned to prepare for a possible pandemic. There are many stories about the boom and bust in real estate, and we are told to brace for more bankruptcies before the market recovers. These stories leave us wondering about the underlying causes of the problems and confused over which policies could lead to improved behavior in the future.

Understanding Dynamic Complexity

Climate change, pandemics, and boom and bust in real estate are complex dynamics that challenge our understanding. We are unable to anticipate the dynamic consequences of policies adopted today, especially when there are long delays between our actions and the system's reactions. Our understanding is also limited by the complexity of the feedback processes that control system behavior. Our actions may be partially erased by the system's internal responses, and the system's apparent resistance to our interventions is confusing. Sorting out the effects of delays and multiple feedbacks is beyond our cognitive abilities, so we look to the past for lessons. But how are we to interpret past patterns in climate change, pandemics, and boom-and-bust cycles? Our understanding of the dynamics of historical patterns is limited by the same complexities that make it difficult to think about the future. There are many interpretations of past behavior, and we are left with limited understanding of both past trends and current problems.

The Premise of This Book

This book is based on the premise that modeling can help us build our understanding of complex problems like those appearing in the news headlines. We build a mathematical model to capture the key interrelationships, and we conduct simulations to see the dynamic pattern. Our cognitive abilities are limited, so we should expect some surprises in the simulations. Indeed, the simulations may turn out to be the very opposite of what we expected. Policies thought to make the system better may make it worse than before. Policies thought to produce winners and losers may turn out to deliver win-win results in the long run. These surprises are the key to improved understanding.

Examples from Many Disciplines

The methods described in this book have been used in a wide range of environmental and business problems, including the problems of climate change (chapter 23), epidemics (chapter 8), and boom and bust in real estate (chapter 19). The methods are explained in introductory chapters and then illustrated with applications to serious problems of the environment. In each case, the problem arises from the way humans interact with the environment. The examples in this book are organized around fundamental patterns, such as exponential growth and oscillations. A panel of six fundamental shapes appears throughout the book. These panels remind us of the premise of systems modeling: the combination of stocks, flows, and feedbacks that explains a dynamic pattern in one system could help us explain the same pattern in another.

This book demonstrates the transferability of systems ideas across many disciplines. You'll see cycles in housing markets and cycles in predator-prey populations. You'll simulate the recovery of threatened water basins and threatened fish populations. And you'll learn about the homeostatic tendencies in physiological systems and in the climate system. These examples are complimented by a diverse collection of examples on the book's website, the BWeb. The BWeb applications deal with anthropology, ecology, economics, genetics, geomorphology, hydrology, limnology, regional planning, and resource exploration.

Interdisciplinary Modeling

The major environmental challenges of our time are inherently interdisciplinary. Models can help us understand the challenges if the models follow the feedback effects that interconnect the environmental, social, and economic systems into a tightly coupled system. This book explains the feedback perspective and the system dynamics method of modeling and simulation. System dynamics is valued for the clarity of representation of the stocks and flows and the feedback processes that control the flows. The approach provides a common language that can be understood by scientists from many disciplines, so it is especially useful in interdisciplinary modeling. System dynamics is also ideally suited for participatory modeling, and it has become a common platform for cooperative modeling of environmental systems.

Modeling the Environment provides opportunities to practice with models in your own discipline and in related disciplines. The best way to begin is to build and verify the models in the book. Then try the exercises at the end of many chapters and appendixes and the additional exercises on the BWeb. A good way to solidify your understanding is to expand and improve one of the models. If you are studying in a one-semester class (i.e., 15 weeks), there should be time to practice with one of the interdisciplinary models. They illustrate the insights that can emerge when contributions from several disciplines are incorporated in a single, highly interconnected model.

Who Should Learn to Model?

Modeling the Environment is suitable for classes from high school to graduate school, and for use in a traditional classroom, in distance-delivery classes, and in the many hybrid combinations of instruction. The target audience for the first edition was college students in undergraduate classes in environmental sciences and regional planning. But the receptive audience has proved to be far larger. Readers of the first edition ranged from students in junior high school to retired business executives. Many readers were able to master the concepts and software without organized instruction. Most readers learn more in groups, especially if you meet together to practice with the modeling software.

Decades of teaching have taught me that students see modeling differently than when I was a student. Their questions have opened my eyes to different perspectives on the use of models. This second edition introduces a student point of view in the form of "questions from Joe" that appear as boxed features in many of the chapters. Think of Joe as a hypothetical student who asks questions that I've often heard in my own classroom. His questions and my responses will help you broaden your own understanding and to appreciate what others may be thinking about modeling.

Your Mathematical Background

Joe's first question is the one I've heard most often at the start of a new class:

Do I need to learn calculus before I can learn to model?

This question is often asked by students who have not taken calculus or who have forgotten calculus from years ago. Some students find mathematics abstract and difficult to understand. They sometimes tell me that modeling is for others to learn. Many students will have a good command of calculus, and they ask if *Modeling the Environment* is their opportunity to put calculus to work on environmental problems.

Calculus may be the first thing on a student's mind, but it is not central to this book. *Modeling the Environment* was written for readers with a wide range of mathematical backgrounds. There is a growing need for modeling projects that help a group of individuals learn together. These projects benefit from experts in several disciplines, from agency staff, and from the stakeholders with firsthand knowledge of the system. This book is written with these individuals in mind. I believe everyone can learn to model, and this book minimizes the mathematical hurdles that block many individuals from trying to do so.

Computer simulation is definitely a form of quantitative analysis, so you need some knowledge of mathematics, and you must be willing to think about the numbers and their units of measure. You probably learned about units in high school, and it is useful to review what you learned (appendix A). It's also helpful to check the units in the equilibrium diagrams (chapter 6). The equilibrium diagrams are also a good way to build your familiarity with the numerical values in your model. I also assume you have learned about graphs. This knowledge is important as there are hundreds of graphs in the book. Be sure to study the graphs carefully, paying particular attention to the vertical scales. And take the time to think of the best combination of variables and scales when you create your own graphs.

I also assume that you have learned introductory algebra. This knowledge is crucial, since the models use algebra to explain the flows. I believe we should aim for clarity, and the clearest models are those whose algebraic equations can be guessed by simple inspection. We should select variable names that are commonly recognized, and we should be able to write the equations with a combination of add, subtract, multiply, or divide. This book aims for clarity in every chapter, and it commits to "friendly" algebra. Indeed, there are only a few instances in which the algebraic equations go beyond add, subtract, multiply, or divide.

This book does not require training in calculus, differential equations, partial differential equations, statistical analysis, or computer programming. Knowledge of these topics is not required, nor is it crucial to your ability to put modeling to use. The crucial requirement is your knowledge of the feedback processes in your system. The models in this book are constructed in a visual manner on the computer. You'll use your knowledge to select the proper combination of stocks, flows, and feedbacks. The tedious job of generating the simulation results is left to the computer. Your challenge is to use the simulation results to build understanding of your dynamic problem.

Website Support for Teachers and Students

Many teachers tell me that they are teaching modeling for the first time, and the first thing they need is access to the figures in the book. Teachers will find these in the instructors' section of the book's website. This section also provides answers to exercises. The BWeb is useful to students as well. They will find exercises beyond those at the end of each chapter, and they will benefit from the separate exercises in fields from anthropology to resource exploration.

Many of the models in the book can be constructed in a step-by-step manner from their descriptions in the chapters. Copies of these models are available in the instructors' section of the BWeb. A few of the models in the book are described in general terms, so readers cannot build them on their own. These models are available to both students and instructors on the BWeb. Case information is also available to all readers. The materials on Mono Lake, the Tucannon salmon population, and the Idagon River simulator are the most extensive as of 2009.

The Stella and Vensim programs are undergoing continual improvements, and the BWeb is a good way to keep pace with the improving software. There will inevitably be typos and other errors in the book, and the BWeb will provide a list of the errors. Check the list if something is erroneous, to see if the error has been reported by a previous reader. If not, report the error, and I will add it to the list. I also welcome comments on the exercises and the answer pages. The website will grow over time, and your feedback will help it grow in useful directions.

New to the Second Edition

This second edition benefits from 10 years of feedback from instructors using the first edition. Indeed, many of the best exercises in the second edition have grown out of the good work by previous instructors and their students. Instructors have also shared their ways of teaching with me, and I have realigned the chapters to take advantage of their experiences.

Many teachers wonder about the mix of time to be spent learning modeling methods versus learning modeling software. The Stella and Vensim programs have become more versatile and easier to learn. Both programs come with excellent online documentation, so the majority of this book is devoted to general concepts, modeling methods, and illustrative applications. However, I believe it is also useful to include some step-by-step instruction on the software. Several chapters have been written as if readers are following along on their own computers. This approach is taken in chapters 2 and 14 and the "Read and Verify" portion of chapter 16. The decade of teaching between the first and second editions of this book has had the important benefit of increased awareness of the technical problems involved in model formulation. These are pitfalls—that is, concealed dangers for the unwary modeler. Chapter 17 describes some of the most common pitfalls identified from conversations with a wide range of instructors, students, and practitioners. Each pitfall is illustrated with a simple model and its problematical behavior. Alternative formulations are presented to show how to avoid the problem.

The previous decade has seen important advances in system dynamics applications. The growth in participatory modeling is especially encouraging (as explained in chapters 13 and 24). The continued use of system dynamics in interdisciplinary applications is also promising. I believe interdisciplinary models have the greatest potential for profound insights, and I have made interdisciplinary modeling the theme of the book.

There have also been important advances in software and advanced methods. The past decade has seen the development of a wide range of icon-based software for stock-and-flow modeling (appendix C). Recent advances in the analysis of uncertainty intervals and the search for key inputs is now within reach of the broad community of modelers (appendix D). The past decade has also seen improved capability to incorporate short-term dynamics within system dynamics models of long-term trends (appendixes E and F). And recent software developments have improved the capability for spatial display of the inputs and outputs of system dynamics models (appendix G).

Author's Perspective

One of my favorite teachers urged all her students to reflect on their theories of how environmental systems work. She argued that these theories dominate the way we think—they shape the questions we ask, the people we listen to, the models we build, the data we seek, and the policies we advocate. She urged all students to give voice to their underlying theories so that teams of students would be more aware of where their teammates are coming from. This was good advice when I was a student, and it is good advice now as I hand this book off to you.

My thinking about environmental systems has been strongly influenced by the ideas published by Jay Forrester and his colleagues at the Sloan School of Management at MIT. Their view of systems and the value of computer simulation underlie much of what you will read in this book. System dynamics has been tremendously useful in my own research and consultancies, and I am impressed by the influential work of many system dynamics practitioners. The method is immensely useful, and we need well-trained people to put it to use on the serious environmental problems of our time.

My thinking on environmental systems was shaped at an earlier age by Garrett Hardin's *Biology: Its Principles and Implications.* His thoughts on homeostasis made a deep impression that holds sway to this day, and his view of the homeostatic plateau is sketched in figure 10.5. You can also see his influence on several of the cases in this book from the summary in table 10.1.

My own thinking on systems has come to focus more and more on the role of delays. This is probably the cumulative result of policy studies in which the ability to simulate the effect of delays emerged as the key to improved understanding. Delays play a prominent role in this book, as you will see in chapter 15 (salmon life cycle delays); chapter 18 (oscillations); chapter 19 (delays in real estate construction); chapter 22 (delays in DDT degradation); and chapter 23 (delays in the effective removal of CO_2 from the atmosphere).

Looking back to an earlier age, I was probably primed to think about delays by a driving test as a teenager. The test was organized by my father on a snowy day in our desert town. (We lived east of the Sierra Nevada, a land of little rain and snow.) My friends and I were learning to drive, and our parents advised us to be careful driving on the slick streets. My father could see that the advice was not getting through to teenagers filled with talk about horsepower and engine size. So he had us drive our cars to the school parking lot, which was set up with rubber barrels in a course requiring a few turns and some braking. The test looked simple enough until the first teenager put his parents' car into a skid that took out two barrels. The rest of us laughed at his error and wondered if he had heard the advice to "be careful." I concluded that my friend was either not listening or not paying attention to the barrels during the test drive. Each of my friends approached the course with confidence, and each of them quickly lost control, sending the barrels flying across the parking lot. I was the last to take the test, so I had the benefit of learning from my friends' failures. I took the wheel expecting to steer through the obstacle course at a somewhat slower speed without any problems. You can probably guess what actually happened—I went into a skid after the first barrel and wiped out the remainder of the course.

We were all embarrassed by our driving, and I was particularly embarrassed since I had failed to learn from my friends. What I remember most is my superficial explanations of my friends' failures. I simply wrote off their experience as something that would not happen to me. The driving test showed that all of us had grossly underestimated the long delay to change a car's direction when the tires have less frictional grip on the surface. We all grossly misjudged the slow speed necessary to drive a car safely on a slippery surface. My father said nothing at the conclusion of the test. But, like Oscar Wilde, he was probably thinking, *Experience is one thing you can't get for nothing.*

I tell this driving test story to dramatize the importance of delays and the difficulty of learning their importance. The driving test reminds us that putting general advice (e.g., "be careful") to good use is extremely difficult. And it reminds us that we do not necessarily learn from previous failures. However, computer simulation modeling can help us overcome the learning obstacles. The existence of a single delay in a key feedback loop may be the confounding factor that makes a system's behavior confusing. System dynamics models are ideally suited to simulate the role of delays. We should strive for realistic models that reveal the system's dynamic problem. Then we should design the models to allow participants to experience the effect of delays through interactive simulation. If we do this work well, we can help to build understanding of environmental systems.

Acknowledgments

Many of the ideas and examples in this book grew out of lessons first presented to me by Dennis and Donella Meadows, two superb professors at Dartmouth College. Their classrooms were a special place to learn, and I am forever in their debt.

My ideas were also shaped by participating in system dynamics classes at the London Business School and the Sloan School of Management at MIT. Thanks to John Morecroft and John Sterman for letting me learn from their approach during my sabbaticals. My most influential sabbatical was spent in the Corporate Planning Department of the Pacific Gas and Electric Company. Thanks to Mason Willrich for letting me learn firsthand about utility planning and modeling.

Many professors have shared their experiences and thoughts on modeling. My appreciation goes to David Andersen, Paula Antunes, Yaman Barlas, Todd BenDor, Paul Campbell, Bob Cavana, Bob Costanza, Henry Coyle, Robert Coyle, Bill Currie, Isaac Dyner, Bill Fleming, Jack Homer, Erik Larsen, Rob MacKay, Marciano Morozowski, Roger Naill, Paul Newton, George Richardson, Nigel Roulet, Khalid Saeed, Laurel Saito, Antonio Samagaio, Fahriye Sancar, Ali Saysel, Thomas Schmickl, Krys Stave, Nuno Videira, Wayne Wakeland, and Kaoru Yamaguchi.

My thinking about modeling has also benefited from conversations and modeling projects with Namsung Ahn, Allyson Beall, Asmeret Bier, Mike Bull, Steve Conrad, Alex Dimitrovski, Richard Dudley, Sy Goldstone, Ron Lohrding, Mike McKay, Jim Mills, Ottie Nabors, Dan Nix, Greg Reis, Ray Rink, Fred Roach, Kevin Tomsovic, Marjan Van Den Belt, and Klaus Vogstad.

Thanks also go to my colleagues at the University of Southern California and at Washington State University who encouraged my teaching of system dynamics and shared their ideas for both the classroom and the book. Special thanks to Peter Gardiner and Bill Budd, the best department chairs one could hope for.

I also wish to thank the folks at isee systems and Ventana Systems for their patience and rapid response to my many software questions. I especially appreciate the help over the years from Karim Chichakly on Stella and from Bob Eberlein on Vensim.

To Todd Baldwin, my editor at Island Press, I express my appreciation for encouraging the second edition. The logistics turned out to be even more formidable than the first edition, and I appreciate the hard work of the Island Press team in dealing with multiple drafts and hundreds of figures. And thanks go to Greg Turner-Rahman from the Department of Art and Design at the University of Idaho for the delightful sketches.

Modeling the Environment grew out of interactions with students at University of Southern California and at Washington State University. The book also benefited from the opportunity to interact with the students from the Bainbridge Graduate Institute and Worcester

Polytechnic University. My thanks to all the students who asked intriguing questions and who strove to make modeling as useful as possible. You made the classroom a rewarding place for me to teach and to learn. This book would not have emerged without your enthusiasm for learning.

I close with words of appreciation to my family, first to my father who helped me to see the world, and to my mother who helped me to reflect on what I saw. And warmest thanks to Amanda and Emilee, two daughters with the unfailing ability to brighten the day and challenge the mind. And to Amy, still my wife and best friend after all these years, thanks for your love, your support, and your sense of what is important.

PART I

INTRODUCTORY MODELING

Chapter 1

Introduction

A model is a substitute for a real system. Models are used when it is easier to work with a substitute than with the actual system. An architect's blueprint, an engineer's wind tunnel, and an economist's graphs are all models. They represent some aspect of a real system—a building, an aircraft, or the nation's economy. They are useful when they help us learn something new about the systems they represent.

Many of us have built and used models. Our first experiences might have involved physical models such as a paper airplane or a cardboard glider. These models were easy to assemble and easy to use. They made it fun to conduct experiments. We tried our experiments, watched the results, and tried again. Along the way, we learned about the dynamics of flight. If your experiences were like mine, you learned that you can't make a paper airplane fly farther by simply throwing it harder. You also learned that each airplane seemed to follow a natural glide path through the air. Through experimentation, you learned the extent to which the plane's natural trajectory could be improved.

This book focuses on mathematical models of environmental systems. The models use equations to represent the interconnections in a system. We will concentrate on a special category of mathematical models that are "simulated" on the computer. They are called *computer simulation models* because the tedious calculations are turned over to the computer. Our job is to think about the best way to construct the model to describe the system. If we do our job well, we can use the model to conduct experiments. We will try our experiments, watch the results, and try again. Along the way, we will improve our understanding of the natural trajectories of environmental systems. Through experimentation, we'll learn the extent to which the system's natural trajectory could be improved.

Informal Models

We use models all the time, but we work mostly with informal models. The images we carry in our minds are simplified representations of complex systems. These are sometimes called *mental models*. Senge (1990, 8) describes mental models as "deeply ingrained assumptions, generalizations, or even pictures or images that influence how we understand the world and how we take action." We use mental models constantly to interpret the world around us, and we may not realize that we are doing so.

To illustrate how quickly and subconsciously we use mental models, take a look at figure 1.1 and explain to yourself what you see. When asked about this image in the classroom,

3

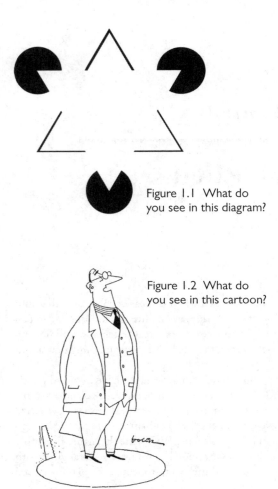

Figure 1.1 What do you see in this diagram?

Figure 1.2 What do you see in this cartoon?

many students report instantly that they see a white triangle on top of another triangle. When asked to explain, they often say that the white triangle has to be there. After all, it's the simplest way to explain the drawing. For example, the white triangle obscures our view of the underlying triangle, and it also obscures portions of the three circles.

Now take a look at figure 1.2 and explain what you see. And what about the curved line on the floor? Is it a circle or an ellipse? When asked about the cartoon in class, many students respond with an elaborate theory about a person hidden from view below the floor. He is playing a trick on the speaker by sawing a hole in the floor. The hole is almost complete, and the pompous speaker will soon fall through the hole. And when asked about the circle or the ellipse, most students say it is a circle. As drawn here, however, the curved line is an ellipse. But you know that it just looks like an ellipse when viewed from our vantage point. This is probably a circle when viewed by the person sawing the hole in the floor. After all, the most efficient way to saw the hole is by sawing in a circle.

Thinking about the Environment

These experiments reveal our ability to form theories to explain the world around us. These theories are mental models. We often form them instantly, and we may not realize that we are doing so. Although we are amazingly clever in thinking about geometric shapes, we are often baffled when thinking about serious problems in the environment. Whether it's a localized problem like urban air pollution or a global problem like the accumulation of greenhouse gases, our mental models seem inadequate to the task. We hear multiple and conflicting explanations of the problem, and we are not sure what to believe. If given the authority to act, we might not know what action to take. And if we follow our instincts, they could lead us in the wrong direction. We may think that the answer is to "push harder" on a particular program, but the system may respond like the paper airplanes from our youth. Pushing harder on an environmental system may be like throwing the airplane harder—it may make the situation worse than before. For instance, you'll see an example of a salmon population in chapter 15. It seems to follow its own natural trajectory over time, and you'll discover that the population can support a large harvest year after year. But if you try to increase the total harvest by pushing harder on the harvest fraction, you'll discover that you lower the harvest in the long run. Your first instinct points in the wrong direction.

Surprise Behavior

You'll see many examples of unexpected behavior in this book. Some systems will show unusually sluggish response. We intervene to change the system for the better, but it continues on the almost the same trajectory as before. The insecticide DDT and carbon dioxide (CO_2) are prime examples of disturbances whose effects are surprising. Chapter 22 shows that DDT would continue to accumulate in ocean fish many decades after efforts to restrict its use. Chapter 23 shows that the CO_2 in the atmosphere could continue to increase for the remainder of the century even after large reductions in emissions. Other systems will show unexpectedly rapid responses to external disturbances. Indeed, the changes may come so fast that the system seems to spin out of control. For instance, in experimenting with fees and rebates to promote the sale of cleaner cars, students see unexpectedly rapid changes in cash flow that can bankrupt the agency in charge of the feebate program (chapter 16).

Surprise behavior takes an entirely different form in Daisyworld, a make-believe world populated by black and white daisies (chapter 11). The world is warmed by solar luminosity, and the planet's temperature influences the spread of the daisies over the surface of the planet. But what if this world were to experience an abrupt increase in the solar luminosity? One would normally think that the planet's temperature will increase, but Daisyworld will react with changes in the mix of black and white flowers and a new temperature that is nearly the same as before. You learn that the interaction of the flowers with their physical environment allows the world to maintain suitable temperatures for flowers across an unusually wide range of values for the incoming solar luminosity.

Unexpected behaviors may be unsettling, especially if you were hoping that a model would lead to proven answers. But we should prepare for and embrace the unexpected results in modeling. Donella Meadows (2009, 87) explains that our mental models "keep track of only a few variables at one time, and we often draw illogical conclusions from accurate assumptions." Surprises are often the key to learning, and they should be viewed as an opportunity for improved understanding, both by individuals and by groups. Unanticipated results in group discussions can be particularly useful in management of large ecosystems. Kai Lee (1993) emphasizes this point in *Compass and Gyroscope*, his book on adaptive management. He describes adaptive management as an approach to natural resource policy that views economic uses of nature as experiments. The key is to learn from such experiments, but learning is a major challenge (Lee 1993, 12), especially in large ecosystems:

> *What makes an ecosystem "large" is not acreage but interdependent use; the large ecosystem is socially constructed. Rivers nurture fish and plants, water fields and cities, provide transport for trade and sometimes hydroelectricity for industry. Multiple uses of a river or other large ecosystem require trading off qualities that are hard to compare, controlled by or benefiting different people. Social constructs can be difficult to alter, and the boundaries between competing claimants to a natural resource have often produced stalemate rather than problem solving. But an adaptive approach can loosen deadlock with surprising outcomes. The social dynamism of learning can undermine socially constructed stalemate.*

System dynamics modeling can speed the learning about large-scale interconnected systems, especially when the modeling includes a group of stakeholders with different perspectives and expertise. My own experience with modeling in groups makes me think that Kai Lee is right about surprises: new insights on how we interact with nature can loosen deadlock and set the stage for exploration of new policies on natural resource management.

Learning by Experimentation

The cases in this book provide opportunities to practice building, testing, and using models. Each new case allows you to learn about the system's natural tendencies by computer experimentation. Learning by experimentation reminds me of the first time I took the controls of the motor at the rear of a fishing boat. My first instinct was to point the motor handle in the direction I wanted to travel. But the boat pivoted and traveled in the opposite direction. I pushed the handle even farther in the intended direction, and the boat headed even farther in the wrong direction. I finally got the message and made corrections, but the lag in the boat's response led to further embarrassment as the path I steered across the lake was a series of arcs to the left, to the right, back to the left, but never in the intended direction. Eventually, with some time and practice, I developed a feel for the system and was able to steer the boat on a true course.

The point of the boating story was my opportunity to practice on a calm lake where errors in thinking could be tolerated. Since I was unfamiliar with boating, my first instinct pointed the boat in the wrong direction. With an opportunity to practice, I learned to control the boat. With environmental systems, however, we seldom have the opportunity to practice. So how might we develop an instinctive feel for the natural tendencies of the system?

The premise of this book is that computer simulation models can help us develop our instincts for managing environmental systems. We build a mathematical model to capture the key interrelationships in the system. Then we conduct experiments with the model. Since the systems are new to us, we should be prepared for surprises along the way. When we check out the reasons for the surprise behavior, we will emerge with new insights on the dynamic behavior. And with new insights and better understanding come better instincts for managing environmental systems.

Experimenting with models is a learning activity that is best pursued in an exploratory manner. Trying lots of simulations is the best way to learn about the simulated tendencies of the system. But the development and testing of computer models also requires a disciplined approach. Good modeling requires us to be clear and explicit about our assumptions. If we do our job well, others will be able to appreciate the assumptions and to understand the conclusions drawn from our experiments. Careful modeling will also permit others to challenge our underlying assumptions, to add more realistic assumptions, to conduct new experiments, and to emerge with new insights on the system behavior.

System Dynamics

The fundamental ideas in this book are not new. They are drawn mainly from the field of system dynamics, which originated in the 1960s with the work of Jay Forrester and his colleagues at the Massachusetts Institute of Technology. Forrester (1961) developed the initial ideas by applying concepts from feedback control theory to the study of industrial systems. One of the best-known applications of the new ideas was Forrester's (1968) *Urban Dynamics*. It explained the rapid population growth and subsequent decline observed in cities like Manhattan, Chicago, and Boston (Schroeder and Strongman 1974, 201). Forrester viewed the city as a system of interacting industries, housing, and people. The city would grow rapidly under favorable conditions. But as its land area filled, the city would shift into stagnation characterized by aging housing and declining industry. During this phase, the city would experience a decline in population. The purpose of *Urban Dynamics* was to help planners deal with the challenges of revitalizing an aging city. As Forrester (1968, 105) wrote: "Reviving blighted areas is a task forced on us by earlier failures in urban management." Through experimenting with the model, he discovered that "pushing harder" on familiar programs could make the situation worse than before. Constructing additional premium housing, for example, could lead to increased stagnation because less land would be available for new industries. He then experimented with a demolition program to remove a fraction of the "slum housing," and the overall result was beneficial. Housing demolition created the space for new industries, permitting a renewal that led to improvements in the mix of industries and workers within the urban boundary.

Forrester recommended that cities adopt demolition programs even though this proposal ran counter to planners' conventional thinking. Critics charged that his recommendation was based on an imperfect model. He replied that all models are imperfect because they are, by design, simplified representations of a system. He acknowledged that the recommendation ran counter to most planners' intuition, but he suggested that our normal way of thinking about complex systems is often limited and misguided. He stated that models are most useful when they lead to "counterintuitive" results that force planners to reexamine their intuitive understanding. In *Toward Global Equilibrium*, Forrester (1973, 5) argued that counterintuitive results are to be expected in all systems, not just in urban systems:

> *The human mind is not adapted to interpreting how social systems behave. Our social systems belong to the class called multiple-loop nonlinear feedback systems. In the long history of human evolution, it has not been necessary for man to understand these systems until very recent historical times. Evolutionary processes have not given us the mental skill needed to interpret properly the dynamic behavior of the systems of which we have now become a part.*

Urban Dynamics highlighted the field's expansion outside the industrial area. The approach came to be known as system dynamics. Perhaps the most widely known application of system dynamics appeared a few years later in a best-selling book titled *The Limits to Growth* (D. H. Meadows et al. 1972). It looked at the prospects for growth in human population and industrial production in the global system over the next century. The model was used to simulate resource production and food supply needed to keep pace with a growing system. It also simulated the generation of persistent pollutants that accumulate and remain in the environment over many decades. The simulations led the authors to conclude that the world system could not support the rates of economic and population growth much beyond the year 2100, if that long, even with advanced technology. The authors concluded that the "most probable result will be a sudden and uncontrollable decline in both population and industrial capacity."

Many interpreted this conclusion as a prediction of "doom and gloom." But *The Limits to Growth* was not about a preordained future. It was about making choices to influence the future. The authors examined simulations with changed attitudes toward population growth and industrial growth. The new simulations led them to conclude that "it is possible to alter these growth trends and to establish a condition of ecological and economic stability that is sustainable far into the future" (D. H. Meadows et al. 1972, 24). They argued that the sooner the world's people begin working toward the goal of a sustainable world, the greater will be their chances of success.

Definition of System Dynamics

System dynamics is a methodology for studying and managing complex systems that change over time. The method uses computer modeling to focus our attention on the information feedback loops that give rise to the dynamic behavior. Computer simulation is particularly useful when it helps us understand the impact of time delays and nonlinearities in the system. Coyle (1977, 2) puts the system dynamics approach in perspective:

> *System dynamics is that branch of control theory which deals with socio-economic systems, and that branch of management science which deals with problems of controllability.*

The emphasis on controllability can be traced to the early work of Forrester (1961) and his background in control engineering. Coyle highlighted controllability again in a highly pragmatic definition:

System dynamics is a method of analyzing problems in which time is an important factor, and which involves the study of how a system can be defended against, or made to benefit from, the shocks which fall upon it from the outside world.

A definition also appears in the online encyclopedia Wikipedia. As of May 2009, Wikipedia defined system dynamics as

an approach to understanding the behaviour of complex systems over time. It deals with internal feedback loops and time delays that affect the behaviour of the entire system. What makes using system dynamics different from other approaches to studying complex systems is the use of feedback loops and stocks and flows. These elements help describe how seemingly simple systems display baffling nonlinearity.

You'll learn about stocks and flows in chapters 2 and 3 and about feedback loops in chapters 9 and 10. And within a few weeks you will be simulating seemingly simple models with baffling nonlinearity.

Six Shapes to Describe Dynamic Behavior

Stocks, flows, and feedbacks are the key to the "system" in system dynamics. But at this stage, it is more useful to start with the "dynamics." The dynamic patterns can take the form of growth, decay, or oscillations. And these same patterns will show up across all environmental systems. I've selected the six shapes in figure 1.3 to focus our attention in this book.

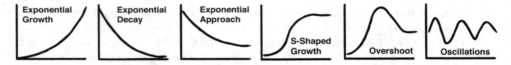

Figure 1.3 Six shapes to represent the dynamic patterns simulated in the book.

Exponential growth is one of the most pervasive and important dynamics in the environment, and you'll simulate this pattern in chapter 2. Exponential growth can be surprisingly rapid as the system doubles in size, then doubles again and again in the same interval of time. Appendix B explains that this interval is known as the *doubling time*. When you understand the doubling time, you'll be in a better position to anticipate the power of exponential growth in environmental systems.

Exponential decay is the inverse of exponential growth. The system will lose half its value in a fixed interval. And then it will lose half of that in the next interval. The fixed interval is called the *half-life* of the system, as explained in appendix B.

The *exponential approach* resembles exponential decay, but the system does not decay all the way to zero. This can happen when the system receives an inflow that counteracts the effect of decay. The system can gradually approach equilibrium. You'll see this pattern in the chapter 5 model of Mono Lake. Figure 1.3 depicts the approach to equilibrium from above. The system can also approach its equilibrium position from below.

S-shaped growth resembles exponential growth in the early stages. But as the system grows larger and larger, it will eventually encounter limitations. The limits could be restrictions on resources, nutrients, breeding habitat, winter habitat, and the like. All systems must deal with a limit of one sort or another. For S-shaped growth to appear, the system must feel the effect of the limits in a manner that will gradually slow the growth and permit the system to reach an equilibrium state that can be sustained year after year. This pattern is fundamental to natural systems, and it will appear in many examples throughout the book.

The *overshoot* pattern starts out like S-shaped growth, but the system does not achieve a smooth accommodation with its limited resources. If there are delays in the reaction to the limits, the growth trend can carry the system beyond sustainable values. The overshoot can be especially pronounced if the resources of the system are damaged during the period of intense crowding. Population overshoots can occur in human systems, as described in *Urban Dynamics* and *The Limits to Growth*. They can also appear in natural systems, as in the deer population overshoot in chapter 21. The overshoot pattern can also occur in coupled human-natural systems (e.g., overfishing of open-access fisheries, described in chapter 15). Understanding the tendency for overshoot is crucial if we are to achieve sustainable management of human and natural systems.

Oscillations appear in many systems, ranging from the beating of the human heart to cycles in the nation's economy. Some systems oscillate in a repetitive and stable manner, one that seems to ensure the longevity of the system. Other systems can enter an oscillatory pattern that is highly unstable and that can threaten the sustainability of the system. Figure 1.3 depicts oscillations that dampen out slowly over time (think of the damped oscillations of a pendulum). But oscillations can take many forms, and they sometimes arise from simple delays in the system (chapter 18). System dynamics is a powerful way to look inside the system for an understanding of oscillations, as you will see with applications to cycles in real estate construction (chapter 19) and in predator-prey populations (chapter 20).

Modeling for Prediction?

The "dynamics" in system dynamics are the fundamental patterns of change, such as growth, decay, and oscillations. System dynamics models are constructed to help us understand why these general patterns occur. They are not constructed to predict the exact value of the system at a specific time in the future. *Predictive models* are quite different. A weather model may be used to predict whether it will rain tomorrow; a stock market model might be used to predict stock prices at the end of the week. These models are sometimes called *forecasting models*. They are designed around a single task—to provide the best possible forecast of the future state of the system. Since their purpose is clearly and narrowly defined, predictive models are easily evaluated. We simply ask how frequently their predictions turn out to be correct. Predictive models are useful in narrowly defined situations in which forecasts are needed and predictive methods can be evaluated. But predictive models are not likely to be generally useful to the student of the environment. For example, an ecologist uses models to understand general properties such as persistence, stability, resilience, or efficiency. Ecosystems are subjected to highly random inputs, so it does not make sense to construct predictive models when basic inputs cannot be measured or predicted. It makes more sense to use models to improve our general understanding and to guide further research.

Modeling for Understanding

System dynamics models are designed for general understanding, but they are often misinterpreted as predictive models, especially if one is looking for a "crystal ball" to forecast the future. Misinterpretation often occurs when a time graph shows an important variable displayed into the future. The time graph might be labeled as a "base case simulation," but readers eager for a forecast will relabel the graph in their mind as "most likely behavior."

To avoid misinterpretation, you should resist drawing any conclusions from one simulation. A single simulation seldom teaches us much about the system. Its purpose is usually to provide a starting point for comparison with additional simulations. Think of a single simulation as one blade in a pair of scissors. Scissors are not designed to cut with one blade working alone. It's only when the two blades work against each other that the scissors serve their intended purpose. System dynamics simulations should also work in pairs. By comparing one simulation against the other, the model will serve its intended purpose. If you find yourself uncomfortable working with pairs of simulations, you are probably looking for a predictive model. In this case, you should turn to forecasting methods to serve your needs.

Modeling for prediction is easy to understand because the predictions can be judged as right or wrong. Modeling for improved understanding is more difficult to understand. If we will not learn how to predict the future, what will we learn to do? If you pose this question to experienced practitioners, they will tell you that managers need useful rules of thumb to help manage complex systems. They will argue that the best use of system dynamics is to help managers develop these rules of thumb. To appreciate what they have in mind, think of the "one car length for every 10 miles per hour" rule of thumb that careful drivers use when driving the freeway. If we are driving at 60 miles per hour, we should strive to leave six car lengths of space between our car and the car in front of us. This rule is a good summary of a complicated calculation involving our reaction time, the momentum of the car, and the braking power of the tires on the road surface. We don't perform this calculation as we drive the freeway. Rather, we follow the "one car length" rule and hope that it will provide sufficient room if we have to hit the brakes. Several of the models in this book will teach us about the momentum built into an environmental system. If we discover that the system is in danger, we may decide to "hit the brakes." The model's simulated response will reveal whether our actions will allow the system to recover from a dangerous situation.

Modeling across Time Scales

Dynamic problems can take many forms, but I believe you can represent most important environmental dynamics by some variation of the six shapes in figure 1.3. Time is on the horizontal axis in each of the graphs in figure 1.3, but what are the units of time? Should we be looking at patterns that unfold in a few minutes, a few hours, a few years? The models in this book simulate dynamics across widely different time scales, as shown in table 1.1. Each model has its own units of time. The shortest unit in this book is seconds (used in simple models of fluid flows). The longest unit is 100 million years. If you've studied geology, you know that this is a suitable measure of time to use in simulating changes in the rock cycle. The majority of the models in the book operate in days, months, or years. The important thing to remember is that the principles of modeling apply across all time scales, so you are free to select the unit of time that best fits the particular dynamic problem you are studying.

Some chapters include two models dealing with the same system, but they use different units of time. For example, chapter 15 describes a model of the salmon smolts' migration that operates in days and a model of the salmon life cycle that operates in months. Don't worry when you see the different units of time used to study the same system. It is often useful to develop several models. Indeed, an organization with responsibility for managing a complicated system might develop a portfolio of models. The portfolio might include a model with time in days, another with time in months, and a long-term model with time in years. The daily model could shed light on the fast-acting dynamics and help in the formulation of the monthly model. The monthly model could shed light on seasonal dynamics and help in the formulation of the long-term model. The construction of three separate models may seem like a lot of work, but it can be worthwhile if it leads to better understanding. But wouldn't it be more efficient to

Table 1.1. Models across time scales.

Time scale	Model
seconds	Joe fills the gas tank (ch 3)
	Water flow through two bottles (ch 6)
minutes	Hikers head up the hill (BWeb)
	Water temperature control (BWeb)
hours	Body temperature control (BWeb)
days	Spread of an epidemic (ch 8)
	Salmon smolts migration to ocean (ch 15)
months	Salmon population life cycle (ch 15)
	The Idagon river simulator (BWeb)
	Genetics and industrial melanism (BWeb)
	Mono Lake brine shrimp population (ch 5)
years	Mono Lake water balance (ch 5)
	Temperature control on daisyworld (ch 11)
	Cleaner cars and feebates (ch 16)
	Cycles in real-estate construction (ch 19)
	Cyles in predator-prey populations (ch 20)
	Overshoot of the Kaibab deer population (ch 21)
	DDT accumulates in the ocean (ch 22)
	CO2 accumulates in the atmosphere (ch 23)
hmyrs	Rock cycle (ch 6), in hundreds of millions of yrs

create a single model with time in days? Then, if you need to simulate dynamics over 50 years, you could simulate the model for 18,250 days. Building one model may seem simpler than building three, but we should resist the temptation to reach for a single model. My experience (both in large organizations and in the classroom) suggests that we will make more progress working with a portfolio of models. The models will be easier to build, to test, and to understand. And they will be easier to explain.

Modeling across Spatial Scales

Environmental systems can be described at many levels and across many scales. For example, rivers or streams might be studied at a small scale (less than 1 meter) if we focus on the gravel and boulder patches in the stream (Frissell et al. 1986). A 1-meter scale would be more appropriate for the study of pool and riffle systems; larger-scale studies could focus on a "reach" (~10 meters), a "segment" (~100 meters), or an entire "stream system" (~1,000 meters). The scale of modeling can also change dramatically in ecological studies. Ecologists may focus their attention at the molecular, physiological, or individual levels. But large-scale scales are appropriate for focusing on populations, communities, and ecosystems. The models in this book deal with widely different scales. In some cases, the model deals with an entire population within a clearly defined spatial boundary (e.g., the deer population of the Kaibab Plateau). In other cases, the spatial boundaries are extremely complicated because of complex life cycle and migration patterns (e.g., in salmon populations).

Some ecologists use models to look at each individual in a population. These models are known as individual-based models, or IBMs (appendix C). You can practice modeling with a simple IBM of hikers heading up a hill (see the book's website, the BWeb). You'll learn that the hikers soon fall into a spatial pattern that leads to relatively uniform speed up the mountain. But most system dynamics modeling is not done at the individual level. The more common practice is to represent the aggregate results of populations. You will see models to simulate millions of smolts migrating down the Snake River and thousands of deer foraging on the Kaibab Plateau. The models are based on average characteristics of the entire population.

The hydrologic models in the book illustrate how the spatial boundaries are set to match the system under study. For example, a basin boundary makes sense for the Mono Lake model in chapter 5. Figure 1.4 depicts the water flows in the Mono Basin on the eastern side of the Sierra Nevada. The large flows from the Sierra can feed the lake or provide export to the south. The largest flow is the evaporation that moves water out of the basin.

Figure 1.4 Water flows in the Mono Basin in the eastern Sierra.

Modeling across Disciplines

The Mono Lake case study will be your first opportunity to see the potential for modeling across disciplines. You'll begin with a hydrologic model of the flows depicted in figure 1.4. The exercises invite you to combine the hydrology model with a biological model of the brine shrimp population. The new model is no longer strictly hydrology or strictly population biology; it is an interdisciplinary combination of both. The new model will allow exploration of export policies that are based on either hydrologic indicators (lake elevation) or biological indicators (shrimp population). You'll see that the system dynamics approach offers the opportunity for interdisciplinary modeling and exploration. This is also nicely illustrated in the Idagon, a model of the hypothetical river system depicted in figure 1.5. The model combines ideas from many disciplines, and it represents different participants' views of the top priorities for managing the river system. The ability to combine perspectives from different disciplines is one of the most useful aspects of the system dynamics approach to environmental systems. We'll illustrate this ability in several of the cases and discuss its value in the concluding chapter of the book.

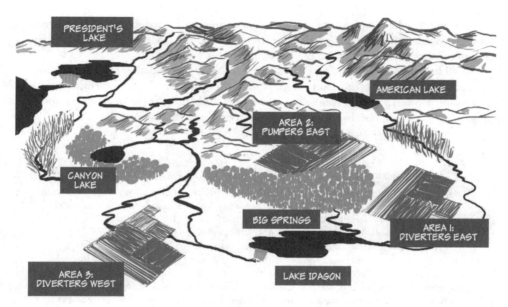

Figure 1.5 Water flows in the Idagon.

The Steps of Modeling

Modeling is an iterative process of trial and error. The model is usually built up in steps of increasing complexity until the simulations show the dynamic pattern under study. A typical modeling project involves the steps listed in table 1.2. Step 1 is to become acquainted with the problem, and step 2 is to be specific about the dynamic nature of your problem. As an example, suppose your system exhibits exponential growth, but you and your colleagues believe the growth is not sufficiently strong. To focus the modeling process, you would draw a graph showing the exponential growth pattern over time. Your graph will have time on the horizontal axis, and you will select the appropriate unit of time. The graph will have an important and easily recognized variable on the vertical axis. The graph can be a simple pencil sketch, like the sketch of exponential growth in figure 1.3. The important thing is to see the general pattern, not the precise numbers. We call this graph the *reference mode*. (*Mode* reminds us that we are simulating a general pattern of behavior; *reference* reminds us that the target pattern provides a point of reference for the modeling project.) You will know that you have a drawn a good

Table 1.2. The steps of modeling.

Step 1.	A is for **Acquainted**	Get acquainted with the system and the problem.
Step 2.	B is for **Be Specific**	Be specific about the dynamic problem.
Step 3.	C is for **Construct**	Construct the stock-and-flow diagram.
Step 4.	D is for **Draw**	Draw the causal loop diagram.
Step 5.	E is for **Estimate**	Estimate the parameter values.
Step 6.	R is for **Run**	Run the model to get the reference mode.
Step 7.	S is for Sensitivity	Conduct a sensitivity analysis.
Step 8.	T is for Test	Test the impact of policies.

reference mode when your colleagues agree that it summarizes the nature of your dynamic problem. If you find that you can't draw the reference mode, you do not have a dynamic problem. In this case, there is no reason to develop a dynamic model.

You'll learn about the remaining steps of modeling as you read about stock-and-flow diagrams, causal loop diagrams, and so on. And you'll read more about the entire modeling process in chapter 13. At this stage, you should turn to chapter 2 to learn about the software programs that make it easy to construct and test dynamic models.

Box 1.1. But what about calculus?

The boxed sidebars in this text use a hypothetical student, Joe, to discuss questions that often arise in class. Joe's first question is whether he needs to review calculus before proceeding to chapter 2. He is also concerned about differential equations. He has not studied differential equations, and he worries that his memory of calculus has faded over time.

If you are like Joe, rest assured—you do not need to know calculus or differential equations to use this book. The book was written on the assumption that you have learned introductory algebra and that you know how to read and interpret graphs. But there is no need to worry if you have not taken calculus or differential equations. Read on, and you'll soon learn that we build system dynamics models in a visual manner on the computer. Your job is to concentrate on the structure of the model, and the tedious job of numerical simulation will be left to the computer. Within a few chapters, you will be building and simulating models that are far beyond the reach of an introductory class in differential equations.

If you have taken calculus, you'll quickly recognize that the software is integrating the effect of flows over time. And if you've taken differential equations, you'll recognize that the models are equivalent to a set of coupled, first-order differential equations. The equations are almost always highly nonlinear, so there is little hope of finding an analytical solution. We will find a numerical solution, as explained in chapter 4.

If you have taken several math courses, your knowledge will be helpful, but only if you give yourself time to assimilate the ideas and models in the book. The models are designed for clarity and ease of understanding. They will not appear as close replicas of the differential equations you have seen in previous classes. Our goal is not to repeat what you have seen before. Rather, we are aiming for a new style of modeling that encourages active participation by individuals with a wide variety of backgrounds.

Further Reading

- Modeling of environmental or ecological systems is explained in a series of books edited by Hannon and Ruth (1997). The book by Deaton and Winebrake (2000) in this series is most similar to the approach in this book.
- Readings on wildlife and ecological modeling are provided by Watt (1968); Odum (1971); Kitching (1983); Grant (1986); Pratt (1995); and Grant, Pedersen, and Marin 1997).
- The use of individual-based models (IBMs) in ecology is reviewed by Grimm (1999).
- Articles on ecological modeling can be found in *Ecological Modelling*.
- Articles on system dynamics modeling can be found in *System Dynamics Review*.
- System dynamics instruction is found mainly in business schools, and several business-oriented texts are available: Coyle (1977, 1996); Richardson and Pugh (1985); Warren (2000); and Morecroft (2007). The definitive text is *Business Dynamics* by Sterman (2000). Both Morecroft and Sterman provide environmental and resource examples alongside of the business applications.
- A unique perspective on systems and their dynamic problems is provided by the posthumous publication of D. H. Meadows's (2009) primer, *Thinking in Systems*.

- Historical perspective on system dynamics is provided by Richardson (1991). He interprets feedback ideas from system dynamics and how they are expressed by prominent scientists from biology, economics, engineering, and the social sciences.

Acknowledgments

- The Kanizsa triangle in figure 1.1 is discussed by Sterman (2000, 17) and by Wikipedia. The image is from Wikipedia.
- The cartoon in figure 1.2 is from Gregory (1998).

Chapter 2

Software: Getting Started with Stella and Vensim

Stella and Vensim are two of the icon-based programs supporting the construction and testing of system dynamics models. I use these particular programs in my own teaching and research, and they are used throughout the book. System dynamics models can also be constructed with Powersim. Like Stella and Vensim, Powersim was designed to provide a user-friendly, icon-based approach to modeling based on the principles first published by Forrester (1961). A variety of other programs is also available to provide icon-based support of dynamic modeling (e.g., Simile, Simulink, and GoldSim). These programs were not designed with the central focus on system dynamics ideas (Forrester 1961). Nevertheless, they have many useful features in common with Stella and Vensim, and it is possible to build models with similar structure and similar results. These and other programs are described in appendix C. Stella and Vensim were selected for this book because of their ease of use and their emergence as popular programs for cooperative modeling of environmental systems. You'll find that the lessons from this book will provide useful guidance regardless of your choice of software.

You will see mostly Stella models in this chapter. The Vensim examples will reveal the close similarity in the programs. I recommend you get started by downloading the trial version of Stella or the learning version of Vensim. They both provide good introductions, many examples, and extensive documentation. Some of my students use both programs; others pick one or the other based on their personal situation and their research interests. You'll see a mix of Stella and Vensim models throughout the book. The book's website (BWeb) provides additional models using both programs. The BWeb also explains my thinking behind the selection of Stella or Vensim for the cases in this book. If you are new to system dynamics, don't worry about which to choose. They are both excellent programs to support dynamic modeling.

Getting Started: Read and Verify

This chapter is different from most chapters in the book in that it is written in workbook style. A good way to practice is to read about each model and immediately verify that you can reproduce the results on your own computer. Each model is described in a step-by-step manner as if you are executing each step with your copy of the software as you read. (As you do so, remember that these results are from version 9.0.3 of Stella and version 5.3 of Vensim PLE, the Personal Learning Edition.) This chapter is written for readers who wish to learn how to build and test models. If you are reading this book for general ideas and concepts, you can skip over the step-by-step instructions. The figures will give a general impression of what can be accomplished with the software.

Exponential Growth in Population

Let's start with an example of exponential growth. Imagine that you have drawn a reference mode similar to the exponential growth sketch in figure 1.3. Population is on the vertical axis, and time is on the horizontal axis. The current population is 100 million, and the population has been growing at 7%/yr in recent years. Suppose that problems are anticipated when the population reaches 800 million, and we want the model to help us anticipate when that will occur.

Figure 2.1 shows a map of a Stella model simulating the population growth. The population is a *stock*, shown by the rectangle in the diagram. The births are a *flow*, shown by the arrow with a valve. This model has one stock and one flow. Stocks and flows are the building blocks of system dynamics, so the software puts them in the first two positions on the panel of icons.

Now imagine you are facing a blank screen and wondering how to start. The best place to start is with a stock. For this example, we will use a stock to represent the population. To create figure 2.1, select a stock from the panel, place it on the screen, and give it the name "population." Then select the flow icon from the panel, click below the stock to establish the cloud, and drag the icon toward the stock until Stella recognizes the connection. Name the flow "births," and you have a map of your first model. If you wish to save your work, click on the File command and drag down to Save.

The map shows the variables and how they interconnect. To write equations, click on the Model tab, and Stella will respond with figure 2.2. The question marks alert us to the variables that need an equation. Let's start with the population. All we need to do is type in the initial value, so we enter 1000000000 since the initial population is 100 million.

But take a closer look at the numbers. Did you see that I typed too many zeros? Such mistakes are common, but they can be avoided if we rethink the units. Let's measure the population in millions of persons and enter the initial value at 100. Close the population equation box, and Stella will show figure 2.3 with one "?" remaining. We could click on "births" and set the value to 7 if we expect births to be 7 million persons/yr.

Figure 2.4 shows that the question marks are gone. The model is ready to simulate. To specify the length of the simulation, go to the Run command and drag down to Run Specs.

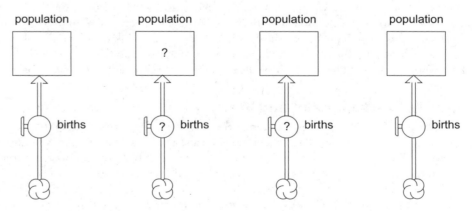

Figure 2.1. Map of   Figure 2.2. Click on   Figure 2.3. Need one   Figure 2.4. The model
the first model.   the Model tab.   more equation.   is ready to simulate.

Time should be in years, starting with 0 and ending with 40. Click on Run, and Stella will simulate the growth in the population for 40 years. Create a graph of the population, and you will see that it grows in linear fashion, increasing by 7 million per year and by 70 million per decade. After 40 years, there will be 380 million persons, well below the 800 million that are said to pose problems.

If you get these results, congratulations on your first simulation. Unfortunately, the model did not deliver the intended results. It gives linear growth, but we are looking for exponential growth. To get exponential growth, the population has to grow faster and faster over time. This can happen if the number of births is proportional to the total population. Let's assume that this is true for our population, and the birth rate is 0.07/yr. But how do we introduce the birth rate into the model?

This is where Stella's converter can help. It is the circle located next to the flow on the panel of icons. The converter's main function is to help explain the flows. Click on the converter, place it next to the flow, and give it the name "birth rate." Then select the red arrow (the action connector) and click on the birth rate. Stella will establish a starting location with a small circle. Drag the arrow to the births and click again when Stella recognizes the connection. Stella will respond with figure 2.5, which indicates that it needs an equation for the births and an equation for the birth rate. The birth rate can be set to 0.07, but what should we write for the births? They depend on the birth rate and on the population. This means we need another connector, as shown in figure 2.6. We now open the flow to write the equation for births.

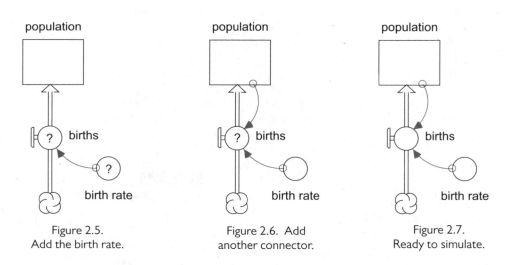

Figure 2.5.
Add the birth rate.

Figure 2.6. Add
another connector.

Figure 2.7.
Ready to simulate.

Some students type in 7 since 7%/yr of 100 million is 7 million/yr. If you give this a try, Stella will complain about unused inputs. It sees the action connectors in figure 2.6, and it wants your equation for births to use the connected variables. Stella doesn't care how you use the two variables, but you know that the number of births is the product of the population and the birth rate. Enter this equation, and Stella will respond with figure 2.7. The model is now ready to simulate.

Before you click the Run command, take the time to write down your guess as to the population at the end of the 40-year simulation. (If you don't know how to guess, turn to appendix B.) If you do recall the doubling-time rule, you'll guess that the population will double to 200 million in the first decade, to 400 million in the second decade, and to 800 million in the third decade. With no limits on the births, the population should continue to grow, reaching 1,600 million by the end of the simulation.

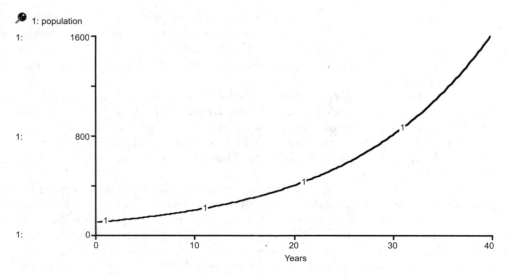

Figure 2.8. Exponential growth in population.

To see these results, click on the graph pad icon and place it on the diagram. You'll then select variables to be graphed and set the scales for the best display of results. In figure 2.8, I selected the population and asked for the vertical axis to be scaled from 0 to 1600.

Figures 2.9 and 2.10 show the corresponding images of the same model in Vensim. Figure 2.9 is a map or diagram of the model. Figure 2.10 highlights the variables whose equations need to be written. Vensim's box variable is used for the population. Vensim calls this a *level*, which is synonymous with a stock. The births add to the size of the population. The icon is a arrow, whose value is depicted by a butterfly valve. Vensim calls this a *rate*, a term that is synonymous with flow. The births depend on the population and the birth rate, as you can tell from the arrows.

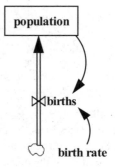

Figure 2.9. Vensim diagram of the population model.

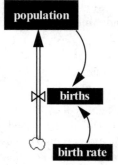

Figure 2.10. Equations are needed before we simulate.

The birth rate is entered by clicking on the VAB (short for variable) icon. Vensim calls this a *variable-auxiliary/constant*. To write the equations, click on the $Y=X^2$ icon, and Vensim will respond with figure 2.10, which shows us the highlighted variables that need equations.

Set the initial value of the population to 100 and the births to be the product of the population and the birth rate. Set the birth rate to 0.07, and Vensim will respond with a diagram free of highlighted variables. The model is ready to simulate. Go to the Model command and drag down to Settings, then ask for time bounds. Set time to be in years, with an initial time of 0 and a final value of 40. Simulate the model with the Model Simulate command. Vensim will ask if you want to assign a name to the data set or if you want to stick with the name "current." If you agree, the results will be stored in a file named current.vdf. Give this a try, and your simulation should show the population reaching 800 million by the 30th year and around 1,600 million by the end of the simulation.

Population Growth with Births and Deaths

The first model simulated the cumulative effect of births, but it ignored deaths. Suppose we are told that the death rate is 2%/yr. If the population has been growing at 7%/yr, we would suspect the birth rate is 9%/yr. Figure 2.11 shows a new version of the model with deaths, a flow that drains the stock of population. To add the flow, select the flow icon, click on the stock to establish the origin of the flow, drag to the right, and release the flow. The new flow will be depicted as removing population from the stock and headed to a cloud. This cloud is sometimes called a *sink*. (The cloud on left side of the births flow is called a *source*.)

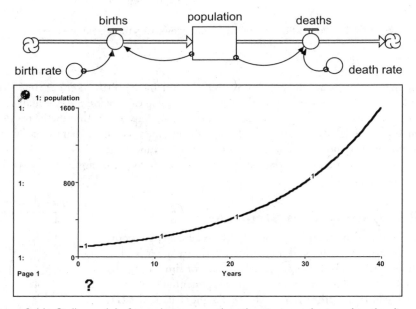

Figure 2.11. Stella model of population growth with a time graph pinned to the diagram.

Figure 2.11 shows the new model with the simulation results in view. The graph is identical to the graph shown in figure 2.8. To create this image, place the graph below the diagram and click the pushpin icon in the upper-left corner. This pins the graph to the diagram, and it will be printed when you print the diagram. The graph confirms the results from before: the population reaches 800 million in 30 years and 1,600 million in 40 years.

Using Converters to Help Understand the Flows

The best way to build a model is to start with the stocks, add the flows, and then use converters to explain the flows. In the previous model, we relied on a birth rate to explain the births and a death rate to explain the deaths. But we do not have to limit ourselves to a single converter to explain each flow. If we add more converters, a clearer picture of the flows may emerge. Figure 2.12 illustrates the previous population model with additional converters. First, notice that the deaths no longer depend on the death rate, which was 0.02/yr in the previous model. This could correspond to an average lifetime of 50 years, and it might be easier for people to understand if the model used the average lifetime. The equation for deaths will be the population divided by the average lifetime.

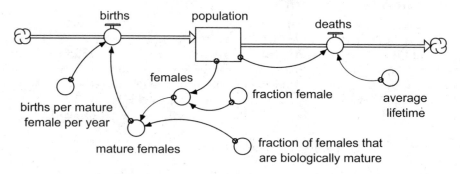

Figure 2.12. Population model with new converters to explain the flows.

The main changes are the extra converters to explain the births. Let's assume that the fraction female is 0.5 and that 36% of the females are biologically mature. And, finally, let's assume that the mature females give birth every other year. This means that the births per mature female per year are 0.5. If you work through the numbers, you'll see that this combination of assumptions corresponds to a birth rate of 9%/yr, the same value as in the previous model. The new model gives the correct results, and it provides a fuller explanation of the exponential growth in population.

Similarity in System Structure

A human population can grow exponentially because a larger population leads to more births, and more births lead to a larger population. Other systems exhibit exponential growth for similar reasons. The balance in your bank account will grow exponentially when the interest added increases the balance, and that higher balance leads to still more interest added in the future.

Figure 2.13 shows a model to see if the balance will grow in exponential fashion. Except for the names, the model is identical to the population model in figure 2.7. Indeed, if you constructed the previous model, all you need to do is change the names, and you will be ready to simulate the bank balance model. It starts with the balance at 100, and the interest added is the balance multiplied by the interest rate. Simulating this model will confirm that the bank balance grows to just over $1,600 in 40 years.

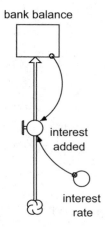

Figure 2.13.
Bank balance model.

Exponential Decay

Exponential decay is the second of the fundamental dynamics sketched in figure 1.3. It can appear when the rate of decay exceeds the rate of growth. The population model will exhibit exponential decay if the death rate exceeds the birth rate.

Let's simulate the model in figure 2.11 with an unusually high death rate of 16%/yr. If we leave the birth rate at 9%/yr, we would expect the population to decline at 7%/yr. You know from appendix B that the half-life for this decay is 10 years. So we would expect to see the population fall by 50% every decade. Figure 2.14 shows a time graph of the population with the vertical axis scaled from 0 to 100. It confirms the expected pattern of decay. For example, there are 50 million people after the first decade and 25 million after the second decade.

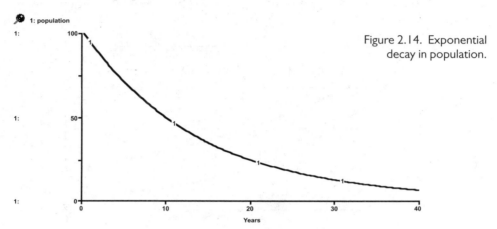

Figure 2.14. Exponential decay in population.

Sensitivity Analysis

To learn more about the importance of the death rate, we can conduct several simulations with different values. Such simulations are called *sensitivity analysis* because they teach us whether the overall results are sensitive to changes in this parameter. To illustrate, let's create simulations with the death rate set to 9%/yr, 12.5%/yr, and 16%/yr . Go to Stella's Run command and drag down to Sensi Specs. Select the death rate, and ask for three simulations with incremental variations from a low of 0.09 to a high of 0.16. Then go to the Run command and drag down to S-Run. Figure 2.15 shows the results in a "comparative" graph. The first simulation ("1" on

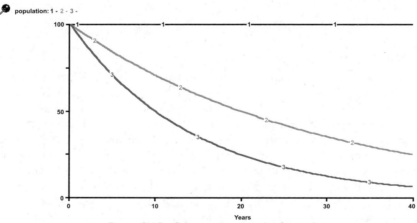

Figure 2.15. Comparative graph of population.

the graph) assumes annual deaths of 9%. This is the same as the birth rate, so the population remains constant at 100 million. The second simulation ("2" on the graph) uses 12.5% for the annual deaths, so we expect the population to decline at 3.5%/yr. You know from appendix B that the half-life will be 20 years. Figure 2.15 confirms that in this second simulation, the population falls to 50 million in the first two decades and then to 25 million in the next two decades. The third simulation ("3" on the graph) uses a death rate of 16%/yr, the value simulated previously.

Figure 2.16 shows the corresponding results of three Vensim simulations of the population model. Vensim allows us to name the results. Rather than sticking with the default name ("current"), I have assigned the names "1st run," "2nd run," and "3rd run." The three values of the death rate are the same as before. A new graph has been created to show the three results, with the vertical axis scaled from 0 to 100.

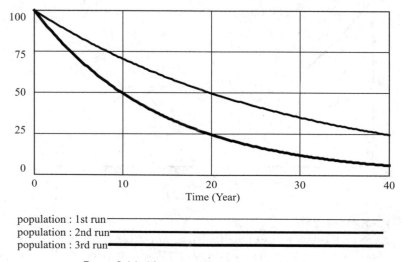

Figure 2.16. Vensim analysis of population decay.

Sensitivity analysis is a standard part of almost all modeling analyses. I recommend sensitivity analysis as the seventh of the eight steps in table 1.2. Models are constructed and tested in an iterative fashion, so sensitivity analyses will be conducted many times during the modeling process. You'll learn more about sensitivity analysis throughout the book, especially in chapter 21 and appendix D.

Nonlinear Relationships

Environmental systems are highly nonlinear, and Stella and Vensim make it easy to include nonlinear relationships. To illustrate, let's assume that the bank pays a higher interest rate if we have a higher balance in our account. The minimum rate is 4%/yr. But if we can build our balance, the interest rate will increase following the pattern in figure 2.17. Readers with a good command of algebra might go to work on a formula that will calculate the new interest rate as a function of the bank balance. But there is no need to devise a complicated formula. The better approach is to take advantage of Stella's graphical function, as depicted in figure 2.18.

To recreate figure 2.18 on your computer, click on the Model tab and open the converter for interest rate. Stella wants you to use the bank balance, so click on the bank balance to make it appear in the equation window. Then click on the "become graphical" function button. Ask for six data points and set the scale on the horizontal axis from 0 to 5000. Set the vertical scale from 0 to 0.08. Then enter six values of the interest rate to match the bank policy. The graphical function provides an opportunity for a visual check on your work by drawing the relationship as the values are entered. The simulation results also provide a check on your work, as you will see in the exercises.

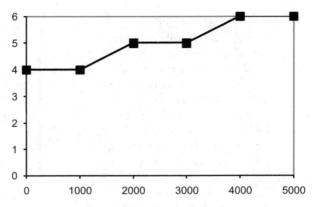

Figure 2.17. Interest rate policy.

These examples (and the exercises that follow) are sufficient to help you get started with the software. We'll return to Stella and Vensim in chapter 14. There you'll find that both programs provide many advanced features to expand the power of modeling beyond what you have seen here.

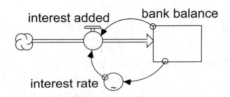

Figure 2.18. Graphical function for the interest rate.

Exercises

Exercise 2.1. Verify

Build the model in figure 2.18 with the new interest rate policy. Set the initial bank balance to 500 and run the model for 80 years. Document your results with a "time series" graph to match the results in figure 2.19.

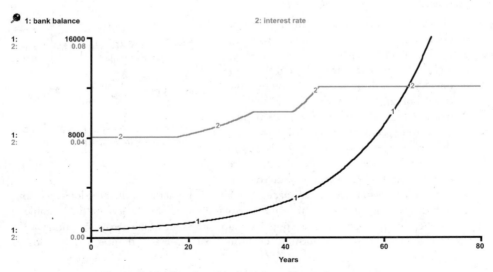

Figure 2.19. Results with the new policy on interest rate.

Exercise 2.2. Scatter graph

Figure 2.20 shows a "scatter graph" that provides a check on the interest rate policy. Select the bank balance to be on the *x* (horizontal) axis and the interest rate to be on the *y* (vertical) axis. Ask for thick lines and set the *x-y* scales to match figure 2.20. Run the model for 80 years, and the results should trace out a pattern of dots to confirm that the interest rate is consistent with the policy shown in figure 2.17.

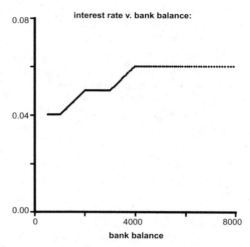

Figure 2.20. Scatter graph.

Exercise 2.3. The graph pad

Stella allows you to store many graphs on the same graph pad. (Think of a pad of paper, with a new graph on each sheet in the pad.) To learn about this feature, open the graph pad from exercise 2.1 and ask for a new page in the pad. Use the new page for the scatter graph in figure 2.20.

Exercise 2.4. Discontinuous graph function

The interest rate policy in figure 2.17 changes gradually with changes in the bank balance. But what if the interest rate were to change abruptly when the balance reaches certain thresholds? Click on the graph icon in the lower-left corner of the graphical function box. Notice that the interest rate changes abruptly at the $2,000 threshold and the $4,000 threshold. Run the new model for 80 years, and document your results with graphs that correspond to figures 2.19 and 2.20.

Exercise 2.5. Vensim graphical lookup

Figure 2.21 shows how the nonlinear relationship for the interest rate policy would be implemented in Vensim. Vensim uses a separate variable to hold the values of the nonlinear relationship. These are assigned the type "lookup," and I normally put "lookup" in their name to remind me of their purpose. You specify the scales for display on the two axes, and Vensim draws the graph as you make the entries. These entries are evenly spaced in this example, but you can

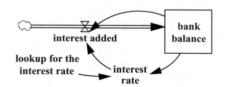

Figure 2.21. Vensim uses a lookup for the interest rate.

define the lookup with uneven spacing on the horizontal axis. When the lookup is completed, open the interest rate to write the equation. Click on the lookup (in "variables"), and Vensim will place the name of the lookup at the start of the equation window. Then enter a left (opening) parenthesis, click on *bank balance*, and enter a right (closing) parenthesis.

Exercise 2.6. Verify that the Vensim lookup works properly

Simulate the model from exercise 2.5 over 80 years, and document your results with the custom graph shown in figure 2.22. Select the bank balance to be scaled from 0 to 16000, and the interest rate to be scaled from 0 to 0.08. And, to match the graph exactly, ask Vensim to use four divisions on both axes.

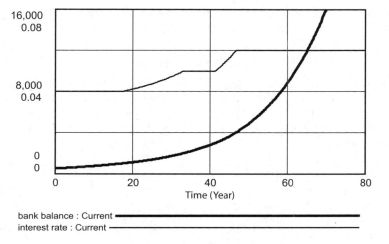

Figure 2.22. Vensim results with a variable interest rate.

Box 2.1. When do I get to write code?

Joe raises his hand to ask about programming. His roommate told him that the top programmers earn big money if they can produce pages of code each day. Joe has done all the exercises so far, but he worries that he has not produced a single page of code.

This question comes up frequently when students have heard about writing code with programs such as Fortran. But system dynamics modeling is not about writing code. Indeed, when you talk to practitioners, they will explain that they spend most of their time away from the computer (e.g., talking with people about the nature of the dynamic problem). When they do go to the computer, they rely on Stella or Vensim to build and test their models. The equations provide the closest correspondence to the code written in programming languages. But Joe has done all the exercises so far, and he hasn't seen a list of equations. That's quite revealing, for it shows that we don't need to see the equations to build and test models. But the equations do exist, as you can verify in the concluding exercises.

Exercise 2.7. View the Stella equations

To see the equations in Stella, click on the Equations tab (located just below the Model tab). You'll recognize the variable names, and you can guess that t stands for time. (The dt stands for the small step in time as we proceed through a numerical simulation, as explained in chapter 4.) Print the equations from exercise 2.1 to verify that they match table 2.1.

Table 2.1. Stella equations from exercise 2.1.

bank_balance (t) = bank_balance(t – dt) + (interest_added) * dt
INIT bank_balance = 500
INFLOWS:
interest_added = bank_balance*interest_rate
interest_rate = GRAPH(bank_balance)
(0.00, 0.04), (1000, 0.04), (2000, 0.05), (3000, 0.05), (4000, 0.06), (5000, 0.06)

Exercise 2.8. Experiment with a poorly formulated model

Figure 2.23 shows a model with an initial population of 800 million people. The births are constant at 100 million/yr. The deaths are constant at 200 million/yr. Build this model and verify the results shown here. (Use the pin in the upper-left corner of the graph to pin the graph to the screen.) The graph shows that the population declines in a linear manner, reaching zero by the end of the eighth year.

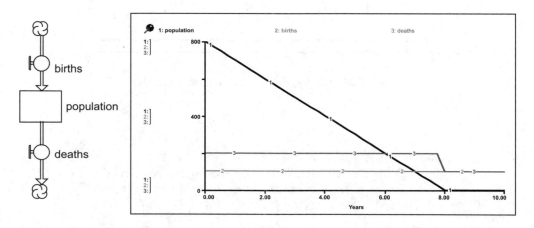

Figure 2.23. Population model with a surprising change in deaths in the eighth year.

At this point, the population should become negative, because the number of deaths is fixed at 200 million/yr, no matter what the size of the population. But the graph shows an abrupt change in deaths at the end of the eighth year. For some reason, the software decides to override our equation and changes the value to 100 million/yr. The population then remains at zero for the rest of the simulation.

Exercise 2.9. Turn off the non-negative option

Click on the population stock and look for the toggle switch for making "population" a non-negative stock. If you see a check mark, the software will not allow the stock to become negative. It does this by ignoring our equation for the outflow, which is about to drive the stock below zero. At first glance, this might seem like a useful thing to do. After all, the population cannot become negative, so isn't it nice of the software developers to have written a rule to replace our equation. In my view, this is the last thing we need. Covering up bad results will simply make the modeling process more difficult. We need to see the bad results as soon as possible so we can reformulate the equation that is driving the stock negative. Click the "non-negative" toggle to turn off this option. Then rerun the model from figure 2.23. Are the bad results visible? What is the population at the end of the simulation?

Exercise 2.10. Add a death rate

Suppose the deaths are based on a death rate of 25%/yr. This rate would create 100 million deaths/yr at the start of the simulation, but deaths would decrease over time, as there are fewer and fewer people remaining. Add a converter for the death rate to the model in figure 2.23, and set the deaths to the product of the death rate and the population. Simulate the model for 20 years. What is the population at the end of the simulation? Will you ever see a negative population in this model?

Chapter 3

Stocks and Flows:
The Building Blocks of System
Dynamics Models

The best way to construct a model is to start with the stocks, add the flows, and then use converters to explain the flows. Population models provide good examples, since we are all familiar with births and deaths. A population model with these two flows is shown in figure 3.1. A single stock represents the size of the population. Births and deaths are the only flows into and out of the population (we are ignoring migration). The flows are represented by double lines that depict the flow of material in and out of the stock. In this case, the material is people, but flows can represent any material that is accumulating in a stock. You'll see flows of water, salmon, insects, carbon, cash, and vehicles, to name a few examples. We use the stocks to represent the present state of the system and the flows to represent the actions that change the state over time. It will take some time for the flows to have their effect on the stocks, so the stocks tend to change more slowly over time. Stocks accumulate the effect of the flows, and they will remain at their current position if there are no flows acting on them. For example, suppose births and deaths were to suddenly go to zero for a month. The population at the end of the month would be the same as at the start of the month.

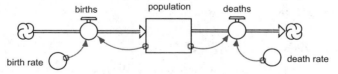

Figure 3.1. Population model with one stock.

The only way a stock may be changed is by the action of the flows. In the case of figure 3.1, the only way the population may change is by the action of births or deaths. You might be wondering about other factors that must have an effect on the population. What about the attitudes on family size? Don't these affect the size of the population? Changing attitudes are certainly important, and they do change the size of the population. Their effect is represented by the way they change the births or deaths. If we thought attitudes were shifting toward

31

smaller family sizes, for example, we would expand figure 3.1 to explain the reduction in the birth rate. This would then lead to a reduction in births and a subsequent impact on the population.

Figure 3.2 shows another population model, but this one does not assign a stock to the total population. Rather, it uses three stocks to keep track of the young, mature, and elderly people. The flows are clear to us because we are familiar with births, deaths, maturation, and aging. This model also calculates the total population, but not as a stock. It is now represented by a converter that is the sum of the three stocks. (The + sign in the diagram means Stella's "summer" adds the three stocks without cluttering the diagram with the information connectors from the three stocks.) The new model includes three stocks to allow for a clearer picture of the age structure.

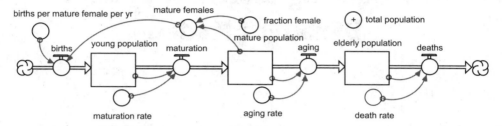

Figure 3.2. Population model with three stocks.

The new model raises the question about the need for many more stocks. It could be useful to assign separate stocks to the males and females and to assign more stocks to keep track of their aging. Perhaps we should use 5-year increments or even 1-year increments. Demographers often show charts of populations in 5-year increments to age 80+ years. A popular form is a bar chart with the number of males on one side and females on the other (see the book's website, the BWeb). Rapidly growing populations have a much higher proportion of young people, and the charts exhibit a "pyramid" shape. It we want to match the pattern in these charts, we would need to add more stocks. With 5-year increments to 80+ years, we would need 17 stocks for the males and another 17 stocks for the females.

So, we have talked about population models with 1 stock, with 3 stocks, and with 34 stocks. How many do we really need? The answer is entirely up to you. You pick the number of stocks to match the dynamic problem under study. My advice is to start with a small number and simulate the model to see what you learn from a simple approach. You can be surprised by the dynamics from a model with only a few stocks.

Why Start with the Stocks?

The stocks are the best place to start because they are the most easily recognized variables in the system. They stand out in plain view. To illustrate, try an experiment. Before reading further, close the book and look at the front cover. Then list three things that are in plain view in the photograph.

If you are like some of my students, your list will include snow in the mountains. You can see the snow in plain view. Figure 3.3 shows snowpack as a stock. It is fed by snowfall and drained by snowmelt. These flows make sense, but take a close look at the photo and ask yourself if you can see these flows. The flows will be familiar to us, but they do not stand out like the stocks. Other students will list the tufa towers, which are prominent in the foreground. The towers are formed by the flow of fresh water from springs, as explained on the book's website (BWeb). Figure 3.4 shows that the mass of tufa towers is reduced by the wave action and undercutting. Once again, the stock variable is in plain view, but the flows are not as apparent.

A third example from the cover photo is the massive Sierra Nevada in the background. The mass of igneous rock could be represented by a stock, as shown in figure 3.5. The process of weathering reduces the mass of rock and adds to the mass of sediments in the photo. Weathering takes place over hundreds of millions of years, so it is not visible in the photo.

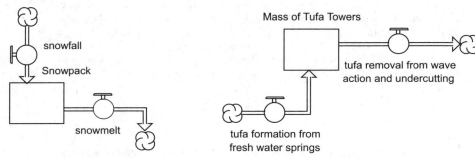

Figure 3.3. Snowpack is a stock.   Figure 3.4. Tufa is a stock.

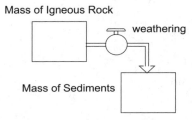

Figure 3.5. Rock and sediments
are stocks.

Stocks are more visible in the world around us, so they are the best place to start. Each stock should be assigned a well-understood name and a well-understood unit of measure. The flows represent the change in the stocks over time, and their units of measure are determined once we select a measure of time. But what if you can't think of the units for the stock? This is a sign that you are not sufficiently familiar with the system. Rather than forging ahead with model construction, it would be better to return to step 1 in table 1.2 and get better acquainted with the system.

The stocks are easily recognized, but the flows may not be in plain view. Understanding the flows requires knowledge of the hydrologic, ecological, or geologic processes. The flows may change rapidly during the time period of interest, but the stocks respond more slowly. They can only change because of the accumulated effect of the flows. If the flows were to go to zero, the stock would remain at its current position. If there were no snowfall or snowmelt, for example, the snowpack would remain at its current value.

You may also think of stocks as providing the storage in a system. D. H. Meadows (2009, 23) notes that storage often acts as a shock absorber or a buffer to protect against rapidly changing flows. Figure 3.6 shows an example of a storage reservoir that may have been constructed to buffer the flows in a river. For example, upstream flows may be highly variable, but the flow at the dam is regulated to create steady flows downstream.

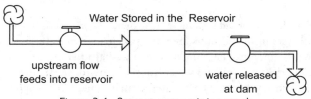

Figure 3.6. Storage reservoir is a stock.

Box 3.1. Joe wonders if long names are a problem

Joe raises his hand to ask about long names like "water released at dam." The long names are troubling him. How can this be mathematical modeling when the diagrams do not resemble what he has seen in previous math classes? Joe remembers almost every math problem started out with a definition of individual letters (e.g., "Let x stand for the amount of water stored in the reservoir"). Sometimes there would be a list of 5 to 10 variables, each with its own letter. And Joe recalls that there were often a lot of Greek letters to represent unknown variables. He's worried that the modeling so far looks nothing like the models in his previous math classes. Joe does not need to worry; the software will have no problem remembering the long names. The names in this book are selected to make the meaning of the stocks and flows clear. We are not working with differential equations, so there is no reason to use individual letters for the variables. To understand difference in naming conventions, look at the table 3.1 description of a differential equation and its analytical solution.

Table 3.1. The model in figure 3.1 represented as a first-order differential equation.

let t = time
let P = population
let P_0 = initial population
let b = births
let d = deaths
let br = birth rate
let dr = death rate
let r = growth rate = br − dr, then dP/dt = b − d = brP − drP

so:

$$dP/dt = (br − dr)P = rP$$

The analytical solution to this equation can be found by trial and error (BWeb):

$$P(t) = P_0 e^{rt}$$

Sources and Sinks

The clouds in the diagrams may be viewed as sources of the flows or sinks for the flows. Systems with sources and sinks are sometimes called *open systems*, since there is flow of material across the boundary. *Closed systems* do not have flow of material across their boundary. The Earth is a prime example. There is certainly energy flow to the Earth, but the Earth is closed in so far as materials and nutrients are concerned. The Earth's materials must be used and recycled again and again to sustain life. This cycling of materials and nutrients is accomplished in biogeochemical cycles such as the carbon cycle and the nitrogen cycle.

The most familiar global cycle is the hydrologic cycle. Figure 3.7 shows this cycle with four stocks to represent the water stored on Earth. The vast majority of the water is stored in the oceans. Much less is stored on land. An extremely small but important amount is stored in the atmosphere.

You can tell that this is a closed system because there are no clouds in the diagram. Most of the environmental systems in this book are open systems. There is both material and energy flowing across their boundary. The human body is perhaps the most familiar example of an open system. We take in air, food, and water from our environment, and we put exhaled air, perspiration, urine, and other matter back into the environment. Although the human body is

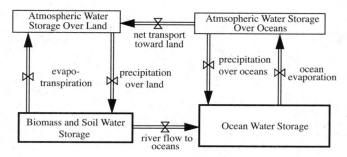

Figure 3.7. The hydrologic cycle.

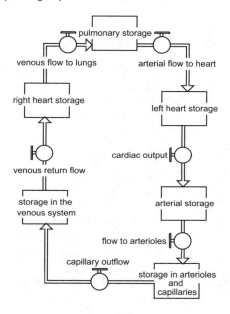

Figure 3.8. The human circulatory system.

an open system, we may elect to represent it as a closed system for the purpose of modeling. Figure 3.8 shows an example for the human circulatory system. The typical body contains around 5 liters of blood, with the majority stored in the venous system. Surprisingly, only a small amount is stored in the arterioles and capillaries, where the nutrient transfer with cells occurs.

How Do We Explain the Flows?

Suppose you have identified the stocks and flows, and you have specified their units of measure. The next step is to explain the flows. At this stage, you should brace yourself for a new challenge: explaining the size of a flow is a lot more difficult than giving it a name and assigning it a unit of measure. Just think of the flows shown so far. How do we explain snowfall and snowmelt? How do we explain the operator's rule for releasing water at the dam? What is nature's rule for evaporation from the ocean? What is the body's rule for controlling cardiac output? Answering such questions requires fundamental knowledge of the processes at work in the system. This book will not provide you with such knowledge. You will bring such knowledge to the process of modeling, and the modeling principles in this book will help you incorporate your knowledge into a new way of viewing the system.

Exercises

Exercise 3.1. Find the errors

Each of the diagrams in figure 3.9 has an error that violates a fundamental principle about stocks and flows. The error is so obvious that Stella will not let you make the error. (I had to use a separate drawing program to create the error.) Identify each error by marking an X through the erroneous part of the diagram.

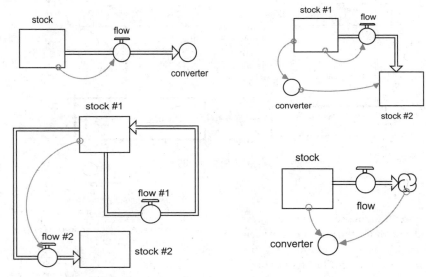

Figure 3.9. Erroneous diagrams.

Exercise 3.2. A good start?

Figure 3.10 shows a model to find the number of students in a college. Stella will let you build this model, and you can add the number of freshmen, sophomores, juniors, and seniors to get the total for the university. Explain whether this is a good way to start a system dynamics model of the student population.

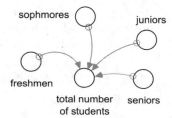

Figure 3.10. A good way to start?

Exercise 3.3. Units for Weathering?

Suppose the mass of igneous rock in figure 3.5 is measured in billions of metric tons. If time is measured in hundreds of millions of years, what are the units for weathering? What are the units for the mass of sediments?

Exercise 3.4. What are the units?

Figure 3.11 shows the flow of petroleum products starting with production in Kuwait and ending with consumption. The stocks act as buffers to help the industry and the consumers ride out the unpredictable variations in supply and demand. The units for time are months, and the units for Kuwait reserves are barrels of crude oil. What are the units for the other variables?

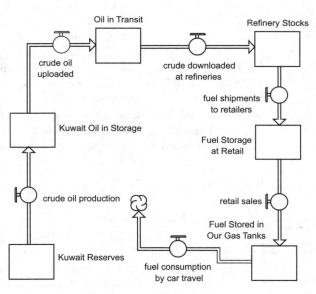

Figure 3.11. What are the units?

Exercise 3.5. The biflow

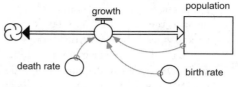

Figure 3.12 shows a biflow, a flow with two arrowheads. The unshaded arrowhead shows the positive direction (when the growth is positive, the population will increase). The black arrowhead shows the negative direction (when the growth is negative, the population will decrease). Write the equation for the growth.

Figure 3.12. Growth is a biflow.

Exercise 3.6. Simulate the biflow model

Build the model in figure 3.12 and set the initial population to 800, the death rate to 0.10, and the birth rate to 0.03. Simulate it for 30 years. Can you guess the size of the population at the end of the simulation? (If not, review appendix B.) Simulate the model and compare with your guess.

Exercise 3.7. Which model is easier to understand?

Build the model in figure 3.1 and assign the same starting value, birth rate, and death rate. Then simulate the model for 30 years. You should see the same results as in exercise 3.6. Which model is easier to understand, figure 3.1 or figure 3.12?

Exercise 3.8. A better way to start

Reorganize the variables in figure 3.10 into a new model of the student population. Use four stocks to keep track of the number of freshmen, sophomores, juniors, and seniors. Include flows to move the students from one stock to another. Then add a converter to keep track of the total number of students.

Exercise 3.9. Flows that drain a stock

Look at the maturation, aging, and deaths flows in figure 3.2. Each flow drains a stock, and there is an "action connector" from the stock to the outgoing flow. These connections tell us that the size of the stock influences the outgoing flow. Explain why this pattern is to be expected in any well-formulated model.

Exercise 3.10. Help Joe finish the model in figure 3.13.

Joe's description of how he knew when his gas tank was full (see box 3.2) may not be precise, but he has provided valuable information that can be used to represent the feedback in this system.

Let's define a "sound index" to take on three values to match Joe's description: *1* stands for free flowing, *2* means some congestion, and *3* stands for running out of space. Now suppose we have controlled experiments to learn that the free-flowing sound occurs with fullness between zero and 80%. Some congestion occurs when the fullness changes from 80% to 90%. And the running-out-of-space sound occurs when the fullness is around 90% to 95%. Suppose Joe says he leaves the nozzle in the full-open position as

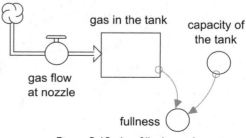

Figure 3.13. Joe fills the tank.

long as he hears the first sound. He cuts the nozzle position approximately in half when he hears the second sound. (This cuts the flow approximately in half.) He cuts the flow to zero when he hears the third sound.

Expand the flow diagram to include the sound index and the changes in the gas flow at the nozzle. Write the equations and simulate the model for 40 seconds. Document your work with (1) the new flow diagram; (2) a list of equations; and (3) a time graph of the sound index, fullness, and gas flow at the nozzle.

Box 3.2. Joe fills the tank

Our friend Joe pulls into the gas station with the tank nearly empty. Figure 3.13 shows his first cut at a model to simulate the gasoline in the tank during refueling. The stock is measured in gallons, and the flow is in gallons/second. The capacity is 20 gallons, and the fullness is the ratio of gas in the tank to the capacity. The maximum flow is 1 gal/sec, so Joe expects the simulation to run for 20 or 30 seconds. The diagram is a good start, but something is missing. There has to be feedback from the fullness to the gas flow at the nozzle. Without information feedback, there would be no control. How would you complete this model? Some students start by looking through all the examples in this chapter, but let's try a different approach. Let's talk to Joe and find out what he was actually doing when he filled the tank. Imagine that you have interviewed Joe, and your notes are below.

My first idea was that Joe watched the meter on the gas pump and lowered the flow when the meter got close to 20 gallons. But he said he couldn't see the meter without his glasses. My next idea was that he used a pump with an automatic shutoff, but he said he used a manual pump. Perhaps he has learned by experience that it takes around 20 seconds, and he counted the seconds before releasing the nozzle. But he shook his head and said he had no idea how long it takes to fill the tank.

Joe saw that we weren't getting anywhere, so he explained that he listened to the sound coming from the gas tank. I pressed for details, but he wasn't very precise about the nature of the sound. There was a "free-flowing" sound when he started. All he could say was that the sound left him with the impression that the gas was flowing easily into the tank. After a time, the sound changed, which made him think there was congestion. And quickly thereafter, the sound changed again, and Joe feared that there was little space left. So he shut off the nozzle, and he was done filling the tank.

Chapter 4

Accumulating the Flows

System dynamics models are constructed as a combination of stocks and flows and then simulated on the computer. The simulation results are generated in a step-by-step fashion by accumulating the effect of the flows. This chapter illustrates the accumulation process with examples of carbon dioxide in the atmosphere and water in a reservoir. These are simple examples, so you will be able to accumulate the flows in a simple table (or in a graph) without the aid of a computer. The accuracy of accumulating the flows in a system dynamics model is illustrated with a population model. You will see that accuracy is easily achieved if we specify a sufficiently short step size for the simulations. The chapter concludes with advice on setting the step size for your model.

Numerical Accumulation of Carbon Dioxide in the Atmosphere

Let's start with a portion of the global carbon cycle. Chapter 23 explains that the growing emissions of carbon dioxide (CO_2) could cause the CO_2 in the atmosphere to double in this century. We'll demonstrate how this can happen by a pencil-and-paper calculation with the model in figure 4.1. The model adds CO_2 to the atmosphere from anthropogenic (human-made) emissions. CO_2 is removed from the atmosphere

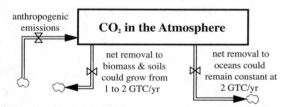

Figure 4.1. A model of CO_2 in the atmosphere.

by the net exchanges between the terrestrial and oceanic systems. The net removals are represented by two flows with long names to remind us of the assumptions adopted for this example. CO_2 in the atmosphere is a stock measured in gigatons of carbon (GTC). (The C stands for the C in CO_2.) The starting value is 750 GTC in the year 2000. Anthropogenic emissions (due mainly to the combustion of fossil fuels) add to the stock of CO_2 in the atmosphere. We start with anthropogenic emissions of 6.4 GTC/yr for the first decade of the exercise.

Emissions increase each decade owing to the growth in fossil fuel combustion. The net removal by the terrestrial system is 1 GTC/yr during the first decade. Some climate modelers predict that net removal will grow over time, so net removal increases in each decade during the first half of the century. Net removal remains at 2 GTC/yr for the second half of the century. Net removal by the oceans is fixed at 2 GTC/yr for the entire century.

Table 4.1. Complete this table to find the CO_2 in the atmosphere at the end of the century.

	Anthropogenic CO_2 emissions (GTC/yr)	Net CO_2 removal by terrestrial (GTC/yr)	Net CO_2 removal by oceans (GTC/yr)	CO_2 in atmosphere (GTC)
2000				750.0
2001–2010	6.4	1.0	2.0	784.0
2011–2020	7.0	1.3	2.0	
2021–2030	8.0	1.6	2.0	
2031–2040	9.0	1.8	2.0	
2041–2050	10.0	1.9	2.0	
2051–2060	11.3	2.0	2.0	
2061–2070	12.7	2.0	2.0	
2071–2080	14.2	2.0	2.0	
2081–2090	16.0	2.0	2.0	
2091–2100	18.0	2.0	2.0	

Table 4.1 is designed for a step-by-step calculation of the cumulative effects of the three flows. We will proceed one decade at a time. Anthropogenic emissions are growing continuously over time owing to the increase in the global combustion of fossil fuels, but the table shows average values for each decade. The second row shows the accumulation during the decade 2001–2010. The average value of emissions is 6.4 GTC/yr, so 64 GTC would be added to the stock during this decade. The net removal is 3 GTC/yr, so 30 GTC would be removed from the atmosphere during this decade. The new value of the stock would be 750 + 64−30, which is 784.

I've entered this value to get you started. To appreciate accumulation, you should complete the rest of the table. If you do the calculations properly, you will arrive at 1,500 GTC of CO_2 in the atmosphere at the end of the table. Atmospheric CO_2 will double in one century. You'll read in chapter 23 that a doubling of CO_2 is predicted by many of the climate modeling teams around the world, and the doubling can arise from the general assumptions shown here. So this is an important projection that we can anticipate by simply accumulating the flows in a step-by-step manner. This process is called *numerical simulation*.

Numerical Simulation and the Step Size

Stella and Vensim simulate models in a step-by-step manner, similar to the calculations in figure 4.1. The calculations didn't take long because you only needed to update the stock every decade. The "step size" of the calculations was 10 years, so you performed the calculations 10 times. If I had provided values for every year, you could follow the same approach with a 1-year step size. Your calculations would be done in 100 steps. If the flows were changing during the course of a single year, a smaller step size would be appropriate. If the step size were a quarter of a year, your calculations would have been completed in 400 steps. Stella and Vensim perform such calculations quickly and accurately. The term *numerical simulation* is used because the results are found by making simple numerical updates in the stocks. Once we have specified the model, our job is easy. We pick the step size and click the "run" button, and the software performs the numerical simulation.

Visual Accumulation of Water in a Reservoir

Figure 4.2 shows a model to accumulate water in a reservoir. The storage is measured in thousands of acre-feet, which we abbreviate as KAF. The model runs in months, so the flows are measured in KAF per month. The purpose of the reservoir is to deliver a constant outflow of 5 KAF/month.

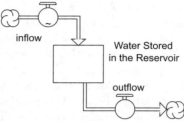

The inflow is highly variable, as shown in the graph in figure 4.3. Inflow peaks in the fifth month and dips to the lowest point in the eighth month. The total inflow over the entire year is 60 KAF, so the average inflow is 5 KAF/month. This is the same as the outflow, so we know the water stored in the reservoir at the end of the year will be the same as at the start of the year. But the reservoir will gain and lose storage within the year. I haven't given the starting value of the storage, and the graph does not give the exact numbers for the inflows. So you cannot perform a numerical calculation as you did in table 4.1. But you should be able to visualize how the reservoir storage will rise and fall over time. Give this a try and explain:

Figure 4.2. Water in a reservoir.

- When will the reservoir hold the most water?
- When will it hold the least water?

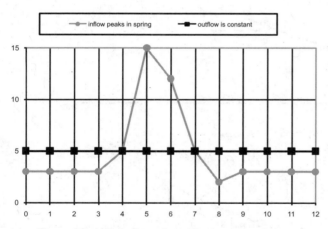

Figure 4.3. Water flows for a 12-month simulation.

The answers may be found on the book's website, the BWeb. Previous students' work on this (and related) exercises has revealed a wide disparity in their ability to think about the accumulation of flows over time. For one reason or another, many students misread the cumulative impact of the flows. One of the most frequent tendencies is to overstate the stock's responsiveness to the changing flows. On the other hand, some students do the accumulations correctly the first time, and many others can teach themselves about accumulation with practice. You can practice with the exercises at the end of the chapter. They will help you improve your understanding of accumulation. And better understanding will put you in a better position to anticipate the results of computer simulations.

Computer Simulation of Population Growth

Figure 4.4 shows a population model to illustrate computer simulation. It starts with 10 million people. The growth rate is 20%/yr, so the population will double to 20 million people in just 3.5 years (see appendix B). We will simulate the model for 5 years with DT (delta time, the increment of time between steps in the simulation) set to 1 year. Our job is to select the value of DT to give an accurate simulation. The population is growing continuously, so we would obtain the most accurate results with an small value of DT. The DT can be set to a fraction of a year, so we might pick 1/8 year to give accurate results. A 5-year simulation would then require 40 steps. To check the accuracy, we could set the DT to 1/16 year, and the new simulation will take 80 steps. We then compare the simulations. If we see essentially the same results, we know the original DT is sufficiently small to give numerical accuracy.

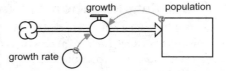

Figure 4.4. Population model.

Figure 4.5 shows the accuracy in simulations with DT set to 1 year in the top graph and 1/2 year in the lower graph. The accurate result is labeled *population* in both graphs. The staircase patterns show the numerical values that are updated with each DT. Each graph has a small circle to mark the spot with 20 million people after 3.5 years. (These are Stella's information buttons. They remain in view if the graph is pinned to the diagram.) It's clear that the population graphs are on target, but the step-by step pattern is lagging behind. The errors are larger in the top graph, and they are growing as the simulation proceeds.

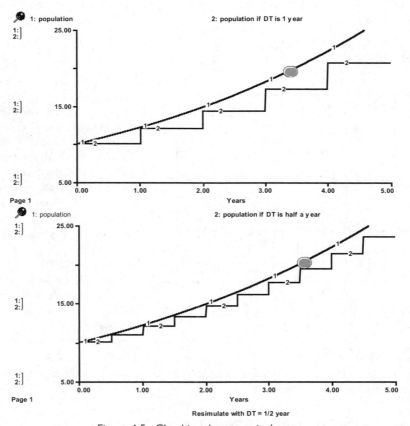

Figure 4.5. Checking the numerical accuracy.

The errors are smaller in the lower graph, but these errors are also growing larger over time. If you repeat these tests, you'll find that you get increasingly accurate results with smaller values of DT. In this example, there are vanishing small errors if DT is 1/16 year.

Advice on DT

The Stella software asks us to specify the units of time and to set the value of DT. The default value of DT is 0.25, with the units for DT the same as the units of time. In the population model, the default DT is 0.25 year. This turns out to be a good selection for most models, but it is not sufficiently small for the rapidly growing population in figure 4.5. With 20% annual growth in population, we need a smaller DT to keep pace with the upward trend. The same reasoning applies if the population were on a steep downward trend. If the change is rapid, we need to update the stocks frequently to keep up with the trend.

The easiest way to pick the value of DT is to draw a graph of the expected pattern over time. (We should have such a graph on hand, as step 2 of modeling [see table 1.2] calls for a "reference mode" graph.) To pick a value of DT, imagine a staircase pattern that follows the ups and downs in your graph. The width of the step in the staircase is a good guess at DT. You simply pick a sufficiently small DT so the staircase pattern follows your graph closely. Run the model with this value of DT. Then cut DT in half and run the model again. If you get essentially the same results, you have numerically accurate results.

Box 4.1. How close is close enough?

Joe understands that the simulations will be increasingly accurate as he cuts DT in half. But he isn't sure about the phrase "essentially the same results." Do all the results have to be close, or just the results at the end of the simulation? And how close is close enough? Do we call the results accurate if they are within 1% of the results in the previous simulation? This is one of those questions that can't be answered in the classroom, but the answer will be clear when you deal with a real problem. You will have specified a reference mode and will be striving for a general understanding of the problem. With this context, you will have no problem deciding whether the new simulation is "essentially the same" as the previous simulation with the larger DT. As you make this judgment, remember that we are modeling for general understanding of dynamic patterns. We are not making point predictions or forecasts of the future state of the system.

Final Suggestion

The final guideline for setting DT is to remember that

DT has no counterpart in the real world.

In other words, we should not be tempted to set the value of DT to match some time interval in the actual system. DT is the step size used in numerical integration; it has nothing to do with a time interval in the real world. When simulating automobile sales, for example, do not be tempted to set DT to 0.25 year if you happen to have quarterly data on sales. If you are simulating a watershed with monthly data on flows, do not feel that DT should be 1 month. DT must be set sufficiently small to ensure numerical accuracy. And to check the accuracy, we cut the value in half and repeat the simulation. We cannot follow these rules if we make the mistake of associating DT with some time interval in the real world.

The temptation to match DT to a real-world time interval occurs among those who envision the world advancing in discrete steps. (Perhaps crops are planted each spring, and the market prices are evaluated when they are harvested the following fall.) If you want a model to match this vision of the world, turn to different modeling methods (BWeb). System dynamics models are simulated on a continuous basis, with time advancing one small step at a time.

The temptation to set DT to match a real-world time interval can also arise when students see the system responding abruptly at a particular time each year. An example could be the sudden appearance of the spring Chinook salmon in the Columbia River each spring. We do not set DT to 1 year to represent such a pattern. It makes much more sense to use conveyor stocks, as explained in chapters 14 and 15.

Number of Steps for Numerical Accuracy

The examples in this chapter require a few hundred steps for the numerical simulations. The CO_2 model in table 4.1 requires only 100 steps if DT is 1 year. It would require 400 steps if DT were 1/4 year. However, suppose that the atmospheric CO_2 were changing much more rapidly, and that we changed DT to 1/16 year. We would now need 1,600 steps. I recommend that we think twice about simulations that require more than 1,000 steps. The BWeb discusses ways to avoid such simulations. It explains that the standard integration method is first-order integration, sometimes called *Euler integration.* The BWeb discusses higher-order integration methods to obtain faster simulations. As a general rule, however, we should not have to turn to higher-order methods. A better approach is to think carefully about the high-turnover stocks that are forcing us to select a small value of DT (as explained in chapter 17).

We'll talk more about DT and numerical accuracy later in the book. But at this point, you know enough to generate numerically accurate simulations. Pick a reasonable value of DT and simulate the model. Cut the DT in half and simulate again. If you get the same results, you have numerical accuracy. Remember that you should not confuse numerical accuracy with model realism. Numerical accuracy simply means that the simulation results are an accurate calculation based on your assumptions.

Exercises

Exercise 4.1. Bees in the hive

Figure 4.6 shows the flow of bees in and out of a hive. Figure 4.7 shows a time graph of the two flows over a 30-minute period.

Questions on the flows: During which minute did the most bees enter the hive? During which minute did the most bees leave the hive? *Questions on the stock*: During which minute were the most bees in the hive? During which minute were the fewest bees in the hive?

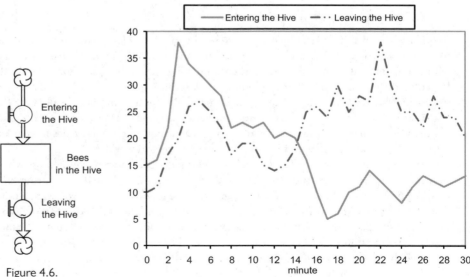

Figure 4.6.
Bees model.

Figure 4.7. Flows of bees in and out of the hive.

Exercise 4.2. Population sketch

Figure 4.8 shows a model of an animal population subject to births and deaths. There are 100 animals at the start of the simulation. Deaths are fixed at 50 animals/yr. The births are 100 animals/yr when time is 0.0 year (figure 4.9). At 2.0 years, births have fallen to 50 animals/yr. The births and deaths are occurring continuously over time. The dot in figure 4.10 marks the starting population at 100. Draw the population curve for the rest of the time interval.

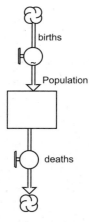

Figure 4.8.
Population model.

Exercise 4.3. Population simulation

Construct the model in figure 4.8. The ~ for births indicates that a graphical function has been used to make the births change over time. Open the births flow and type *time*. Then click the "become graphical" function button, and you will see a graphical box. Enter values for births to match the values in figure 4.9. Then simulate the model for 16 years with DT = 0.25 year. Does the population match your sketch?

Exercise 4.4. Stella's sketchable graph

Sketchable graphs are a device to encourage us to think the likely results before we run the model. Add a new graph to the model in exercise 4.3 and click on the sketchable option. This will create a graph of a selected variable that may be compared to your sketch of the expected behavior. Select population to appear on a scale from 0 to 200. Then drag the cursor across the graph to give a rough approximation to the pencil sketch from figure 4.10. Then simulate the model. You will see the results along with your sketch.

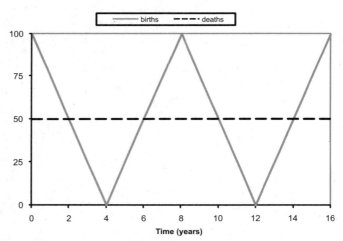

Figure 4.9. Births and deaths for accumulation exercise.

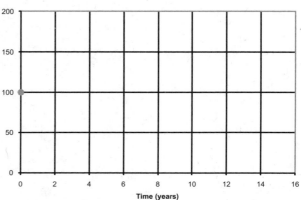

Figure 4.10. Sketch the population results.

Chapter 5

Water Flows in the Mono Basin

Mono Lake is an ancient inland sea on the eastern side of the Sierra Nevada, a land of stark contrasts and spectacular vistas. Volcanic islands rise in the middle of the lake, and tufa towers ring the edges. It is also an oasis for wildlife in the high-desert country. Migratory birds use the lake as a stopover, and nesting birds raise their chicks on the islands. The birds are drawn by a simple, but extraordinary ecosystem. Microscopic algae thrive in the lake, providing the food supply for brine shrimp and alkali flies. Brine shrimp and alkali flies are astoundingly prolific organisms that can provide virtually limitless food for birds under proper conditions.

Mono Lake was selected for the first case study because it offers important lessons for policy making as well as for modeling. From a policy point of view, Mono Lake is a story of how

> a handful of people began a campaign to save a dying lake, taking on not only the City of Los Angeles, but the entire state government by challenging the way we think about water. Their fight seemed doomed in the beginning, but long years of grassroots education and effort finally paid off in 1994, when the California Water Resources Control Board ruled that Los Angeles use of Mono Basin waters be restricted. Over time, the lake will return to a healthy condition. . . . The battle over Mono Lake is one of the longest and most fiercely contested conservation battles in U.S. history, and that rare one with a happy ending. (Hart 1996)

From a modeling point of view, Mono Lake is well suited to demonstrate the power of stock-and-flow modeling. We will be simulating the flows and accumulation of water in Mono Lake, so the stock-and-flow concepts will be easy to understand. By the end of this chapter, you will see a model that may be used to simulate changes in the lake level with different policies on the water exported to Los Angeles.

Background

The Mono Lake story began early in the century in Los Angeles, a city of around 100,000 people whose water supply was mostly from local wells. As the city grew, its water managers looked north to the Owens Valley for new sources of water. The Los Angeles Aqueduct was completed in 1913 under the direction of William Mulholland. But by 1930, the Los Angeles population had reached 1 million, and the city looked to even more distant water sources. By 1941, the Colorado River Aqueduct was completed, and the Los Angeles Aqueduct was extended north to the Mono Basin.

Diversions from the Mono Basin began in 1941. Stream flows that would normally reach the lake were diverted into a tunnel running beneath the Mono Craters and released into the basin to the south. From there, they flowed by gravity and siphons for nearly 400 miles to Los Angeles. By the 1970s, the diversions averaged around 100,000 acre-feet/yr (100 KAF/yr). The lake level was 6,417 feet above sea level in 1941 and held around 4.3 million acre-feet of water. Salinity, a crucial factor for the alkali flies and brine shrimp, was around 55 grams/liter (55 g/L). Then the diversions began, and the lake level declined steadily during the next four decades. By 1981, the lake's volume had been cut approximately in half, and its salinity had climbed to around 100 g/L. The lake level stood at 6,372 feet above sea level, 45 feet below its height when the diversions began.

Increased salinity can reduce algae production and lower the survivability of alkali flies and brine shrimp. When these herbivores decline in number, the nesting birds may not find adequate food to raise their chicks. A declining lake level poses other dangers as well. When the lake receded to 6,375 feet above sea level, for example, a land bridge emerged connecting Negit Island to the mainland (see the book's website, the BWeb), making a once secure nesting habitat vulnerable to predators.

Reversing the Course

Students from the University of California studying the lake during the 1970s were alarmed at what they found. They feared that higher salinity could lead to serious declines in the brine shrimp population and a subsequent loss of suitable habitat for the bird populations. In 1978, a group of students formed the Mono Lake Committee, a grassroots education and advocacy group. Mono Lake also drew the attention of the National Audubon Society, which filed suit against the City of Los Angeles in 1979. The California Supreme Court responded in 1983. It held that the public trust mandated reconsideration of the city's water rights. The court noted that Mono Lake is a scenic and ecological treasure of national significance and that the lake's value was diminished by a receding water level. The court issued an injunction to limit the city's diversions while the State Water Resources Control Board reviewed the city's water rights.

The control board considered a variety of alternatives. One extreme was the "no restriction" alternative, in which the city would be free to divert water as in the past. With no restrictions, the control board expected the lake to decline for another 50 to 100 years and reach a dynamic equilibrium at around 6,355 feet above sea level. The opposite extreme was the "no diversion" alternative. If all Mono Basin's streams were allowed to flow uninterrupted to the lake, the lake level would climb over a period of 100 years, eventually reaching 6,425 feet. The court decision was issued in 1994. It concluded that the appropriate balance between the city's water rights and the public trust would be served by allowing the lake to rebuild to the elevation of 6,392 feet. A small plaque next to the boardwalk to the northern shore marks this target elevation (BWeb). By May 2009, the lake level had climbed to over 6,382 feet, less than 10 feet below the target elevation.

Water Flows in the Basin

The first step in the modeling process is to become acquainted with the system and the problem. The BWeb provides additional information to supplement the short summary in this chapter. Figure 5.1 is a sketch of the main water flows in the basin. The principal flow into the basin is the runoff in five gauged streams that drain the Sierra Nevada.

Rush Creek (photo 5.1) is the largest of the five gauged streams. It delivers 60 KAF/yr in an average year. The other streams deliver a combined total of 90 KAF/yr. The City of Los Angeles operates diversion points to direct water that would normally flow to the lake into the Lee Vining Conduit for export from the basin. The diversions averaged 100 KAF/yr in the 1970s, leaving 50 KAF/yr to reach Mono Lake.

Figure 5.1 Water flows
in the Mono Basin.

Precipitation and evaporation can vary from year to year depending on the weather and the size of the lake. Evaporation is the largest flow out of the basin. The two small arrows at the bottom of figure 5.1 represent a collection of inflows and outflows that are small and that may not depend on the size of the lake. The overall balance of all of the flows in the basin is a net loss of approximately 50 KAF/yr.

Now imagine what would happen if this annual loss were sustained year after year. Diversions began in 1941. By 1961, the lake would have lost 1 million acre-feet. By 1981, the lake would have lost 2 million acre-feet. This simple arithmetic is sufficient to illustrate how a lake with around 4.3 million acre-feet when the diversions began could be cut approximately in half by the early 1980s. Let's turn now to computer simulation modeling to look at this problem in more detail.

Photo 5.1 Rush Creek flow.

Purpose of the Model

The second step in the modeling process is to specify the purpose of the model as clearly as possible. For this chapter, our purpose is to project the future size of the lake given different assumptions about the amount of water exported to Los Angeles. To be more specific, we should draw a sketch of an important variable over time. Let's start with a sketch that shows the volume of water if export were allowed to continue at the high values from the 1970s. Figure 5.2 is a sketch of the expected trend. It serves as a target pattern or reference mode for the modeling. The general shape is what is important, and this particular shape corresponds to the exponential approach, the third of the six dynamic patterns in chapter 1. Figure 5.2 begins in the year 1990 and extends to the year 2090. This long time period seems reasonable given the State Water Resources Control Board's assessment of the time needed for the lake to reach dynamic equilibrium.

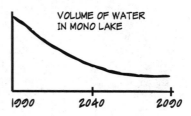

Figure 5.2 The reference mode.

The length of time on the horizontal axis is called the *time horizon* of the model. With a 100-year time horizon, we can decide on what to include and what to exclude from the model. Since we are looking out 100 years, we can ignore daily or monthly variations in the lake volume. Notice that the line shows a smooth decline in the volume of water in the lake. This means we are going to ignore seasonal variations. The shape of the reference mode also reminds us that we are ignoring year-to-year variations in the weather. One year may be particularly wet, the next particularly dry. These variations could be important in the short term, but we will ignore them in this model. For now, we will assume that every year is an average year as far as the hydrology is concerned.

One final feature of figure 5.2 should catch your attention—there are no numbers on the vertical axis. You might have wondered why we don't show the lake volume at around 2 million acre-feet at the start and then an estimate of the final volume in the year 2090. You should feel free to include or exclude numerical detail when sketching the reference mode. If the numbers are available and they improve communication among members of the team, include

---

Box 5.1 Joe asks why we need a target pattern

Joe doesn't like the idea of specifying a target pattern, and he wonders why we don't build the model first and see what the computer tells us about the dynamics. He fears we are biasing the results with a predetermined pattern. This is a natural reaction, and you may be feeling anxious as well. Don't let this feeling prevent you from this important step in the modeling process. You will make more progress if you start with a predetermined image of the expected dynamics. To illustrate with a model from our youth, think about the first time you built a paper airplane. You didn't build it blindly, thinking that you would learn its likely pattern of behavior once you threw it in the air. You built it with a predetermined image of its glide path through the air. Building a system dynamics model is similar: you start with an image of the dynamics, and that image guides your decisions on which variables to include and which to exclude from the model.

I have learned this lesson the hard way because my colleagues and I often jumped into model construction without drawing a reference mode. These experiences left us with large, overly complicated models. We eventually came to the disheartening conclusion that we didn't know why we were building the model. We concluded that modeling without a target pattern is like packing for a trip when you don't know where you are going: brace yourself for a lot of excess baggage.

them in the sketch. If not, feel free to leave them out. It's the general shape of the diagram that counts at this stage of the process.

List a Policy and Anticipate Its Impact

Specifying a policy variable at the outset is useful because it guides decisions on what to exclude from the model. For our purposes, there is one policy variable: the amount of water exported from the Mono Basin. Our purpose is to learn how the size of the lake varies with variations in water export. A time graph of the likely response to a policy change is helpful at this stage. Figure 5.3 shows an example. Let's suppose that the lake volume declines for 50 years owing to high export. And at this point, the problems of low volume and high salinity present overwhelming evidence of a threatened ecosystem. The dashed line represents the volume at which the salinity problems are fully evident. Now imagine that export is cut to zero, and that we wonder about the likely response of the lake. Figure 5.3 shows one student's pencil sketch of the possible response. She feared that the lake's downward momentum would lead to a continued decline for several years after the change in export. The sketch shows a delay before the volume changes in an upward direction and exceeds the danger level. The student worried that the lake might need 5 to 10 years to recover to a safe level, and she argued that this sluggish response means exports should be reduced in advance of the lake reaching the threshold. The model in this chapter will help us understand whether the lake would exhibit such a sluggish response.

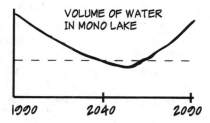

Figure 5.3. Sketch of possible response to cutting export to zero in 2040.

Iterative Modeling

Modeling is an iterative process of trial and error. The best approach is to begin with the simplest possible model that could explain the reference mode. You should build, test, and reflect on the first model before moving to a more complicated model. The worst thing you can do is try to build the perfect model from the start. You will make more progress in the long run if you start simply and learn as much as you can from each new model. Now, before reading ahead, ask yourself what you would do next. What are the stock variables? And what are the flows that directly influence those stocks? What units would you use to measure the stocks and flows? Then think of the converters you would use to explain the flows.

First Model of Mono Lake

Let's keep the first model simple by limiting ourselves to a single stock. Stocks represent storage, and the vast majority of the water in the basin is stored in Mono Lake. So it makes sense to assign a stock to the volume of water in the lake. The volume will be measured in thousands of acre-feet (KAF). The sketch in figure 5.1 shows three flows reaching the lake and two flows draining the lake. These flows are connected to the stock shown in figure 5.4. The model runs in years, so the units for each of the flows must be KAF/year. The area of the lake would be measured in thousands of acres. The precipitation rate and the evaporation rate will be measured in feet per year. The main flow into the basin is the runoff from the Sierra streams. The converter is named "Sierra gauged runoff" to match the name used by Vorster (1985). The export is the amount of water diverted into the tunnel under the Mono Craters to leave the basin. We subtract export from the gauged runoff to get the flow past the diversion points, the flow that actually reaches Mono Lake. The evaporation is the lake's surface area multiplied by

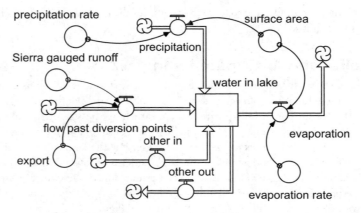

Figure 5.4. First model of Mono Lake.

the evaporation rate; the precipitation is the surface area multiplied by the precipitation rate. Let's set the area to 39,000 acres (39 K acres), the area in 1990. If the evaporation rate is 3.75 feet/yr, the evaporation from the surface of the lake would be just over 146 KAF/yr. The precipitation rate is 0.67 feet/yr, as shown in table 5.1. The final parameters are the "other in" and "other out" flows, which are explained on the BWeb.

Now, what do you expect to see from this initial model? Will it generate the reference mode, the pattern in figure 5.2? Figure 5.5 shows the simulation result. Flow past the diversion points is constant at 50 KAF/yr, and the evaporation is

Table 5.1. Parameter values.

Parameter	Value
Initial water in lake	2,228 KAF
Sierra gauged runoff	150 KAF/yr
Export	100 KAF/yr
Other in	47.6 KAF/yr
Other out	33.6 KAF/yr
Precipitation rate	0.67 feet/yr
Evaporation rate	3.75 feet/yr
Surface area	39 K acres

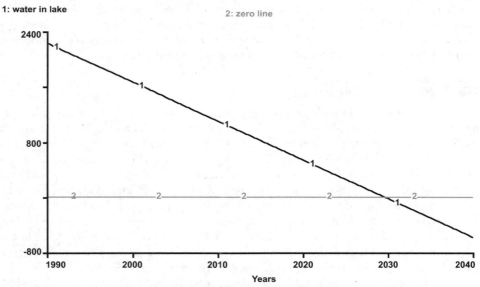

Figure 5.5. Simulation results from the first model.

constant at 146 KAF/yr. The water in the lake declines in a linear fashion throughout the simulation. Indeed, the volume reaches zero by the year 2030, and it continues to decline, going well below the "zero line" (included in the graph for emphasis). It makes no sense to see the water in the lake become negative, and some students are tempted to invoke the non-negative option on the stock variable. But you know from exercise 2.9 that this would be covering up the bad results.

At this stage, you might wonder if this clearly erroneous behavior is the result of the parameter values in table 5.1. Perhaps we should try a simulation with a different estimate of the precipitation rate or the evaporation rate. Do you think these experiments would change the linear pattern? Would they ever allow the model to generate the reference mode? Give these experiments a try, and you'll discover that this model will never generate the target pattern. No matter what parameters you try, the pattern will turn out to be either linear decline or linear increase. The problem with this first model is not the input parameters; it's the structure. It's time to expand the stock-and-flow diagram to improve the structure.

Second Model of Mono Lake

If Mono Lake were shaped like a cylinder, its surface area would remain constant as the volume declines. But the basin looks much more like a shallow cup than a cylinder. The lake's surface area tends to shrink as the volume declines. Let's change the surface area to an internal variable that will decline with a decline in the volume of water in the lake. The challenge is to represent the surface area as a function of the volume. The BWeb shows the surface area of the lake as it appears in a geographic information system (GIS) map of the Mono Basin. The GIS makes it clear that the surface area does not conform to a simple geometric shape (like a circle or an ellipse). There are irregular shorelines, islands, and other irregularities that mean that we cannot express the area through a simple geometric expression. Fortunately, geologic and bathymetric surveys (Vorster 1985, 261) provide the areas associated with different volumes of water in Mono Lake. The surveys are reported in table 5.2, with volume measured in thousands of acre-feet (KAF). The table also provides the elevation of the lake associated with each volume. At one extreme, the bottom of the lake is at 6,224 feet above sea

Table 5.2. Survey results.

Volume of water (KAF)	Surface area (K acres)	Elevation (feet above sea level)
0	0.0	6,224
1,000	24.7	6,335
2,000	35.3	6,369
3,000	48.6	6,392
4,000	54.3	6,412
5,000	57.2	6,430
6,000	61.6	6,447
7,000	66.0	6,463
8,000	69.8	6,477

level. At the other extreme, the lake would be at 6,477 feet if it were to hold 8,000 KAF. The surface area is measured in thousands of acres (K acres). With a volume of 2,000 KAF, for example, the surface area would be 35.3 K acres.

The elevation information is extremely useful since scientists use elevation as a proxy for a wide variety of conditions at the lake, as shown in figure 5.6. This diagram shows an elevation gauge of the lake level with expectations for different impacts marked along the sides. The formation of the Negit Island land bridge, for example, occurs when the elevation falls below 6,375 feet. The gauge shows the historical range at the top. The lightly shaded range marks

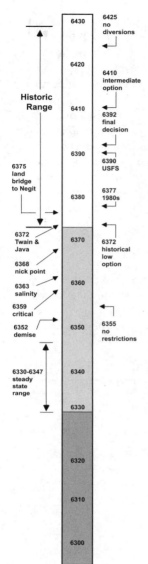

Figure 5.6. Mono Lake
elevation gauge.

elevations that would occur if high exports were to continue into the future. The dark shaded portion at the bottom represents elevations that are not expected because annual inflows would prevent the lake from falling into this zone. The short labels alert us to key elevations and the Control Board alternatives. More detailed information is provided on the BWeb.

Figure 5.7 shows a new model of Mono Lake to incorporate the survey results. We add an information connector from the water in the lake to the surface area. We then invoke "become a graph" with water in the lake on the horizontal axis. We ask for nine points on the horizontal axis and enter the nine values of surface area from table 5.2. We then add a new converter for the elevation, add the connection from water in the lake, and follow the same approach. Stella will assign the ~ symbols to remind us of the nonlinear graphs.

The results from the second model are shown in figure 5.8. They are certainly much different from the results of the first model. The surface area begins the simulation at 39 K acres and declines as the volume of water declines. The new pattern is just what we are looking for. There is a gradual decline of water in the lake, eventually reaching dynamic equilibrium. The decline in elevation is somewhat irregular owing to bathymetric irregularities (BWeb). By the end of the simulation, Mono Lake holds less than 1 million acre-feet of water, and its elevation is below 6,320 feet.

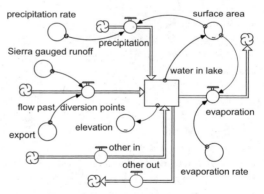

Figure 5.7. Second model of Mono Lake.

Checking the Model

One way to check the model is to set the model inputs to match time series information during the four decades of diversions. Our confidence in the model would be bolstered if it could simulate the 45-foot decline in lake elevation that occurred between 1941 and 1981. Another test would be to check yearly variations in the simulated size of the lake. These historical comparisons have been performed by Vorster (1985) to check his water balance model for the Mono Basin. His thesis is exemplary because of the careful documentation, the independent estimates of each flow, and the historical comparisons. Many of the parameters in this chapter are taken from Vorster's thesis, so we should expect to see similar results when our model is simulated under similar conditions. Vorster expected the lake to equilibrate at 6,335

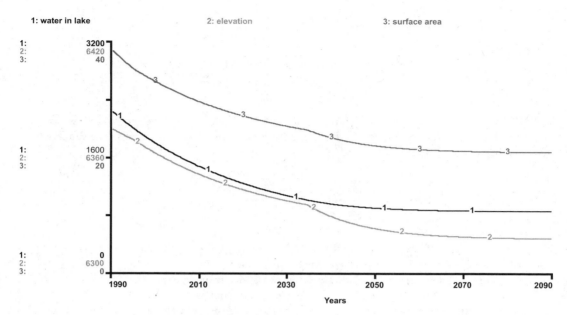

Figure 5.8. The second model matches the reference mode.

feet if the export remains at 100 KAF/yr. This result is more than 15 feet higher than the elevation at the end of the simulation in figure 5.8. Something is missing from the second model.

Third Model: Changing the Evaporation Rate

The missing factor is the change in the rate of evaporation as Mono Lake's water becomes more and more dense over time. The increased density arises from a fixed amount of dissolved solids held in solution in a shrinking volume of water. Highly saline waters tend to evaporate more slowly than fresh water owing to a reduction in the vapor pressure difference between the surface of the water and the overlying air. We can represent the change in evaporation by taking advantage of evaporation studies. These studies are sometimes called pan studies since pans of water (with different salinities) are subject to evaporation. The results are documented by Vorster (1985, 90) and tabulated in table 5.3. Column 1 of the table uses specific gravity to measure the density of the water. (A value of 1.1 means the lake water is 10% heavier than fresh water.) Column 2 reports the "effect" of higher salinity on evaporation. If Mono Lake's water is 10% heavier than fresh water, for example, its evaporation rate would be 92.6% of the rate for fresh water.

Table 5.3. Effect of specific gravity on the evaporation rate.

Specific gravity	Effect on the evaporation rate
1.00	1.000
1.05	0.963
1.10	0.926
1.15	0.880
1.20	0.833
1.25	0.785
1.30	0.737
1.35	0.688
1.40	0.640

The specific gravity is given by:

$$\text{specific\_gravity} = \frac{(\text{water\_in\_lake}*\text{density\_of\_fresh\_water} + \text{total\_dissolved\_solids})}{\text{water\_in\_lake}*\text{density\_of\_fresh\_water}}$$

Let's work through the numbers for the start of the simulation. The density of fresh water is 1.359 million tons per KAF, and the total dissolved solids is 230 million tons. The simulation begins with 2,228 KAF of water in the lake, so we expect to see:

$$\text{specific\_gravity} = \frac{2{,}228\_\text{KAF}*1.359\_\text{million\_tons} / \text{KAF} + 230\_\text{millions\_tons}}{2{,}228\_\text{KAF}*1.359\_\text{million\_tons} / \text{KAF}}$$

which turns out to be 1.076. Interpolating from table 5.3, we know that the evaporation rate will be around 94% of the evaporation rate for fresh water.

Figure 5.9 shows that the specific gravity is used to determine the specific gravity effect on the evaporation rate (~). The effect is 1.0 when the specific gravity is at 1.0, so the model would multiply the freshwater evaporation rate by 1.0 to get the actual evaporation rate. As the specific gravity increases, however, the effect will decline, as indicated in table 5.3. The evaporation rate is the product of this effect and the freshwater evaporation rate.

Figure 5.10 shows the results of the new model with export held constant at 100 KAF/yr. The time graph shows volume, surface area, elevation, and specific gravity. The specific gravity begins the simulation at around 1.076, as we expect. It ends the simulation at 1.163. The pattern for elevation is the same as before—a gradual decline, eventually reaching dynamic equilibrium. By the end of the simulation, the elevation is 6,336 feet above sea level, 16 feet higher than in the previous model and within a foot of the corresponding simulation by Vorster (1985, 225).

Box 5.2. The Friendly Algebra Pledge

Joe raises his hand to say he likes the look of the equation for specific gravity. It reminds him of the math he has seen before, and he asks why we haven't seen such formulas in any previous models. Does Stella have a problem with complicated algebra?

Stella and Vensim have no problem, but I recommend we avoid complicated expressions because they often get in the way of understanding. With simple algebra, everyone knows what the model is doing, and everyone can contribute to its development. The simple algebra is not just for the beginning chapters while you are learning; it's good practice for all modeling, especially modeling in groups. To encourage this practice, I adopt the Friendly Algebra Pledge for the entire book: the converters and flows will be a simple combination of the inputs; either add, subtract, multiply, or divide.

All good rules will be broken from time to time, and the first exception to this rule is the equation for specific gravity. The information button in figure 5.9 notes this exception to the rule. If you were viewing this model on the computer, you could click this button, and Stella would open a window with information on the equation and the reason for using a more complicated equation. In this particular case, the equation was written to match the equation for specific gravity used by Vorster (1985).

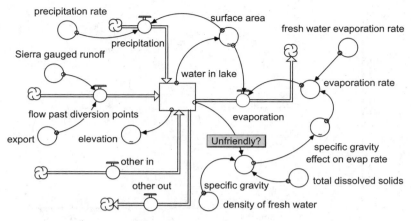

Figure 5.9. Third model of Mono Lake.

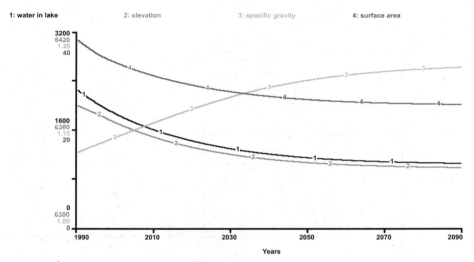

Figure 5.10. Results from the third model.

Policy Testing

Let's turn now to the policy question raised early in the chapter: How rapidly can Mono Lake change direction if we decide to cut the export? Figure 5.11 shows the decline in elevation over a 50-year period with export fixed at 100 KAF/yr. The elevation will be at 6,342 feet in 2040. For purposes of this test, let's suppose that this elevation is a dangerous or threshold level (BWeb). An extra variable is added to the graph to draw our attention to this hypothetical threshold. Now, what do we expect to see if the export is cut immediately to zero? From the figure 5.3 sketch at the outset of the modeling, we anticipated that some downward momentum will carry the elevation below the threshold. Perhaps it would be 5 to 10 years before the lake could recover to safe values again?

Figure 5.12 shows the second half of the policy test. This graph reveals that the lake would respond rapidly to the change in export. There is no downward momentum once the export is changed to zero, and the elevation climbs back more rapidly than the descent prior to the

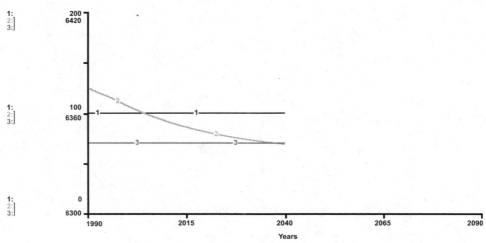

Figure 5.11. Export is fixed at 100 KAF/yr in the first half of a policy test.

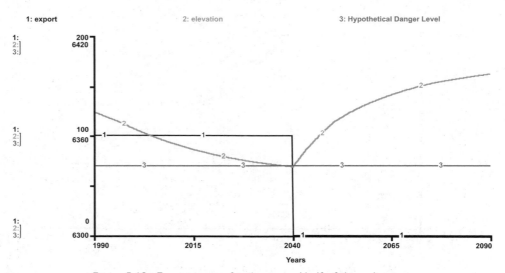

Figure 5.12. Export is zero for the second half of the policy test.

change in export. The lake requires only around 15 years to recover to the starting elevation. By the year 2090, the elevation would be well above the elevation at the start of the simulation.

Where Is the Downward Momentum?

The policy test came as a surprise to the first of my students to simulate the dynamics of Mono Lake. She had read about the long delays in getting environmental systems to change course, and she had included such delays in her original sketch of the policy response (figure 5.3). She found herself wondering what to trust, her instincts or the model? The surprise results can be a sign that the model is poorly formulated, so it's important to try new simulations with different parameter values. If you try this with the third model, you will see the same result again and again: Mono Lake's elevation responds immediately and rapidly when the export is cut to zero. The lake reaches a threshold in the year 2040 with the flows nearly in balance. In the next year, the lake is receiving an extra 100 KAF/yr of water, and it responds immediately with a rapid increase in elevation.

So, why is there no downward momentum? You've read in chapter 2 that the stocks sometimes account for the momentum in a system. The model of Mono Lake has only one stock, the volume of water stored in Mono Lake. Perhaps we will see some downward momentum if additional stocks are included in the model. Water is also stored in other lakes, in the snowpack, and in the groundwater in the basin. Perhaps if these were added, a new model would show a different policy response. You can experiment by expanding the hydrological aspects of the model (BWeb), and perhaps your new model will show major changes in the lake's response to a change in export. For example, you might expand the model to include snowpack or groundwater storage. If these stocks hold a relatively large amount of water, you might see some downward momentum that was not seen in figure 5.12.

Why Not Include the Extra Stocks from the Start?

If you are like many students, you thought of snowpack and groundwater storage when we went to work on the first model. You might have wondered why these stocks were not included at the very beginning. I started with only one stock because it's always best to start simply and take the time to learn from the simple model. When you reach a model that reproduces the reference mode and delivers some insights on policy testing, you will be in a much better position to make meaningful additions. To illustrate, think of what you would do next if you were to add the snowpack to the third model. The goal is now clear: Would the addition of the snowpack change the rapid response pattern shown in figure 5.12? This question gives focus to further work and helps one decide how much detail to include in the new stocks and flows. If the additional detail does not change the policy results, then the new variables should only be retained if they add a sense of realism to the model.

Moving beyond Fortran

The third model is an excellent model of the water flows in the Mono Basin. Its main advantage is simplicity and clarity. The diagram (figure 5.9) occupies less than half a page, but that is sufficient to show every variable and every interconnection. The names are simple, and the units are clear. The simulations show the reference model, and the results have been checked against a more detailed model. The model may be used to test the impact of different export policies.

On the other hand, we have checked the response of the third model by comparison with the water balance model by Vorster (1985). Our own modeling was made possible by Vorster's hard work in designing and checking his model. Vorster's model was implemented in Fortran, a procedural programming language popular among water resource analysts in previous decades. The Fortran code makes calculations similar to the dynamic results shown in this book, so what is the value of the Stella model? The appeal of the Stella model is the clarity of design; the model is designed to communicate as well as to calculate. The calculated elevations are similar to the elevations estimated by Vorster, but the communication of the stocks and flows is much more powerful with the Stella model.

Moving beyond Hydrology

The Stella model also provides a launching pad to move beyond hydrological issues. Perhaps there is significant downward momentum associated with the ecology of Mono Lake. If so, ecological considerations may call for taking action in advance of reaching a dangerous threshold. Figure 5.13 shows portions of a BWeb exercise that explores the interconnections between the hydrology and the brine shrimp population, a key element in the ecological system. This diagram shows all the stocks and flows and some of the information connections. The model may be used to examine a fundamental policy question—should we control the export from the elevation or from the adult shrimp count? The lower portion of the model shows three stocks to keep track of the life cycle of the brine shrimp and the influence of salinity on the cyst deposition. With higher salinity, we expect fewer overwintering cysts, fewer births, and fewer adults each summer. If this were to occur, the manager might react by lowering the export. This would allow the lake to climb to a higher elevation and the salinity to fall to a lower level. With lower salinity, we might see a rebound in the brine shrimp population to the levels needed to support the bird populations.

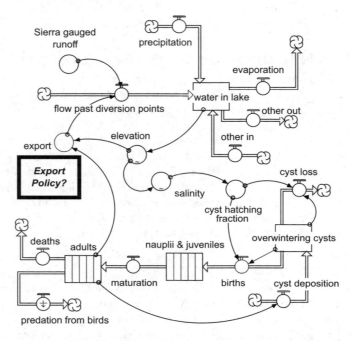

Figure 5.13. Combining Mono Lake hydrology and population biology.

The BWeb exercise provides an opportunity to explore alternative management strategies with an interdisciplinary model. The model in figure 5.13 is a combination of hydrology and biology that is seldom achieved because investigators too often confine their studies within strict disciplinary boundaries. The system dynamics method allows us to reach across these boundaries to create a model for rapid exploration of different management strategies. In my opinion, this is the ultimate value of system dynamics modeling of environmental systems.

Do We Need to Cut Export in Advance?

Let's close the Mono Lake case with the policy question raised at the outset: Does it make sense to cut exports in advance of the lake falling to a particular threshold elevation? The hydrological simulations suggest that a "wait and see" export policy may work fine. If you conduct the model merger exercises (BWeb), you may see a case for acting in advance because of the interconnected nature of the hydrology and the population biology. In my opinion, however, the key delays in this system do not dwell in the hydrological or ecological factors that have been discussed so far. I believe the key delays are more likely to be found in the political and legal aspects of policy making in the Mono Basin.

Exercises

Exercise 5.1. Verify the first model

Build the model in figure 5.4 and verify the results in figure 5.5. (Remember to turn off the non-negative option for the stock.)

Exercise 5.2. First model with bad results covered up

Run the model from exercise 5.1 with the non-negative option turned on. Which flow does Stella change to prevent the stock from going negative?

Exercise 5.3. Verify the third model

Build the model in figure 5.9 and verify that you get the results in figure 5.10. (Be sure to set the vertical scales to match those in figure 5.10.)

Exercise 5.4. Rapid response in 2000 as well?

Perhaps the rapid response in figure 5.12 is due to the fact that we waited until 2040 to change the export. Rerun the model with the pause interval set to 5 years and with initial export equal to 100 KAF/yr. When you reach the year 2000, the elevation should be 6,364 feet. This is a more realistic threshold, as you can tell from the elevation gauge (figure 5.6). Cut export to zero in the year 2000 and complete the rest of the simulation. Document your results with a graph similar to figure 5.12. Does the elevation show a rapid response similar to figure 5.12?

Exercise 5.5. Simulate a buffer policy

Modify the model to allow it to simulate an export policy similar to a proposal from the Mono Lake Committee. The idea is to specify a target range for the lake elevation where you are confident that the ecosystem is safe as long as the elevation is within the range. If the elevation is 6,380 feet or lower, no export is allowed. If the elevation is 6,390 feet or higher, 100 KAF/yr is allowed. If the elevation is within the buffer zone, the export will change in a linear manner, as shown in figure 5.14. Simulate the model from 1990 to 2010 and document your work with a time graph of elevation and exports. Pick the scales on the vertical axis to make it easy to check that your export policy is working correctly.

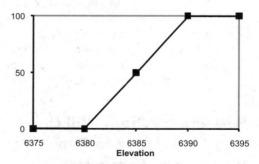

Figure 5.14. Export depends on the elevation.

Exercise 5.6. Control board policy

The California Water Resources Control Board issued its export policy in 1994. The overall goal was to build the lake toward an elevation of 6,392 feet. This would be done in stages, as described by Hart (1996, 171). No export was allowed until the elevation reaches 6,377 feet. Then the allowed exports would be 4.5 KAF/yr until the elevation reaches 6,390 feet; 16 KAF/yr until the elevation reaches 6,391 feet; and 30.8 KAF/yr if elevation exceeds 6,391 feet. Simulate this export policy over a sufficiently long time period for the lake to reach dynamic equilibrium. Document your results with a time graph of elevation and exports.

Exercise 5.7. Sensitivity

Use Stella's sensitivity specs command to study the sensitivity of Mono Lake's elevation to changes in export. Ask for five simulations with export set to 0, 25, 50, 75, and 100 KAF/yr. Show the results in a comparative time graph with the scales shown in figure 5.15.

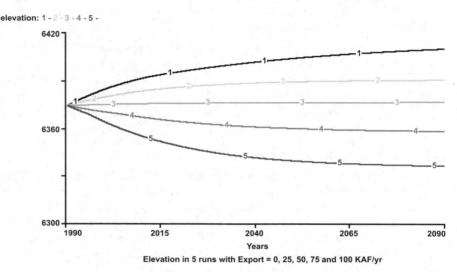

Figure 5.15. Sensitivity analysis of Mono Lake elevation.

Exercise 5.8. Explain the other flows

Table 5.4 lists the various smaller flows that make up the other inflows and outflows in the third model. The important assumption is that these flows do not change with changes in the size of Mono Lake. Expand your model from exercise 5.3 to include each of these flows as converters. Then add the converters to give the other flows. Run the new model and verify you get the same results as exercise 5.3.

Table 5.4. Other inflows and outflows.

Other inflows (KAF/yr)
Ungauged Sierra runoff = 17
Non-Sierra runoff = 20
Net land surface precipitation = 9
Diversion inflows = 1.6

Other outflows (KAF/yr)
Exposed lake bottom evaporation = 12
Evapotranspiration = 13
Groundwater export = 7.3
Net Grant Lake evaporation = 1.3

Postscript: Where Will the City Find Water to Replace Mono Basin Exports?

The focus of this chapter is the Mono Basin and the need for reduced diversions in order to rebuild the lake. But many readers will wonder about the wider implications of reducing the diversions. Will the extra water come from the Colorado River, new reservoirs in northern California, or perhaps desalinization plants along the ocean? Many believe that the most attractive supply of new water is hidden in the homes and businesses in the city of Los Angeles. Eliminating inefficiencies in water use is an important low-cost supply of water, and the city has made serious investments in water conservation and recycling (BWeb).

Further Reading

- Information on Mono Lake is given in several reports by academics, all of which are listed on the BWeb. The BWeb also provides supporting information about the hydrology and ecology of the basin.
- The most extensive single collection of information on the Mono Basin is the environmental impact report prepared for the State Water Control Board's review of the Mono Basin Water Rights of the City of Los Angeles (EIR 1993).

Acknowledgments

- Photo 5.1 is used courtesy of the Mono Lake Committee (BWeb).
- Table 5.2 is based on Vorster (1985, 261).

Chapter 6

Equilibrium Diagrams

System dynamics is the study of how systems change over time. Our focus is on dynamics, but dynamics are often difficult to understand because of the many forces at work in the system. This chapter describes the equilibrium diagram, a simple way to portray the balance of forces. An *equilibrium diagram* is a stock-and-flow diagram with the numerical values and the units for each variable written directly onto the diagram. This diagram provides a snapshot of the system at one point in the simulation. If we understand the snapshot, we will be in a better position to understand the moving picture, the dynamics of change in the system. You can take a snapshot at any time in the simulation that is worthy of study. This chapter focuses on the moments when the system is in equilibrium. These are special moments because the flows into and out of each stock must be in balance. An equilibrium diagram is simple way to check how this balance is achieved.

Mono Lake in Equilibrium

The model of Mono Lake (figure 5.9) shows that the lake could rise or fall over time depending on the value of export. If we watch the simulations long enough, the lake eventually reaches a certain size and remains there for the remainder of the simulation. The lake may be said to be in dynamic equilibrium. *Equilibrium* means that the lake's conditions remain constant over time. *Dynamic* means that there are flows in and out of the system. Figure 6.1 shows the balancing flows in a simulation with the export held constant at 100 KAF/yr.

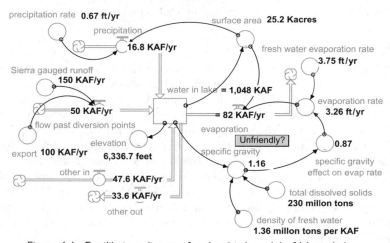

Figure 6.1. Equilibrium diagram for the third model of Mono Lake.

The lake volume is 1,048 KAF. For the stock to remain at this value, the inflows must balance the outflows. If you sum the three flows into the stock, you will discover that the total inflow is 114.4 KAF/yr. And if you sum the two outflows, you will learn that the total outflow is 115.6 KAF/yr. So the volume of water in the lake is slowly declining at this point in the simulation. This slow decline will lead to a smaller volume, a smaller surface area, and a somewhat smaller evaporation. If we were to run the model until the year 2190, the lake elevation would decline by an additional 10 inches. At the somewhat smaller size, the inflows and outflows would be in balance, and the lake has found its equilibrium position.

Checking the Units

Equilibrium diagrams are a simple way to check the balance of flows that create dynamic equilibrium. They are also a simple way to help us check the units in the model.

Figure 6.1 shows the units for every variable, including the dimensionless "specific gravity effect on evap rate." Since we are using simple algebra, it is easy to check the units by inspection. For example, the evaporation rate is a dimensionless effect times the freshwater evaporation rate measured in ft/yr. This means the evaporation rate must also be measured in ft/yr. The evaporation flow is the product of surface area in K acres and the evaporation rate in ft/yr, so the units of evaporation must be KAF/yr. The units of the stocks and flows are also consistent. The stock is in KAF, and each of the flows is in KAF/yr. The units for the specific gravity are not immediately obvious, but the "unfriendly button" reminds us that the algebraic expression is more complicated than simple addition, subtraction, multiplication, or division. (In other words, this particular equation is one of the few equations in the book that violates the pledge to adhere to friendly algebra.)

The units for the variables with nonlinear graphs (~) may appear puzzling. For example, how do we get elevation measured in feet when the only input is the water in the lake, measured in KAF? The reason for the confusion is that we are accustomed to applying algebra to the units (i.e., if we multiply the numbers, then we multiply the units.) But there is no algebra with the ~ variables. We use the nonlinear graphs to give the elevation of the lake as a nonlinear function of the volume of water in the lake. This particular relationship was based on survey results and not an algebraic formula whose units could be checked for consistency.

Appendix A provides a review of units and the options for checking consistency within the Stella and Vensim programs. Many of my colleagues make use of these programs, but I find it more instructive to check the units by inspection of the equilibrium diagram. You will learn more about the units by checking them yourself than by letting the software check them for you. And you will also become more familiar with the numbers.

Checking the Numbers

System dynamics models help us understand dynamic behavior. We focus on the general patterns of behavior because our goal is to understand the general trends, not to make forecasts of a numerical result at a specific point in time. Successful modelers do not lose sight of this goal. At the same time, however, successful modelers take the time to become familiar with the numbers. We should be aware of their approximate size and the interrelationships, but we do not need to become overly concerned with the precise numerical value. If we were modeling gasoline use by automobiles, for example, we should have sufficient numerical literacy to know that a fuel efficiency of 25 miles/gallon is plausible, but an efficiency of 250 miles/gallon needs to be checked. Of course, anyone who drives a gasoline-burning car would be sufficiently familiar these numbers to take a second look at 250 miles/gallon. But we do not always have such familiarity at the outset of an environmental study, and an equilibrium diagram can speed the process of familiarization.

The Mono Lake equilibrium diagram adds to our understanding of a model that has already been constructed and tested. The diagrams can also be used to help us understand a new model that we may be thinking of constructing. The diagrams help us anticipate the flows needed for dynamic equilibrium and to become familiar with the numbers. This value is illustrated with examples of biogeochemical cycles.

Equilibrium Diagrams and the Biogeochemical Cycles

Figure 6.2 shows the first illustration of a biogeochemical cycle. This is the global hydrologic cycle, with water storage in cubic kilometers (ckm) and water flows in ckm/yr. The diagram reveals the huge differences in storage. The vast majority of the water is stored in the ocean, but only a tiny portion is stored in the atmosphere. The diagram reveals that atmospheric storage is a high-turnover stock, one that supports high flows with a small

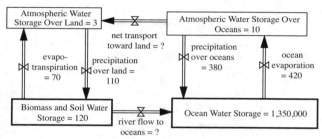

Figure 6.2. Equilibrium diagram for the global hydrologic cycle.

amount of storage. Now, suppose you are thinking of building a model of the hydrologic cycle and you have become familiar with most of the flows, but you can't find estimates of the net transport toward land and the river flow to oceans. What would be the values of these flows if the hydrologic cycle is in dynamic equilibrium?

The next example is the global rock cycle depicted in figure 6.3. The units of storage are millions of Pg (petagram, or billion metric tons). Time is measured in 100 million years. (By this measure, 1 unit of time corresponds to the time from the Cretaceous period to the present.) A simulation for 40 time units would allow one to represent the oldest rocks found on earth (see the book's website, the BWeb). Figure 6.3 shows the mass of rock in the asthenosphere portion of the mantle at the base of the diagram. A long name is assigned to the ocean crust production to make the units clear: 7,000 Pg is produced every 100 million years. All the flows are measured in these units. Notice that the ocean crust subduction is also 7,000 Pg/100 million years. This means the oceanic igneous rock is in dynamic equilibrium. Continental igneous rocks include granite and basalt. This stock is drained by weathering and metamorphism and is increased by continental crust production and by melting of metamorphic rock. Estimates of melting are difficult to come by (BWeb), but you should be able to estimate the melting if the system is in dynamic equilibrium.

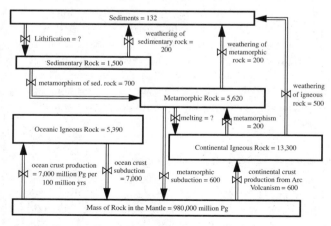

Figure 6.3. What are the missing flows in the global rock cycle?

The upper portion of the diagram shows the accumulation of sediments and the formation of sedimentary rock. These include shale, coal, limestone, and sandstone. Sedimentary rocks are formed by lithification, a process of compaction and cementation. Estimates of lithification are also difficult to come by, but you can estimate the lithification if the sedimentary rock is in dynamic equilibrium.

The third example of a biogeochemical cycle is the global phosphorus cycle. Phosphorus (P) is one of the essential elements of life, and it is often the limiting nutrient for plant growth. The top of figure 6.4 shows that land plants store 3,000 billion kilograms (bkg) of P. The plant uptake is not shown, but you can estimate the uptake if you assume that the stock of P in land plants is in equilibrium.

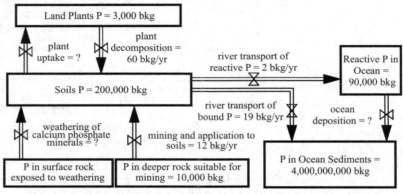

Figure 6.4. What are the missing flows in the global phosphorus cycle?

The storage of P in the soils is 200,000 bkg, around seven times larger than the P in plants. Nearly all of the phosphorus in terrestrial systems was originally derived from the weathering of calcium phosphate minerals. This flow is marked by a question mark in figure 6.4, but you should be able to estimate its value if soils P is in equilibrium. Figure 6.4 makes it clear that the entire P cycle is not in equilibrium. The flows from surface rock and deeper rock (suitable for mining) are not replenished over time, and there is a net transport of phosphorus from these rocks to the ocean sediments. The majority is transported in bound form, so it quickly adds to the P in ocean sediments. A small amount is transported in reactive form and is then subject to deposition. The ocean deposition flow is shown by a question mark in figure 6.4, but you know it must be 2 bkg/yr if reactive P in the ocean is in equilibrium.

These examples show the flow of material in the biogeochemical cycles. Did you notice that there were no clouds in these three equilibrium diagrams? That's because these cycles work with a fixed amount of material in a closed system. The models simulate the changing form of these materials over time. Land use models are similar; they simulate the changing categories of land use over time. Equilibrium diagrams can help us become familiar with cycles in land use as well.

Land Use on Daisyworld

Figure 6.5 shows a model of Daisyworld, an invented world with only three categories of land use. The land can be empty ground, or it can be covered by white daisies or by black daisies. Figure 6.5 shows a typical 1,000 acres of the Daisyworld landscape. The area is occupied by 403 acres of white daisies and 271 acres of black daisies. The remaining 326 acres is bare ground. There are no clouds in the diagram, so you know that no new land is created and

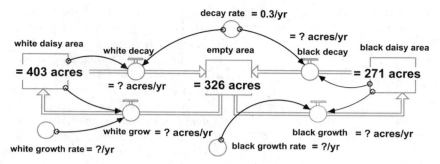

Figure 6.5. What are the missing values for Daisyworld?

none of the existing land is lost. The white daisy area will increase owing to white growth and will decline owing to white decay. The process of decay is a change in land use: land that was previously covered by white daisies is now empty ground. Both the flows influencing the white daisy area are marked with a question mark in figure 6.5. You can estimate their values if you are told that the white daisies are in dynamic equilibrium. And if you turn your attention to the right side of the diagram, you should be able to estimate the black decay and black growth. And, finally, you should be able to specify the white growth rate and black growth rate.

You'll learn more about Daisyworld in chapter 11. At this stage, think of how easily you were able to fill in the missing numbers in the diagram. The job was easy because you understand that the flows into and out of each stock must be in balance and because you know that each equation uses friendly algebra. You guessed each equation based on the names and units and then determined the numerical value for equilibrium.

Land Use and Forest Management

The next equilibrium example involves the management of forested land. Figure 6.6 shows a model to simulate the harvesting policy of the landowner. Six stocks are used to keep track of the land described by the type of trees on the land. Table 6.1 lists the time intervals used to calculate the succession flows for each stock. We assume that 8 years are required for seedlings to become saplings and another 8 years for the saplings to turn into trees that the landowner would classify as suitable for small pulp. Five more years are needed to see large pulp and another 5 years to see small logs.

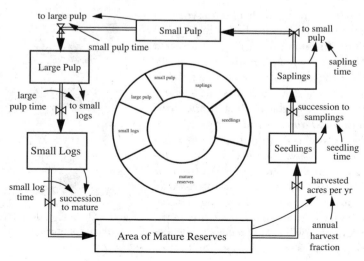

Figure 6.6. Model of forest land use.

Table 6.1. Time intervals for succession.

Seedlings	8 yrs
Saplings	8 yrs
Small pulp	5 yrs
Large pulp	5 yrs
Small logs	8 yrs

The final time interval is the 8 years for the small log trees to grow sufficiently to be classified as mature reserves.

The image in the center of figure 6.6 alerts us to the relative size of the land categories that would be expected from the current harvesting policy. (Inserting graphic images into a model diagram is explained on the BWeb.) The harvest fraction is the policy variable. It is currently 0.0375/yr, and it applies to 42,667 acres of mature reserves. So, 1,600 acres of mature reserves would be harvested each year. The company plants seedlings as the trees are removed. From this action alone, the seedling area would increase by 1,600 acres/yr. But the seedling land that was planted 8 years earlier will be graduating to the saplings category. If the forest is in dynamic equilibrium, the total seedling area would remain constant. For this to happen, the succession to saplings must be 1,600 acres/yr.

Suppose the company's harvesting policy has been followed for several decades and the forest is in dynamic equilibrium. From the information given so far, you should be able to estimate the amount of land in each of the six categories. Starting on the right side of the figure 6.6, you know that the equation for succession to saplings is the seedling area divided by the seedling time. (Remember: simple algebra and unit consistency.) So you know that the seedling area needs to be 8 times larger than the flow of 1,600 acres/yr. This tells us that the seedling area is 12,800 acres. Similar reasoning will allow you to estimate the acres in each remaining category. Give this a try and add up the acreage. You should get a total of 97,067 acres.

This model ignores some of the challenges of managing a real forest. There are no complications from tree loss due to fire or infestations and there are no variations in the time intervals. Also, the harvesting is concentrated at the 4 o'clock position in figure 6.6, which allows us to easily visualize how the land uses would change over time. (Think of the harvesting proceeding in a clockwise direction over time.) This model is intentionally simple to allow you to guess the values of each of the flows and each of the stocks. At first glance, the model may appear too simple to teach us much about the dynamics of harvesting. But dynamics are hard to simulate in our heads. Sometimes simple models can deliver surprising results, as you will see when you try the harvesting exercises on the BWeb.

Stability Conditions

Equilibrium diagrams help us see the forces that keep a system in dynamic equilibrium. But they don't tell us whether the equilibrium is stable. The stability of equilibrium positions is depicted in figure 6.7. This drawing shows a marble resting on three different surfaces. The marble is at rest in all three examples, so it is in equilibrium. But you know at a glance that these are very different types of equilibriums. The marble at the bottom of the cup is an example of *stable equilibrium*. If you were to deflect it from the resting position, it would return to the original position. The middle example is an *unstable equilibrium*. A small disturbance is all that is needed for the marble to depart from the equilibrium position. Once disturbed, gravitational forces would take the marble farther and farther from the original position. The third example is called *neutral equilibrium*. If the marble is moved to the right or left, it will simply remain at the new position. These stability conditions are obvious to you because

Figure 6.7. Stable, unstable, and neutral equilibrium.

you have spent your whole life becoming familiar with gravitational forces. The stability of environmental systems is not so easy to discern. We can, however, test for stability through computer simulation.

Testing for Stability

A simple test for stability is analogous to disturbing the marble on the cup. We introduce a change in the system and watch the system react. If it returns to its original position, we know the equilibrium is stable. Figure 6.8 shows a simple way to introduce the disturbance. The stock variable is any stock in the model that you wish to test. The regular flow in and the regular flow out represent flows that are part of your existing model. The new flow is the disturbance flow that is added for the test. The ~ in the flow icon tells us that we are using a graphical function. The fact that there are no inputs to the disturbance means time is on the horizontal axis of the graph function. We set the disturbance flow to zero until we reach a point in time when we wish to disturb the stock. Then we change the flow to a value sufficient to disturb the stock from its equilibrium position. The disturbance flow should return to zero in the next unit of time. This will allow us to watch the system react to the disturbance. If the stock returns to the original value, we know the equilibrium is stable. If the stock moves farther and farther away from the equilibrium, we have an unstable equilibrium.

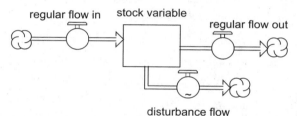

Figure 6.8. Addition of a disturbance flow to test for stability.

Stability tests are useful to check our understanding of the system. We can often guess whether the equilibrium is stable. For example, if you notice that the system seems to find the equilibrium position on its own, then the equilibrium is probably stable. The Mono Lake model is illustrative. When the export was fixed at 100 KAF/yr, the lake found its equilibrium value on its own, as shown in figure 5.10. This is a sign that the equilibrium is stable, which you can confirm in the exercises.

The Value of Equilibrium Diagrams

Equilibrium diagrams are a simple way to see the balance of flows in the system. They help us understand the equilibrium conditions, and they also help us to become more familiar with the numbers and the units in the model. They are useful when checking an existing model or when we are thinking about the flows in a new model. They are one of the easiest ways to aid the development of good models and to improve our understanding of environmental systems.

Exercises

Exercise 6.1. Missing Numbers for Silver Lake

Figure 6.9 shows an equilibrium diagram for a model of Silver Lake. The lake is fed by two creeks and drained by Silver Creek flow. The net evaporation represents evaporation minus precipitation. Net evaporation is the product of the net evaporation rate and the surface area. What is the net evaporation? What is the surface area?

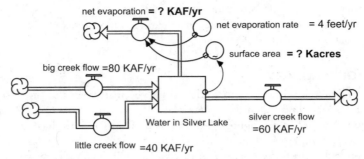

Figure 6.9. Complete the equilibrium diagram for Silver Lake.

Exercise 6.2. Stability of Mono Lake by water removal

Introduce a disturbance in the model from exercise 5.3. Set time to run from 1990 to 2150. Set the disturbance flow to remove 20 KAF/yr in the year 2090. Document your test with a time graph showing the water in the lake on a scale from 0 to 3200.

Exercise 6.3. Stability of Mono Lake by water addition

Repeat the stability test from exercise 6.2, except the disturbance flow should add 20 KAF/yr to the stock of water in Mono Lake in the year 2090.

Exercise 6.4. Importance of addition or removal

Can you think of a situation in which the stability test gives a different result if you disturb the stock by adding rather than removing material?

Exercise 6.5. Missing numbers in forest carbon model

Figure 6.10 describes a model of carbon flows in a tropical forest. The carbon flowing into the system is fixed at 30 GT/yr (gigatons per year). The carbon influx is split in four directions with 20% going to the roots, 30% to the leaves, 20% to the branches, and 30% to the stems. The transfer rates are all expressed as a fraction/yr. The diagram shows the flows of carbon to litter on the forest floor, to the humus stage, and eventually to charcoal. All stocks are in dynamic equilibrium except for one—the stock of carbon stored as charcoal. It continues to grow slowly over time. Will the growth be exponential or linear? The six question marks indicate missing entries in this equilibrium diagram. Give the numbers and the units for the missing entries.

Exercise 6.6. Verify forest equilibrium

Build the model in figure 6.10 with the initial value of all stocks at zero. Simulate the model long enough for the forest to reach equilibrium. (This is sometimes called *spinning-up* the model.) Does the model confirm your estimates in exercise 6.5?

Exercise 6.7. Stability of the forest equilibrium

Exercise 6.6 works if the model finds the equilibrium on its own. Do you think the equilibrium is stable or unstable? Check your answer with a disturbance to remove 2 GT of carbon from the stock of carbon in the litter.

Exercise 6.8. Circulatory system

Figure 6.11 shows the blood storage in the human circulatory system. Total storage is 4,800 ml, just under the 5 liters of blood in the average person. Each flow is the size of the stock divided by an average time for that stock. For example, the left heart time is

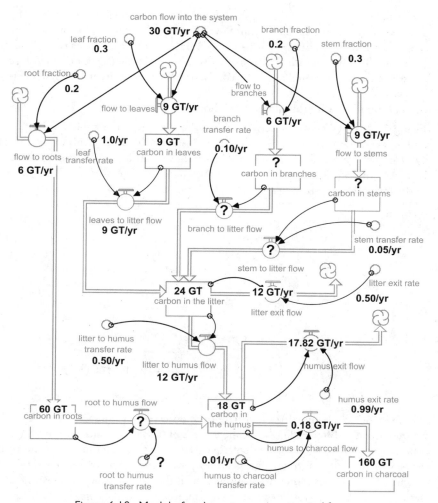

Figure 6.10. Model of carbon storage in a tropical forest.

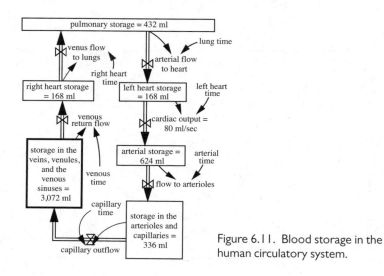

Figure 6.11. Blood storage in the human circulatory system.

2.1 seconds, and the outflow from the left heart is 168 ml divided by 2.1 seconds. This gives 80 ml/sec for the cardiac output flow. What are the values of the other five flows if the system is in equilibrium? The time interval for the left heart storage is 2.1 seconds. What are the time intervals for the other five stocks? (To check your estimates, add the six time intervals. It should take 60 seconds to complete the circulatory system cycle.)

Exercise 6.9. Equilibrium values for the two bottles

Figure 6.12 shows a two-bottle system. Storage is measured in cubic centimeters (cc), and flows are measured in cc/sec. Flow 1 is constant at 5 cc/sec. It fills the first bottle, a cylinder with an area of 5 square cm. The overflow from the first bottle drains into the second bottle, a cylinder with an area of 10 square cm. The overflow from the second bottle leaves the system.

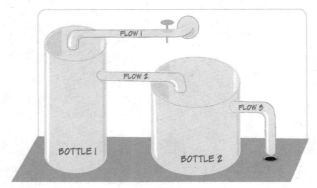

Figure 6.12. Sketch of the two-bottles system.

Table 6.2 shows overflows that have been observed for the two bottles. The values are reported for different heights in increments of 0.5 cm. (Overflows for intermediate values can be found by linear interpolation.) The height of water in the first bottle must exceed 14 cm if there is to be some overflow. The overflow will be 1 cc/sec if the height reaches 14.5 cm, and it increases substantially if the water reaches greater heights. The second bottle has no overflow until the height exceeds 9 cm. There is 1 cc/sec overflow if the height reaches 9.5 cm, and it increases substantially thereafter. Figure 6.13 shows a first cut at a model to simulate the filling of the two bottles over time. We begin with

Table 6.2. Overflows.

Height of water in bottle 1 (cm)	Overflow from bottle 1 (cc/sec)	Height of water in bottle 2 (cm)	Overflow from bottle 2 (cc/sec)
13.0	0	8.0	0
13.5	0	8.5	0
14.0	0	9.0	0
14.5	1	9.5	1
15.0	5	10.0	5
15.5	10	10.5	10
16.0	20	11.0	20
16.5	64	11.5	64
17.0	90	12.0	90

empty bottles, so the stocks are initialized at zero. The inflow is set to 5 cc/sec, but the question marks show that we don't have equations for the overflows. It might help us to formulate these equations if we think about the equilibrium situation. What would be the values of the two overflows in dynamic equilibrium? What would be the volume of water in the first bottle? In the second bottle?

Exercise 6.10. Simulate the two-bottles system

Complete the model in figure 6.13 by formulating equations for the overflows. Be sure to add as many converters as needed to make your theory of the overflows clear. Simulate for 80 seconds with a DT (delta time—the step size of the numerical simulation) at 0.25 second. Document your results with a time graph of the two volumes. Does your model reach equilibrium? Do the equilibrium results match your answers in exercise 6.9?

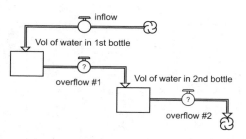

Figure 6.13. Getting started on a model.

Exercise 6.11. Stability of the two-bottles system

Introduce a disturbance in the second stock by removing 25 cc of water in the 40th second. Run the model for 80 seconds and document the results with a time graph of the volume in the second bottle.

Acknowledgments

- The global water cycle is adapted from Schlesinger (1991).
- The rock cycle is adapted from the teaching website by David Bice (BWeb).
- The phosphorus cycle is described by Schlesinger (1991) and by Botkin and Keller (1998).
- The forest land use model is adapted from the teaching notes by Boyce (1991)
- The forest carbon model is adapted from Huggett (1993).
- The human circulatory system is described by Guyton and Hall (1996).

Chapter 7

S-Shaped Growth

S-shaped growth is shown in the panel of dynamic patterns in figure 7.1. It is one of the most important patterns in natural systems, and it appears in a variety of forms in the book. This chapter describes the growth of a flowered area and the growth of a sales company. These seemingly dissimilar systems exhibit identical behavior over time. Both systems are able to grow when they are small, and their growth is exponential. Like all systems, they must deal with limits of one sort or another, and these two systems are able to come into accommodation with their limits and make the transition to equilibrium in a smooth, controlled manner. And, finally, both systems reach an equilibrium state that can be sustained year after year. This is a remarkable pattern of behavior, one that can be examined with computer simulation.

Figure 7.1. S-shaped growth, the fourth of the six dynamic patterns.

Flowered Area

Suppose we wish to simulate the spread of flowers across a 1,000-acre landscape of suitable land. You've seen stocks and flows to deal with flowers in chapter 6, so you would probably be drawn to a model like the one shown in figure 7.2.

Let's set the initial value of the flowered area to 10 acres and the empty area to 990 acres. The total area uses Stella's summer feature. Its value will be constant at 1,000 acres. Suppose we set the decay rate at 20%/yr, and we set the decay to the product of the flowered area and the decay rate. The growth in flowered area is the product of the flowered area and the growth rate, and let's suppose the growth rate is 100%/yr. There are only 10 acres of flowers at the

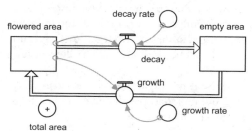

Figure 7.2. First model of the flowered area.

77

start, so there will be minimal limitations on the flowers' ability to spread. (The seeds will find plenty of open space in all directions.) With these assumptions, the flowered area would expand by 80%/yr.

The growth rate with no resource limitations is sometimes called the *intrinsic growth rate*. The intrinsic growth rate applies in our example when the flowers occupy only a small fraction of the area. But as they spread across the landscape, there will be less empty area available, and germination will be increasingly difficult. The actual growth rate will be less than the intrinsic growth rate. Figure 7.3 shows a growth rate multiplier to lower the actual growth rate over time. The multiplier uses Stella's graphical function, with the fraction occupied on the horizontal axis. The multiplier is 1 if the fraction occupied is zero. This ensures that the actual growth rate is the same as the intrinsic growth rate. As the fraction occupied increases, the multiplier will decline. But the shape of the decline is not obvious. Selecting a realistic shape requires an understanding of the process of germination and expansion. For this example, let's adopt the assumption that the growth rate is reduced in a linear manner, as shown in figure 7.4. The graph shows the multiplier at 0 if the fraction occupied reaches 1. This makes sense as there would be no room for the flowered area to expand.

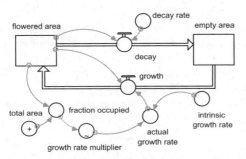

Figure 7.3. Second model of the flowered area.   Figure 7.4. Growth rate multiplier.

You should be able to build and simulate this model from the information provided so far. What would you expect the flowered area to do over time? At the beginning of the simulation, the flowers occupy only 10 acres; the net growth will be 80%/yr; and the doubling time is less than a year. Think of how rapidly the flowered area will spread if the doubling time is fixed. The acreage would double to 20 acres in less than a year. Shortly thereafter, we would have 40 acres, then 80 acres, then 160 acres, then 320 acres, then 640 acres. Then, in one more doubling, the area would exceed the 1,000 acres of suitable land. But you know that the growth will slow owing to the effects of the growth rate multiplier. Can you do the simulation in your head? If so, how many years are required for the flowers to cover the entire area?

Figure 7.5 shows that the flowers will spread remarkably quickly in the first few years, but they will never cover the entire area. The simulation starts with 10 acres and reaches 500 acres by the 6th year. The 6th year shows the highest growth in flowered area because the growth exceeds the decay by the largest margin. The upward trend slows noticeably in the following years. We see a decline in the growth and further increases in the decay. These flows come into balance around the 12th year, and the system is in dynamic equilibrium. The flowers cover 800 of the 1,000 acres.

You might be wondering why the flowered area did not expand to fill the entire 1,000 acres. An equilibrium diagram helps here. Figure 7.6 shows that growth and decay are in balance at 160 acres/yr. The growth rate and decay rate are balanced as well.

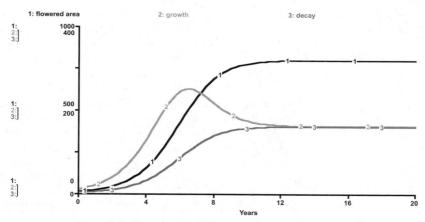

Figure 7.5. Flowered area shows S-shaped growth to 800 acres.

The decay rate is fixed at 0.2/yr throughout the simulation. The flowers spread across the landscape until their growth rate is reduced to 0.2/yr, exactly the value needed for growth and decay to be in balance. This balancing point is found when the flowers occupy 800 acres and the growth rate multiplier falls to 0.2. The system finds the equilibrium situation on its own, so we expect this to be a stable equilibrium.

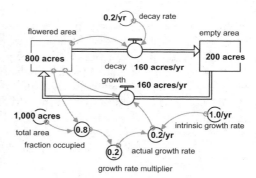

Figure 7.6. Equilibrium diagram for the flowers model.

Variations in the Pattern of S-Shaped Growth

The growth rate multiplier is an example of a density-dependent relationship. It moves from 1 to 0 as the density of the flowers increases from 0 to 100% of the area. The linear graph was selected for the first simulation, but there is no need to limit ourselves to a linear relationship. Figure 7.7 shows three relationships. All three curves adopt the same starting and ending points. They ensure that the intrinsic growth rate applies when there are only a few acres of flowers. And they all ensure that there is zero growth if the flowers were to cover the entire area. But their intermediate values are quite different. The top curve would be suitable if the flowers do not feel a significant effect of limited space until they cover 50% of the area. The lower curve would be used if the reduction in growth is quite pronounced as soon as the flowers fill 20% of the space. The linear curve make sense if the reduction in growth rate is proportional to the fraction occupied.

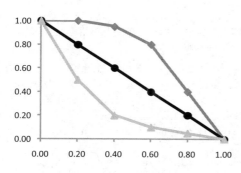

Figure 7.7. Growth rate multipliers.

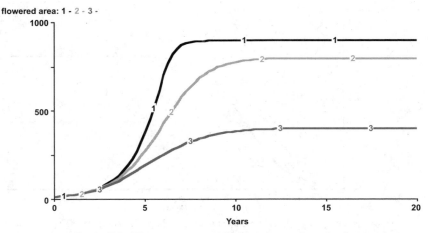

Figure 7.8. Three versions of S-shaped growth in the flowered area.

Figure 7.8 shows that the variations in the density-dependent relationship lead to major changes in the eventual spread of the flowers across the landscape. The three simulations are quite similar for the first three years because the growth is dominated by the intrinsic growth rate. The first simulation shows that the flowers would rapidly reach an equilibrium of 900 acres. The second simulation is the same as in the previous example (figure 7.5); the flowers grow to 800 acres. The third simulation shows the growth slowing down considerably by the 5th year, and the flowers reach dynamic equilibrium with only 400 acres occupied. In all three cases, the flowered area finds an equilibrium state with the growth rate at 20%/yr, exactly the amount needed to counter the rate of decay. All three examples show that the equilibrium can be sustained year after year.

The Logistic Equation and Its Limitations

The middle trajectory in figure 7.8 will be familiar if you have studied population biology. This curve is often called a *logistic curve* or the *logistic equation*:

$$A(t) = A_0 e^{rt}/(1+(A_0/K)*(e^{rt}-1))$$

where:

$A(t)$ = area of flowers as a function of time
t = time in years
A_0 = 10 acres at the start of the simulation
r = net growth rate at the start = intrinsic growth rate − decay rate = 0.8/yr
K = 800 acres, the area shown at the end of the simulation

This equation is the analytical solution to the differential equation for the flowered area (see the book's website, the BWeb). I introduce it here because of its wide use in ecology and population biology. Ricklefs (1990, 329) explains that it was the "first equation in ecology to generate research that would provide the data to test a mathematical model." It was used in models of yeast, fruit flies, water fleas, flour beetles, and even humans, and it "took a permanent place in population biology."

You can verify that the flowered area model matches the logistic equation in the exercises at the end of the chapter. It's reassuring to confirm the match between the model and a familiar equation. But the logistic equation is only one of many possible versions of S-shaped growth, and the equation works only if we expect the forces that slow the growth to apply themselves in a linear manner. Odum (1971, 185) explains that this assumption probably only works for organisms with simple life histories (e.g., yeast in a limited space). There is also the difficult question of how to interpret the K in the logistic equation. It has become customary to call K the *carrying capacity*, but this term can be confusing. In the flowers example, K is 800 acres, the acreage covered by the flowers at the end of the simulation. The term *carrying capacity* leaves an impression of a physical limit. In the flowers example, the landscape can physically accommodate 1,000 acres. But K is 800 acres, not 1,000 acres. For the logistic equation to work, K must be defined as the final acreage when the flowers reach dynamic equilibrium. But how do we know in advance that the flowered area will reach 800 acres? Some would respond that curve fitting gives us the answer. You can experiment with BWeb exercises to learn the challenges in estimating K from time series data. By the time you have sufficient data to determine the value of K, the S-shaped pattern will be complete. (There will be no need for a forecast.) So, if we can't guess the value of K, what should we do to understand S-shaped growth?

Suggestions for Studying S-Shaped Growth

The logistic equation has earned "a permanent place in population biology" because it provided a convenient mathematical expression to allow scientists to report their empirical evaluations during an era prior to the use of computer simulation techniques. As you study systems that exhibit S-shaped growth, you should move beyond the logistic equation. You will make more progress if you follow the approach with the flowers example. Use stocks and flows to represent the structure of the system; use numerical simulation to learn the dynamic behavior over time; and use equilibrium diagrams to verify that the equilibrium conditions make sense.

Growth in a Sales Company

These suggestions apply to any system that is likely to exhibit S-shaped growth, not just to ecological systems. To illustrate the generality of the approach, let's turn to an example of a growing company. Figure 7.9 shows a Vensim model to simulate the growth in the sales force of a company selling widgets. (A widget is a hypothetical manufactured product. Manufacturing is easy for this company; the challenge is selling the widgets.) The initial sales force is 50 people, and they are subject to a 20% annual exit rate. The company recruits new people based on the budgeted size of the sales force. The department budget is 50% of the company's annual revenues from the sale of widgets. Widget sales depend on the number of salespeople and their individual effectiveness.

Let's work through the numbers to become familiar with what a sales force of 50 people can accomplish. Each person can sell 2 widgets/day for 365 days/yr (sorry, no vacation). Total sales is 36,500 widgets. At $100 per widget, the company's annual revenue is $3,650,000. It allocates half to the sales department, so the

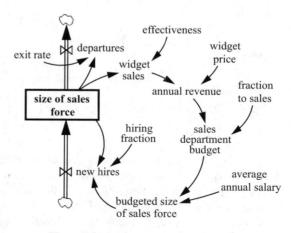

Figure 7.9. First version of a sales model.

sales manager has a budget of $1,825,000. The average annual salary is $25,000, so the manager can afford a sales force of 73 persons.

The current sales force is 50 persons, so the difference between the budgeted number and the current number is 23 persons. The hiring fraction is 1/yr, so the company would hire 23 persons in the year. Meanwhile, 20% of the current sales force are departing, so 10 people leave the company. The net effect is a growth of 13 people per year, or 26%/yr. At this rate, the company would double every 2.7 years. In 27 years, there would be 10 doublings. The sales force would be over 50,000 people.

There are limits to the growth in all systems. A sales company's limits could be the capacity of the factory (the ability to produce and deliver widgets on schedule) or the saturation of the sales territory (how many widgets do customers really need?). We'll assume that saturation is the limit for this company, as shown in figure 7.10. The diagram shows that effectiveness now depends on the size of the sales force. The diagram highlights the sales growth loop discussed previously: more salespeople lead to more sales, more revenues, a larger budget, and the hiring of still more people. But the effectiveness will change in the new model as more and more people are working the same territory. Their effectiveness is now a function of the size of the sales force. If we have only 50 persons, the effectiveness will be 2 widgets/day. But as more and more people are knocking on the same doors, the individual effectiveness declines. This decline is implemented using Vensim's "lookup for effectiveness" based on the values shown in table 7.1. Each person can sell 2 widgets/day as long as there are no more than 400 salespersons. As the sales force grows beyond 400, each individual's effectiveness declines. For example, each member of a 800-person sales force would be able to sell 1.6 widgets/day. If the sales force were to grow still larger, individual effectiveness would decline still further.

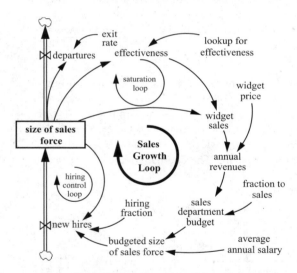

Figure 7.10. Second version of the sales model.

Table 7.1. Effectiveness depends on the size of the sales force.

Size of sales force	Effectiveness
0	2.0
200	2.0
400	2.0
600	1.8
800	1.6
1000	0.8
1200	0.4

You should be able to build and simulate this model from the information provided so far. Let's start with 50 persons, and we expect to see growth of 26%/yr due to the power of the sales growth loop highlighted in figure 7.10. But as the sales force climbs past 400 persons, there is saturation of the market, and individual effectiveness declines. Can you simulate this system in your head? Do you think the sales force will grow to a size that can be sustained year after year?

Figure 7.11 shows a Vensim graph of the simulation. The top curve is the size of the sales force. It starts at 50, reaches 400 in the 8th year, and is growing rapidly around the 9th year. Then the growth slows, and the company reaches an equilibrium of just over 750 salespeople. The lower curves are the new hires and the departures, each measured in people per year. New hires peak at just below 200 people/yr in the 9th year. From that point forward, new hires decline while departures continue to increase. These two flows come into equilibrium by the 16th year. The end result is a company with just over 750 people. The equilibrium is dynamic, with around 150 new hires and 150 departures each year, and it can be sustained year after year.

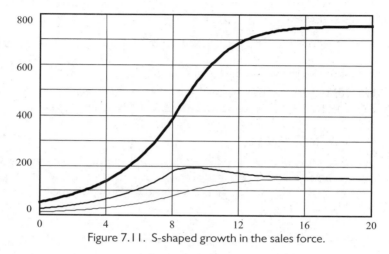

Figure 7.11. S-shaped growth in the sales force.

The Essentials of S-Shaped Growth

At first glance, a sales company seems entirely different from a field of flowers. But the simulations show that these systems exhibit essentially the same dynamic behavior. Both systems start small and grow in exponential fashion. The growth in flowers is made possible by a high intrinsic growth rate. The growth in the sales force is made possible by a high financial return to each person's efforts. Both systems face limits, so they cannot grow forever. What is remarkable is how they come into accommodation with their limits. Both systems feel the effects of their limits as they fill the space available, and they find the size at which the dynamic forces are in balance.

And, finally, it is important to note what we did not see in the simulations. Both systems avoided the behavior in the final two panels of figure 7.1. Both systems avoided the overshoot pattern because they reached an equilibrium state that was sustained year after year. And both systems avoided oscillations. We'll see examples of oscillations and overshoot later in the book. For now, turn to chapter 8 to see S-shaped growth in the spread of an infectious disease in a human population.

Exercises

Exercise 7.1. Verify the flowers model

Build the model in figure 7.3. Simulate it with DT = 0.125 years and verify the behavior in figure 7.5.

Exercise 7.2. Stability of the flower equilibrium

Introduce a disturbance to remove 20% of the flowers in the 15th year of the simulation. Run the model for 30 years to show whether the equilibrium is stable.

Exercise 7.3. Importance of the initial conditions

Create a comparative time graph of the area of flowers from four simulations with different starting values for the flowered area. Set the starting values to 5, 10, 15, and 20 acres. (Don't forget to adjust the starting values of the empty area accordingly.)

Exercise 7.4. Verify the sales force model

Build the model in figure 7.10 and verify that it gives the results in figure 7.11.

Exercise 7.5. Stability of the sales model

Introduce a disturbance flow in the 16th year to remove 200 people from the sales force. Run the model for 30 years to show whether the equilibrium is stable.

Exercise 7.6. Build a bigger company

Use the sales force model from exercise 7.4 to learn if you can build a bigger company by devoting a larger fraction of company revenues to the sales department budget. Conduct a sensitivity test with the fraction of revenues allocated to the sales department set at 45%, 50%, and 55%.

Exercise 7.7. Verify the match with the logistic equation

Add some new converters to the model in figure 7.3 to match the variable names in the logistic equation. Set r to 0.8; A_0 to 50; and K to 800. Then add a converter "A of t," which is defined by the logistic equation. (You will need to use Stella's exponential function.) Simulate the new model with DT = 1/4 year, and verify that you get the results shown in figure 7.12.

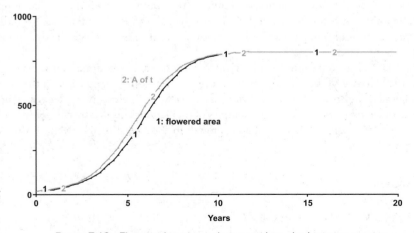

Figure 7.12. Flowered area is a close match to the logistic equation.

Exercise 7.8. Obtain a better match with the logistic equation

The small differences in figure 7.12 are due to the numerical accuracy of the simulation. Cut DT to 1/8 year and show the comparison. Do the differences disappear? If not, cut DT in half again and repeat the comparison.

Further Reading

- Sterman (2000, 616) describes a sales force model with sales effectiveness influenced by delivery delays. He also describes the different forms of S-shaped growth and their widespread usefulness in business and epidemiology.
- The logistic equation in biology and ecology is described by Ricklefs (1990); Hastings (1997); and Murray (2002).
- Odum (1971, 185) describes the logistic equation and gives good advice against the use of curve-fitting techniques to estimate the value of K.
- Odum's advice is reinforced by Sterman's (2000, 330) explanation of the futility of econometric estimates of K in business settings. By the time we have sufficient time series data on new product sales, there is no need for a forecasting model.

Chapter 8

Epidemic Dynamics

The rapid spread of infectious disease is an epidemic, and the history of epidemics includes staggering accounts of human misery. For instance, around a third of the population in 14th-century Europe may have died from the Black Death (bubonic plague). The worldwide flu epidemic in 1918–1919 infected almost half the world's population and killed around 20 to 30 million people. Infectious diseases create massive human misery today, especially in the tropical countries, where some 250–300 million people are infected with malaria. Some scientists warn of a global rise in infectious diseases, both old and new. Old diseases once believed to be controlled have appeared in new regions of the world. The new diseases include HIV/AIDS and a new form of tuberculosis. Many epidemiologists warn that it is only a matter of time before a pandemic on the scale of the Spanish influenza of 1918–1919 sweeps across the globe (Walters 2003, 147).

This chapter uses system dynamics to help us understand the rapid spread of an infectious disease. We will simulate a hypothetical disease that is nonfatal and spreads rapidly, and the infected individuals recover in only a few days. Thus we would expect the epidemic to run its course in 20 days. Our first step is to draw a reference mode (a target pattern of behavior). We expect to see exponential growth in the number of affected persons in the early stages as more infected people lead to more infections and still more infected people. The infections will eventually slow as there are fewer vulnerable people remaining. This pattern should sound familiar from the examples of S-shaped growth in chapter 7. The S-shaped pattern is the closest match with our situation. Reliable data on infections are often difficult to obtain, but some well-documented cases suggest S-shaped growth in the number of affected persons (Sterman 2000; Murray 2002).

A First Attempt at an S-I-R Model

One of the most common models in epidemiology classifies people as either susceptible to the disease, infected with the disease, or recovered from the disease. This is sometimes called the *S-I-R model*, as the affected people move from susceptible to infected to recovered. These categories correspond to the stocks in figure 8.1. The model simulates a population of 10,000 people, beginning with 9,998 susceptible and 2 infected. The duration of infection is 2 days, and we define the recoveries as the infected population divided by the duration.

All that remains is to explain the infections. With only one question mark in the diagram, we might think model construction is largely done. After all, we've written equations for 6 of the 7 variables. But explaining the infections is the big challenge. This disease is passed from person to person, so the key is the interaction between the infected people and the susceptible

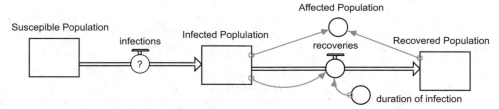

Figure 8.1. Stocks and flows in an S-I-R model of an epidemic.

people. Let's suppose that each infected person comes into contact with 6 people per day and that 25% of these contacts are sufficiently close to cause infection. The 2 infected persons will have 12 contacts in the first day, and 3 of these will lead to infections. The stock of infected people will increase by the 3 infected people. Meanwhile, the process of recovery is at work during the first day. There are 2 infected people, and we assume they are evenly distributed by time of infection. One person is only recently infected; the other was infected on the previous day. By the end of the first day, the second person will have recovered. After 1 day, the stock of infected people is increased by 3 and reduced by 1. It would grow from 2 to 4 persons in a single day.

A doubling time of 1 day will lead to extremely rapid growth, so we are on the right track to simulate the first stage of S-shaped growth. Let's implement this theory as shown in figure 8.2. We assume 12 contacts per day per infected person, and we set the infectivity to 0.25. The infections are the product of the infectivity and the total contacts per day. And, to eliminate unnecessary connectors, the affected population is calculated with Stella's summer.

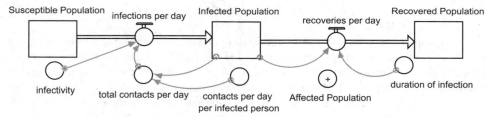

Figure 8.2. Explaining the infections in an S-I-R model.

This model is simulated over 20 days to give the results in figure 8.3. The infected population grows rapidly, reaching 5,000 people partway through the 8th day. Shortly thereafter, the infected population peaks at over 6,600 people. Then it suddenly changes direction, declines rapidly, and reaches zero around the 16th day.

The rapid growth in the infected population at the start of the simulation is a good start, but the abrupt change in the 8th day was not expected. Do you think this unexpected result is leading us to a potential insight? Or is it telling us that we need to rethink the model? The answer involves the susceptible population that reaches zero by the 8th day. We cannot have a negative number of susceptible people, so the curve remains at zero for the remainder of the simulation. But our theory of infections calls for more people to be infected. There are over 6,000 infected people who have 12 contacts per day, and 25% of these should lead to infections. But this could not happen as there are no susceptible people remaining in the population.

There are two problems with this model. First, the equation allows infections to occur when there are no susceptible people remaining. The second problem is that the non-negative

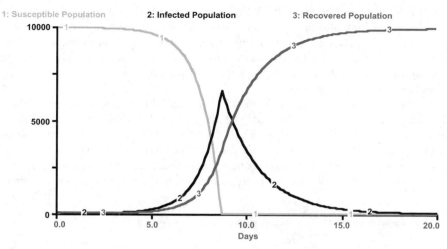

Figure 8.3. Simulation of the first version of an S-I-R model.

option applies to the stock of susceptible people. You know from exercise 2.9 that this option should be turned off. However, for this example, the stock was intentionally in the non-
negative position to remind us of this feature. I know from firsthand experience that it is easy
to forget to turn off this option. The end result can be real confusion. In this case, our equation for infections is wrong, but Stella covers up the mistake by writing its own equation after
the 8th day. We are left with figure 8.3, which gives the false impression that we are getting
close to an explanation of the epidemic. We are actually far from explaining the infections in a
plausible manner. To see how poorly the first model really behaves, turn off the non-negative
option and repeat the simulation. Infections will grow forever, creating millions of infected
people in a town with only 10,000 people.

Expanding the Model

Figure 8.4 shows how we could expand the model to account for the slowdown in the infections as the stock of susceptible people is depleted over time. The new model retains the assumption of 12 contacts per day for each infected person. But it is only the dangerous contacts
that could lead to infection. These are the contacts between an infected person and a susceptible person. The fraction of contacts with a susceptible person is identical to the fraction of

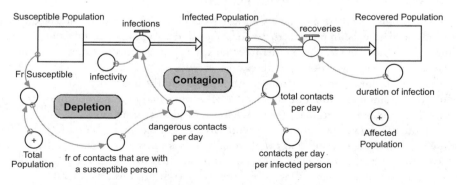

Figure 8.4. Representing both contagion and depletion.

the population that is susceptible. (If 75% of the total population is susceptible, then 75% of the contacts are with a susceptible person.) This assumption makes sense if there is uniform mixing of people. The buttons in the diagram draw our attention to the contagion and depletion processes that are simulated within the model. Contagion is responsible for the rapid growth at the outset of the epidemic; depletion is responsible for slowing the growth as the number of susceptibles are reduced.

The new simulation is shown in figure 8.5. The infected population peaks in the 9th day with over 30% of the population infected. The infected population falls during the second half of the simulation as recoveries exceed new infections. Over 90% of the population has been affected by the disease by the end of the simulation. The results match the reference mode specified at the outset, so we have reached an important milestone in the modeling process (see table 1.2). At this point, it's useful to test the sensitivity of the results to changes in the parameter values.

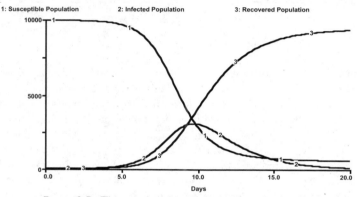

Figure 8.5. The simulated epidemic runs its course in 20 days.

Sensitivity Analysis

Let's use the model to explore different outbreaks with changes in assumptions. We will retain the general stock-and-flow structure and the assumption of uniform mixing The sensitivity analysis will focus on the three parameters:

- Contacts per day per infected person = 6
- Duration of infection = 2 days
- Infectivity = 0.25 infections created per dangerous contact

The product of these parameters indicates the potential for a newly infected person to spread the disease. With 6 contacts/day for 2 days and 25% leading to infection, each newly infected person would lead to 3 other infections. The product is sometimes called the *contact number*, and we would expect an outbreak if it exceeds 1.

Figure 8.6 shows three simulations with variations in the infectivity (50%, 25%, 12.5%). The contact numbers would be 6, 3, and 1.5, so we would expect an outbreak in all three cases. The model confirms that outbreaks happen in all three situations.

Figure 8.7 shows three simulations with variations in the duration of infection. The first simulation assumes that infected people require 4 days to recover. The second case assumes 2 days; the third case, 1 day. The contact numbers would be 6, 3, and 1.5, so we again expect an outbreak in all three cases. Once again, the model confirms this expectation.

These tests show that the S-shaped pattern persists across a wide range of conditions, and they show the usefulness of the *contact number* in anticipating the outbreak of the epidemic.

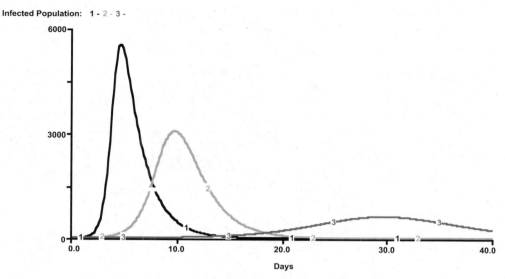

Figure 8.6. Infected population with the infectivity at 50%, 25%, and 12.5%.

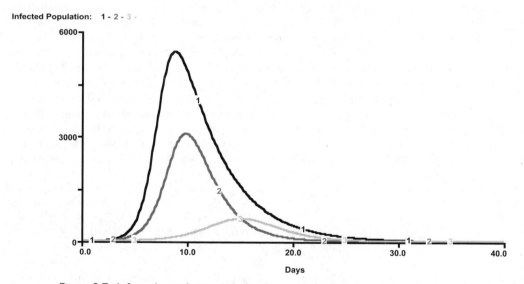

Figure 8.7. Infected population with the duration of infection at 4 days, 2 days, and 1 day.

Policy Analysis

Let's put the model to use to study the impact of a policy to reduce the contacts between the infected population and the susceptible population. The current model assumes that the infection spreads rapidly and that there is no change in the behavior of the infected people during the 20-day time period. Each member of the infected population has 6 contacts with other people each day at the start, during the middle, and at the end of the simulation. But an educational campaign (in advance of an epidemic) might make people more aware, and there could be a reduction in the contacts per day once an epidemic is underway.

Figure 8.8 shows a new model with the daily contacts as a nonlinear (~) function of the infected population.

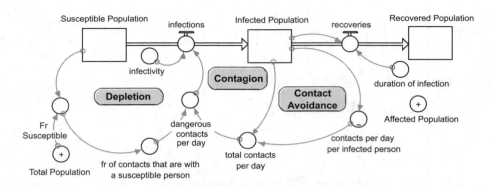

Figure 8.8. Expanding the model to represent contact avoidance.

Figure 8.9 shows the simulated impact of contact avoidance. The first simulation is the reference case shown previously in figure 8.5. The daily contacts are fixed at 6, and the infected population peaks at just over 3,000 in the 11th day of the simulation. The second simulation assumes that daily contacts will be cut in half when the situation is sufficiently serious to warrant the change in behavior. This is assumed to occur when the infected population reaches 2,000. This contact avoidance has no effect until around the 10th day. It lowers the peak number, thus leaving more people in the susceptible category. This leads to somewhat greater number of infected people later in the simulation. The third simulation assumes that the public has been sufficiently educated to get the word out sooner. This simulation assumes that infected people will cut daily contacts in half when their numbers exceed 1,000. The peak is now around 1,500 and appears in the 9th day. This simulation shows a somewhat higher infected population after the 15th day because of the people remaining longer in the susceptible category.

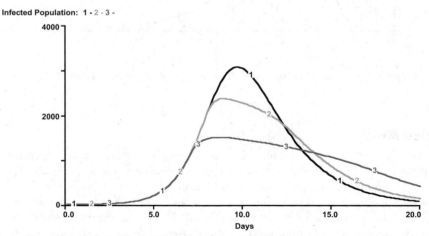

Figure 8.9. Infected population in three simulations to study contact avoidance.

Figure 8.9 shows that contact avoidance can reduce infections in the short term but increase them in the long term. Figure 8.10 shows the relative importance of these counteracting effects by showing the total affected population. The first simulation is the reference case with no change in daily contacts: 94% of the population is affected by the epidemic by the 20th day. The second simulation has contacts cut in half when the infected population exceeds 2,000. The third simulation cuts the contacts in half when the infected population exceeds 1,000. This third case shows substantially lower numbers in the 12th day. But the numbers continue to grow as susceptible people who avoided infections earlier become infected later in the time period. By the 30th day of the simulation, 87% of the population has been affected. Contact avoidance (as simulated here) leads to only a modest reduction in the affected population.

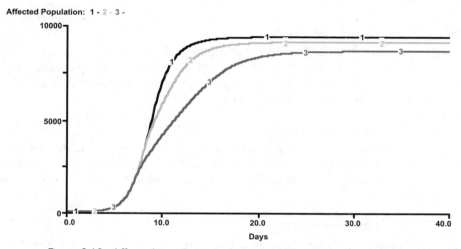

Figure 8.10. Affected population in three simulations to study contact avoidance.

Number of Stocks for Insight

This simple model illustrates the application of system dynamics to epidemiology. It demonstrates that a model with only three stocks can deliver important insights about the outbreak of an epidemic. Sometimes, a model with four stocks is more descriptive because it is important to represent the exposed state before people become infectious. Such models are sometimes abbreviated as S-E-I-R because they represent the transitions from susceptible to exposed to infected and finally to recovered. And in some cases, we use many more stocks to provide a useful representation of the epidemic. For example, we'll break the community of 10,000 people down into five separate groups in one of the exercises, and the model will have 15 stocks. The book's website, the BWeb, includes a model of the spread of yellow fever. This disease involves a complicated interaction between susceptible and infected humans and susceptible and infected mosquitoes. We will use 10 stocks to simulate the outbreak of yellow fever.

These examples indicate that the number of stocks will change depending on the type of epidemic under study. But regardless of the number of stocks, all epidemic models will simulate the processes of contagion and depletion. Contagion is responsible for the rapid growth at the start of the outbreak, while depletion is responsible for the slowdown in the final stage of S-shaped growth. These processes are highlighted by the buttons in figure 8.4 because they are key to simulating the S-shaped growth in the number of people affected by the epidemic.

Similarity of Feedback Structure

If we revisit the sales company example from chapter 7, we'll see corresponding processes. For example, the "sales growth loop" in figure 7.10 corresponds to the contagion loop in an epidemic. The sales company eventually leveled off owing to saturation of the market. The saturation of sales in figure 7.10 corresponds to the depletion process in an epidemic. The similarity in the loop structure of these systems is the reason they have similar dynamic behavior. The diagrams provide a button or a label to draw our attention to the key loops. Since the loops are key to understanding the dynamic behavior, it is useful to examine their structure more closely. We can do so with causal loop diagrams, as explained in the next chapter.

Exercises

Exercise 8.1. Confirm friendly algebra

There are no buttons in figure 8.4 to alert us to unfriendly algebra, so you know that the equations are simple add, subtract, multiply, or divide. Write the equations for all the flows and converters in figure 8.4. (You should be able to do this by thinking about the variable names in the diagram.)

Exercise 8.2. Equilibrium diagram

Figure 8.5 shows the infected population just past its peak around the 10th day. Build this model and verify the results in figure 8.5. Then run it again with the pause interval set to 10 days. Create an equilibrium diagram (snapshot) of the situation in the 10th day.

Exercise 8.3. Similar sensitivity test

Conduct three simulations with the contacts per day set to 12, 6, and 3. Your results should be identical to the results in figure 8.6.

Exercise 8.4. More aggressive contact avoidance

The contact avoidance simulation cuts the contacts per day per infected person from 6 to 3 partway through the simulation. Even with these fewer contacts, the contact number still exceeds 1. So, let's try a more aggressive policy. Repeat the third simulation in figure 8.9 with the contacts cut from 6 to 2 after the infected population exceeds 1,000.

Exercise 8.5. Earlier contact avoidance

Repeat the third simulation from figure 8.9, where daily contacts are reduced from 6 to 3. But this time, make the change in contacts after the infected population exceeds 500. Compare this policy with exercise 8.4 by showing a comparative time graph of the total affected population. Which policy leads to a lower number of affected persons by the end of the simulation?

Exercise 8.6. Dynamics of recovery

The recoveries are defined as the stock of infected people divided by the duration of infection. The duration is set at 2 days, but that does not mean that each person is infected for 2 days. Read ahead (chapter 14) to learn about the *conveyor stock*, one that will hold the material in position for a specified time interval. Then change the infected population to a conveyor stock as shown in figure 8.11, and set the "transit time" to the duration of infection. Set the duration at 2 days, and the new model will hold each person in the infected category for exactly 2 days.

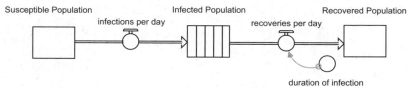

Figure 8.11. Infected population is represented by a conveyor stock.

Then simulate the new model to learn the importance of the new representation of the recovery process. Document your results with the comparative graph shown in figure 8.12. The first curve is the infected population shown previously in figure 8.5. The second curve shows the number of infected people with a conveyor stock using 2 days as the transit time. It should peak at over 5,600 in the 8th day.

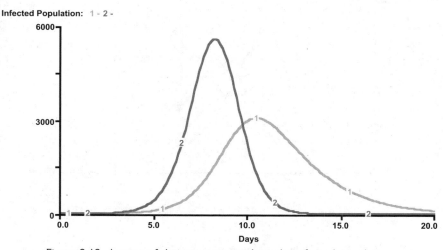

Figure 8.12. Impact of the conveyor stock on the infected population.

Exercise 8.7. Include five groups of people using arrays

Read ahead in chapter 14 to learn about arrays. Then define a dimension G to take on elements G1, G2, G3, G4, and G5 to represent five groups of people. Build the model in figure 8.13, and set the initial values of the susceptible populations to 1,999 for all five groups. The initial values of the infected populations will be 1. The same equations apply to all five groups. The key difference between the groups is the contacts per day per infected person. Set these parameters to 4, 5, 6, 7, and 8. You should then be ready to simulate a population of 10,000 people broken down into five equal-sized groups with different contact frequencies. The model simulates the infections within each group, so the basic assumption is that the groups are highly segregated from one another. (Infections can only spread between members of the same group.) Simulate the new model, and verify that you get the results in figure 8.14. This graph shows that the epidemic will spread most rapidly through the 5th group, the group with 8 contacts per day.

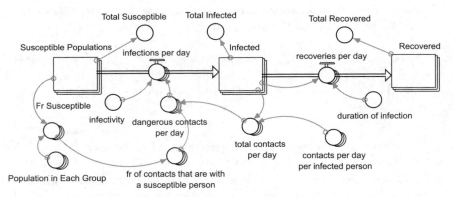

Figure 8.13. Epidemic model with arrays.

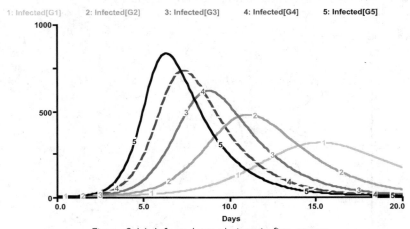

Figure 8.14. Infected populations in five groups.

Exercise 8.8. Vensim version of epidemic model

The diagram in figure 8.15 shows the Vensim version of the epidemic model. Build this model and verify that it gives the results shown in figure 8.5.

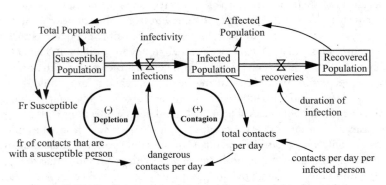

Figure 8.15 Vensim version of the epidemic model in figure 8.4.

Exercise 8.9. Include five groups of people using Vensim subscripts

Read ahead in chapter 14 to learn about subscripts and arrays. Define a subscript G (using DSS version of Vensim) to take on elements G1, G2, G3, G4, and G5 to represent the five groups of people. Build the model in figure 8.16 using the same assumptions as in exercise 8.7. Simulate the new model, and verify that you get the results shown in figure 8.17.

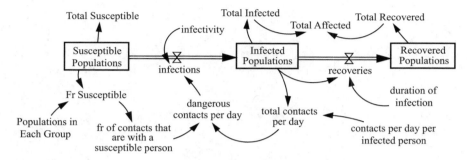

Figure 8.16. Vensim version of an epidemic model with five groups of people.

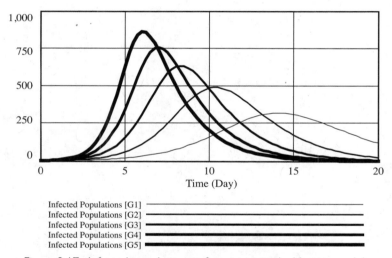

Figure 8.17. Infected populations in five groups in the Vensim model.

Further Reading

- Botkin and Keller (1998) describe the world's most deadly infectious diseases and the epidemiological transition as countries industrialize.
- Murray (2002) provides some history of epidemics and presents differential equation models of epidemic dynamics.
- Sterman (2000) describes system dynamics modeling of infectious diseases with applications to immunization policies.
- Dangerfield et al. (2001) describe a system dynamics model of HIV/AIDS in Great Britain.

- Ritchie-Dunham and Galvan (1999) describe a model of dengue fever in Mexico.
- Rahmandad and Sterman (2008) use epidemic modeling to compare system dynamics modeling with agent-based modeling.
- Walters (2003) describes six modern plagues and how we are causing them.

Chapter 9

Information Feedback and Causal Loop Diagrams

System dynamics helps us to analyze complex systems, with special emphasis on the role of information feedback. This chapter describes causal loop diagrams to portray the information feedback at work in a system. The word *causal* refers to cause-and-effect relationships. The word *loop* refers to a closed chain of cause and effect that creates the feedback.

Population Example

A population model is shown in figure 9.1, and the corresponding causal loop diagram is shown in figure 9.2. The words represent the variables in the system. The arrows represent causal connections. They are drawn in a circular manner in figure 9.2 to show the closed chain of cause and effect. The arrow from births to the population stands for the fact that births add to the size of the population. The arrow from the population to the births stands for the fact that a larger population will tend to have more births in the future. The important thing is the closed chain of cause and effect: a larger population leads to a higher number of births, and higher births lead to a higher population. This is an example of *positive feedback*. (The term *positive* comes from control engineering and has no value connotation.) We label positive feedback loops with a plus sign (+) in the middle of the loop.

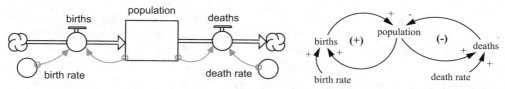

Figure 9.1. Population flow diagram.   Figure 9.2. Population causal loop diagram.

The loop on the right in figure 9.2 shows the arrows between deaths and population and between population and deaths. The arrow from deaths to the population is labeled with a minus sign (–) at the tip of the arrow. This means the deaths will reduce the size of the population. The arrow from the population to the deaths is a positive arrow. It stands for the fact that a larger population will tend to have greater number of deaths. The closed chain of cause and effect creates *negative feedback*. (The term *negative* is also from the field of control engineering; it has no value connotation.) We label negative feedback loops with a minus sign (–) in the middle of the loop.

Feedback Control in a Home Heating System

Since the feedback terminology originated in control engineering, it's useful to consider your home heating system, an engineered system designed for temperature control. If you live in cooler regions, you're familiar with the thermostat control system that governs furnace operation. You haven't seen a model of this system, but you can still appreciate the causal relationships in figure 9.3. The arrow from energy content to air temperature is positive because these variables change in the same direction. (If energy content increases, the air temperature increases.) Moving around the loop on the right, the next arrow connects the air temperature to the heat loss. This arrow stands for the fact that a higher indoor temperature causes greater heat loss through the walls and windows of the house. The next arrow connects heat loss to the energy content. This arrow is labeled as negative because the heat loss will reduce the BTUs of energy stored in the house. The closed chain of cause-and-effect relationships forms a negative feedback loop. If left to act on its own, this loop will work to remove the energy from the house until the air temperature falls into balance with the outdoor temperature.

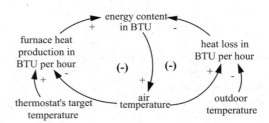

Figure 9.3. Feedback loops in a home heating system.

The loop on the left shows the heat added to the room by the furnace. It depends on the capacity of the furnace and the fraction of the time that the furnace is operating. The operation is controlled by the temperature of the room at the location of the thermostat. When the air temperature goes up, the thermostat turns the furnace off. In other words, a higher air temperature means less heat production from the furnace. The furnace loop is a negative feedback loop that strives to move the inside air temperature close to the target temperature.

Now, can we anticipate the combined effect of the two loops? Think of your home heating system as two coupled negative feedback loops that are striving to reach different goals. If the house is equipped with a large furnace, the heat production loop will dominate, and it will succeed in driving the indoor air temperature to the target. However, if the furnace is inadequate to the task, the heat loss loop will dominate, and the indoor air temperature will fall until it matches the value of the outdoor temperature.

Positive and Negative Feedback

Positive feedback is a closed chain of cause and effect that can lead to growth in the system. A population can grow when more births lead to more people, and more people lead to more births in the future. You can identify a loop as positive if it appears to perform the same function as the population loop in figure 9.2. If it leads to a larger and larger system over time, it's probably positive feedback. In some cases, the function of the loop will not be obvious. When this happens, you may simply count the number of negative arrows around the loop. If there are no negative arrows, it's a positive feedback loop. If you count two, four, six, or any even number of negative arrows, it's also positive feedback.

The feedback loop involving deaths in figure 9.2 is typical of negative loops that wear down the system over time. We may see a population declining owing to deaths, a workforce declining owing to retirements, or a lake shrinking owing to evaporation. If you encounter a loop that appears to be wearing down a system over time, it's negative feedback. There will be cases in which you won't be able to discern the function of the loop. When this happens, you can simply count the number of negative signs around the loop. If there is only one negative arrow, it's a negative feedback loop. If you count three, five, seven, or any odd number of negative signs, it's also negative feedback.

Now consider the furnace feedback loop in the home heating system. Its purpose is to control the energy content of the room so that the temperature approaches a goal. The heat loss loop is also goal oriented. Its goal is to bring the indoor air temperature into balance with the outdoor temperature. When you encounter a loop that appears to be striving for a goal, it's negative feedback. As before, you can check your intuition by verifying that there is an odd number of negative signs around the loop.

Guidelines for Labeling Each Arrow

The arrows in causal loop diagrams are labeled + or − depending on whether the causal influence is positive or negative. We use the + to represent a cause-and-effect relationship where the two variables change in the same direction. In figure 9.4, for example, the arrow can mean that an increase in A causes an increase in B. It could also mean that a decrease in A causes a decrease in B. This is the meaning of the + arrow connecting the birth rate and the births in figure 9.2, for example. A higher birth rate means higher births; a lower birth rate means lower births. The two variables change in the same direction. A positive arrow can also stand for the effect of a flow on the stock that accumulates the inflow. In figure 9.1, for example, the positive arrow from births to population stands for a flow into the stock.

Figure 9.4. Causal arrow with a + sign.

Figure 9.5. Causal arrow with a − sign.

Figure 9.5 shows an arrow marked with a − sign. This is used when the two variables change in the opposite direction. The arrow in figure 9.5 can mean that an increase in X causes a decrease in Y. It can also mean that a decrease in X causes an increase in Y. This is the meaning of the − arrow connecting the outdoor temperature and the heat loss in figure 9.3. A higher outdoor temperature means less heat loss; a lower outdoor temperature means higher heat loss. The two variables change in the opposite directions. A negative arrow can also stand for the causal link between a flow and the stock that is drained by the flow, as in figure 9.2, where the deaths cause a reduction in the population.

When assigning the + or − to each arrow, focus your attention on one causal connection at a time. Try to imagine a situation where every other variable is held constant; only the variable under consideration is allowed to vary up or down. Suppose we are considering whether to assign a + or − to the arrow linking C to X in figure 9.6. We should consider the impact of C on X if the values of A and B are held constant. (You will sometimes see the Latin *ceteris paribus* ["other things being equal"] used to remind us to think of the impact of C on X with *other variables held constant.*)

It's sometimes difficult to force ourselves to hold other variables constant because we immediately see the consequences of a change in our mind's eye. Consider what goes through your mind when labeling the arrow from target temperature to heat production in figure 9.3. You are told to think of the temperature

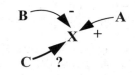

Figure 9.6. Thinking about the effect of C on X.

inside the house as constant. Then you are to ask if an increase in the target temperature would cause an increase in the heat production. Since it does, you would label that arrow with a +. But you would immediately recognize that higher heat production leads to more energy content and a higher temperature inside the house. This may seem like a contradiction—wasn't the inside air temperature supposed to be constant? Don't let this apparent contradiction bother you. We are simply holding variables constant for purposes of labeling a diagram. The labels are a communication device to help us identify the feedback in the system. The actual simulation of whether temperatures go up or go down is reserved for the computer model.

Examples of Coupled Loops

You've seen two examples of loops working in tandem with one another. The two loops in figure 9.2 show the interplay of births and deaths working through population as a common variable. Figure 9.3 shows the interplay of heat production and heat loss. Figure 9.7 shows a third example from economics.

The diagram shows two loops interacting through the market price of an automobile. Car production builds the inventory of cars at the dealers. A higher inventory leads to a lower market price, and lower market prices cause less car production in the future. There is only one negative arrow in the production loop, so we know that this is negative feedback. This loop describes the effect of the supply curve from your study of economics. The loop on the right side of the diagram involves the demand curve from introductory economics. If the price were to increase, the retail sale of cars would tend to fall (all else held equal). Retail sales drain the in-

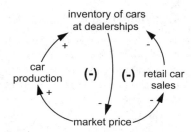

Figure 9.7. Coupled loops in an automobile market.

ventory of cars held in stock at the dealership. And a decline in the inventory will cause the dealers to raise their prices in the future. There are three negative arrows in the demand loop, so we know that it also generates negative feedback.

Negative Loops Act to Negate Outside Disturbances

These examples reveal a basic property of negative loops—they act over time to negate the impact of outside disturbances. In the home heating system, for example, the furnace loop acts to negate the impact of changes in outdoor temperature. If it gets colder outside, there is more heat loss and a drop in the indoor air temperature. But the furnace then comes on more frequently, and indoor air temperature is returned to the target value. In the automobile market, both the production and sales loops act to negate outside disturbances. Let's suppose that a hailstorm destroys 20% of the cars held in inventory. The market price would increase; higher prices would spur more production; and higher production would build inventory. Higher prices would also reduce demand, which would also help to build inventory. Once again, the negative loops act to negate the impact of the original disturbance.

The hailstorm story illustrates a useful way to think about the feedback. Pick any variable in the loop, invent a disturbance, and ask yourself if the closed chain of cause and effect will end up negating the impact of the disturbance. If it does, you have negative feedback. Positive feedback loops have the opposite effect. They lead to changes over time that amplify the impact of the disturbance. Suppose, as an example, the population in figure 9.1 receives 1 million people who migrate into the country. This adds to the population, which means there will be more births and a still-higher population in the future. The effect of the positive feedback is to magnify or reinforce the impact of the outside disturbance.

These examples confirm what was said previously about *positive* and *negative feedback*. These terms do not carry any judgmental meaning. Positive loops are not called positive because they deliver good results. Negative loops are not expected to deliver poor results. The terms originate from the field of control engineering where *positive feedback* could refer to runaway growth (e.g., a poorly designed microphone loudspeaker system that creates an ear-shattering sound). Control engineers use the term *negative feedback* when designing devices to achieve stable control of electromechanical systems. A common example is the servo-mechanism.

Servomechanism

A servomechanism is a self-acting machine to control the operation of a larger machine. One of the most famous servomechanisms is the "centrifugal governor" or "flyball governor" shown in figure 9.8. This machine was first used in flour mills to control millstone operation, and it was adapted for automatic control of steam engines by James Watt. He wanted a self-acting machine to control the speed of steam engines so they would be a safer part of a larger machine such as a steamship. The illustration shows steam from the boiler approaching a chamber where the valve position controls the amount of steam vented to the atmosphere. Most of the steam is allowed to reach the turbine, which will spin the shaft and deliver power to the ship. But the same shaft is connected to the governor. Now, imagine what would happen if the steam engine were running "too hot." Too much steam would flow toward the chamber, and some of it would reach the turbine, spinning it more rapidly. But the governor would also spin more rapidly, and the centrifugal force drives the flyballs farther out and lifts the valve in the chamber. The rising valve vents more steam to the atmosphere, and less steam reaches the turbine. The rotation speed of the shaft that drives the ship is thereby controlled by the design of the governor, not by the steam that happens to be produced at the boiler.

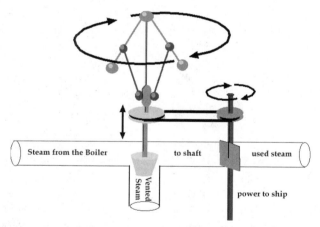

Figure 9.8. The centrifugal (or flyball) governor.

The centrifugal governor operates through the negative feedback loop shown in figure 9.9. Steam generation is an input to the system, but only a fraction of the steam is allowed to reach the turbine. This fraction is controlled by the height of the valve in the chamber. Let's work our way around the loop with a "thought experiment" in which the shaft is spinning too slowly. A low speed means lower speed of the governor, less centrifugal force on the balls, and a lower height of the valve in the chamber. A lower height lowers the fraction of steam vented to the atmosphere. This means that more steam will reach the turbine, and more steam to the turbine will correct for the shaft spinning too slowly.

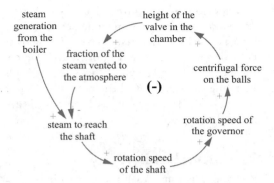

Figure 9.9. Negative feedback in the centrifugal governor.

Look at the Big Picture

Feedback ideas may be new to you, so you may not immediately recognize the feedback loops that control the systems around you. You will come to see the loops more readily if you ask yourself about the bigger picture. To illustrate, I ask you to imagine that you are in charge of a power plant project that may require 5 years for construction. You are an expert on power plant projects, and you anticipate a need for 1,000 workers at the peak of construction. The workers will look for housing in the nearby town with only 2,000 people. The town has housing and public facilities for a small population, so it will be inundated when the new workers and their families arrive. How will the town cope with all the new people?

Figure 9.10 shows the housing part of a diagram to address this question. The starting variable is the number of construction workers at the project site. This is what you know best; there could be around 100 construction workers in the early stages of site preparation, but their number could then reach 1,000 during the peak of construction. The construction workers would add to the total workers in town and to the housing demand. The surge in construction workers would lead to a surge in the demand for housing, a drop in the vacancy rate, and a drop in the construction worker satisfaction.

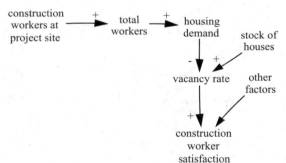

Figure 9.10. Housing in a boomtown.

This diagram represents one part of the bigger picture. It focuses on the number of workers and their dissatisfaction when they cannot find adequate housing. One way to open our eyes to the bigger picture is to think of the inputs and ask if the inputs vary independently from what is happening in our system. If you were running the power plant construction project, you should ask yourself about the expected number of construction workers at the site. Would they really peak at 1,000? Or would their numbers depend on conditions in the town?

Figure 9.11 shows what could emerge when you answer this question. It includes a normal need for construction workers, the numbers that you know well. But the actual number of workers depends on their productivity. (If the workers are half as productive, you could need twice as many to do the same job.) Worker productivity will fall if there is a decline in worker satisfaction owing to inadequate housing. The bold arrows draw our attention to the positive feedback loop in the boomtown. Suppose we start at the top with an increase in the normal

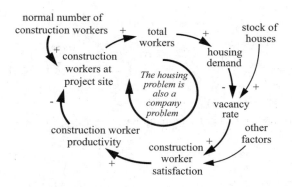

Figure 9.11. The bigger picture of the boomtown housing problem.

number of workers. This will lead to an increase in the workers at the site, an increased demand for housing, and a reduction in the vacancy rate. Lower vacancies would lead to lower worker satisfaction and lower productivity. If you were in charge of this project, you would then have to hire still more workers to get the job done. This vicious circle acts to make the situation worse and worse over time. The end result could be a massive increase in the cumulative cost of your construction project.

The positive feedback loop is named to remind us of the new perspective that comes from looking at the bigger picture: *the housing problem is also a company problem.* If you have to manage a workforce with high turnover and low productivity, the cost of the power plant could increase by hundreds of millions of dollars. The housing problem is a threat to your company, and you need to help the town deal with the problems during the construction boom.

Now, imagine what could happen if you make the housing argument to your colleagues at the power plant company. They are experts in power plants, not in housing. They will ask what you know about local zoning rules, infrastructure provision, and the profitability of housing. These are all new to you, and you will be tempted to retreat to your comfort zone. Perhaps it would be safer to ignore the factors in figure 9.11. And if you are accustomed to models to help plan the construction project, perhaps you should focus on what you know best. Building a model of the system in figure 9.11 might be the quickest way to expose yourself to criticism.

The point of the boomtown story is that it takes courage to look at the bigger picture. A causal loop diagram can go a long way to give us a new perspective on the problem. We should then follow up with a simulation model to test our theory about the larger system. You'll need a modeling method that focuses on the feedback structure of systems. System dynamics provides the method; you must provide the courage to put it to use on the big picture.

Creating Causal Loop Diagrams from Flow Diagrams

This book began with examples of flow diagrams before we looked at the causal loop diagrams, and my recommendation is that we use stock-and-flow diagrams prior to drawing the causal loop diagrams (see table 2.1). This is a pragmatic suggestion based on the observation that most people "see" the stocks and flows more easily than they "see" the loops. Now suppose we have a good flow diagram, and we wish to draw the causal loop diagram. How can we make sure we find all the loops? And how can we avoid drawing a diagram that ends up looking like a plate of spaghetti?

The best staring point is the same as when we first constructed the stock-and-flow diagrams: *start with the stocks!* The best way to begin is to write down the names of the stocks and leave room for the other variables to appear later. Then add the arrows to represent each flow. Inflows are labeled with a +; outflows are labeled with a –. Then work your way from one flow to another adding the necessary arrows to explain each flow. By the time you're done, you will have found every loop in the model.

Let's try this approach with the population diagram in figure 9.12. Figure 9.13 illustrates the first step toward a causal loop diagram; it shows the three stocks and the four flows. The

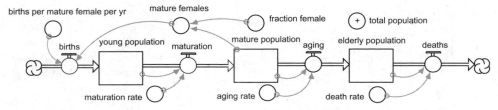

Figure 9.12. Flow diagram of a population model.

Figure 9.13. Start with the stocks and flows.

arrows represent the impact of each flow on the stocks. At this stage, you won't see a single feedback loop in the diagram. They appear later when we add the arrows to explain each of the flows.

Figure 9.14 shows how the diagram would appear after we explain the deaths, aging, and maturation flows. Deaths are shown to depend on the size of the elderly population and the death rate. Adding these arrows reveals a negative feedback loop on the right side of the diagram. The next flow is aging. It depends on the size of the mature population and the aging rate. Adding these arrows reveals a second negative loop in the middle of the diagram. The maturation flow depends on the young population and the maturation rate. Adding these arrows closes the third negative loop in figure 9.14. At this point, you'll notice a trend developing—each of the loops works its way through one of the stocks.

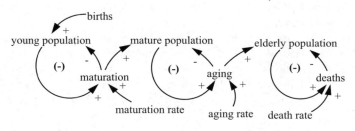

Figure 9.14. Explaining three of the flows.

Figure 9.15 completes the causal loop diagram by explaining births, the final flow in the model. Births depend on the number of mature females and the births per mature female per year. The number of mature females depends on the size of the mature population and the fraction of the population that is female. Adding these arrows allows us to see the

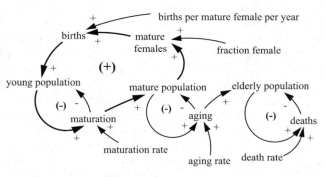

Figure 9.15. Explaining births reveals the positive feedback loop.

positive feedback loop in the system. This loop involves two of the stocks—the stock of young people and the stock of mature people.

Figure 9.15 shows all the feedback loops in the system. The one positive loop gives the system the power to grow. The three negative loops act to wear out the system over time. The only variable from the original flow diagram that does not appear in the final diagram is the total population. If you pencil in the total population, you will discover that its inclusion adds no new feedback to the system. At this point, you may choose to include the total population or leave it out, depending on the purpose of the diagram. The normal purpose is to discover the feedback loop structure, so it would normally be better to ignore the total population to avoid cluttering up the diagram.

Software Support for Loop Identification

You can turn to the software to support your investigation of the feedback loops once you've constructed the stock-and-flow diagram. For example, Vensim's loops tool will provide a listing of all the loops along with the names of the variables appearing in the loop. The Stella loop pad tool is illustrated in figure 9.16. It begins by asking which of the stock variables is of interest. We select a stock, and Stella responds with a list of loops that work their way through that particular stock. Figure 9.16 shows an example for the population model. This diagram was obtained by selecting the young population as the relevant stock and asking for the loop that involves births. Stella responds with all the loop's variables interconnected with arrows in a circle. It then asks you to label the loop with the letter *C* or the letter *R*. We use *C* if the loop acts to counteract change; we use the letter *R* if the loop acts to reinforce change. In other words, *C* stands for negative feedback, and *R* stands for positive feedback.

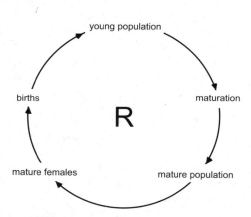

Figure 9.16. Stella's Loop Pad tool.

A Stock in Every Loop

The loop tools are organized around the principle that there must be at least one stock in every loop. You might be wondering why this is true. Clearly, the rule applies to the population model, but what about the loops in the automobile market model in figure 9.7? In this case, you would probably assign a stock to the inventory of cars at dealers. And what about the heat control model in figure 9.3? Is there at least one stock in each loop? In this case, the stock is probably the energy content. It would increase owing to furnace heat production and be reduced by the heat loss.

Of course, three examples do not prove a fundamental principle. Suppose you were to try an experiment to contradict the principle. Construct the population flow diagram in figure 9.12, and attempt to close a feedback loop that does not involve one of the stocks. Imagine, for the sake of argument, that the births per mature female depends on the fraction female. Then, to close a loop without involving a stock, suppose that the fraction female depends on the births per mature female per year. Give this a try. You'll discover that you can't make the closing connection because Stella will respond with: *Sorry, but that would create a circular connection.*

This warning may seem surprising since the system dynamics approach is predicated on the need to understand circularity in systems. Why should we be afraid of a circular connection? The purpose of this warning is to prevent lazy thinking about circularity. If you encounter this warning, take it as a challenge to think more deeply about the system. Previous students typically encounter this warning when attempting to mimic a set of simultaneous algebraic equations from a textbook description of a system. When you find yourself in this predicament, think about the stocks and flows that are hidden behind the textbook description. In many cases, you will add a stock, and you will be able to close the loop in a more realistic manner. In some cases, the flows act almost instantaneously to adjust the stock, so the new stock is not a good addition to the model. In these situations, your best approach is to solve the simultaneous algebraic equations and insert the solution as a converter variable.

Feedback Loops for S-Shaped Growth

The main reason for drawing causal loop diagrams is to see the feedback loops that can help us explain dynamic behavior. And when we see systems with similar dynamic behavior, we expect to see a similar combination of feedback loops. Let's explore this idea with the examples of S-shaped growth from chapter 7.

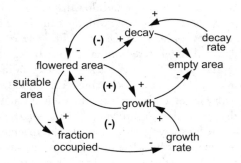

Figure 9.17. Feedback loops in the flowers model.

Figure 9.17 shows the feedback loops in the flowers model in chapter 7. There is a single positive loop that gives the flowers the power to spread early in the simulation. There is a single negative loop involving the decay of flowered area over time. The third loop acts to slow the rate of growth as the flowers occupy a larger fraction of the suitable area.

Figure 9.18 shows the feedback loops in the sales company model in chapter 7. The positive loop involves widget sales, revenues, and the budgeted size of the sales force. It gives the system the power to grow early in the simulation because each salesperson is able to generate a large contribution to the departmental budget. Figure 9.18 shows a negative feedback loop involving the departures of employees. It is similar to the decay of the flowered area, and it acts

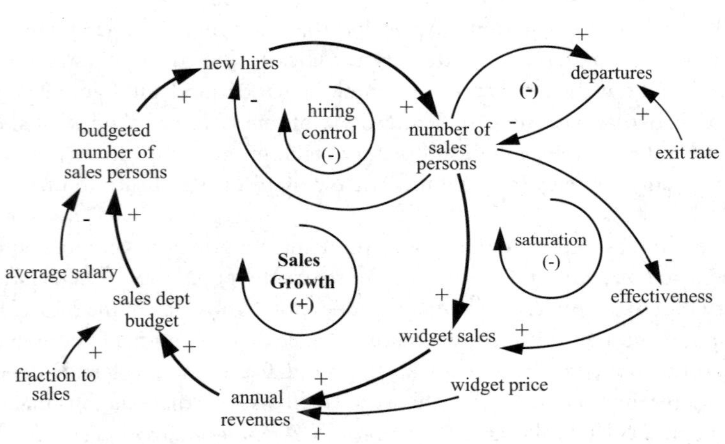

Figure 9.18. Feedback loops in the sales model.

to wear out the system at a constant rate over time. The saturation loop shows the decline in effectiveness of each salesperson as the number of salespeople grows over time. This is negative feedback that acts to weaken the strength of the positive loop as the salespeople saturate the area. Eventually, each salesperson will only be sufficiently effective to contribute enough to the departmental budget to cover his or her own salary, and the company will come into equilibrium.

The comparison of the flower and sales systems confirms that their similarity in behavior is not a coincidence. These two systems have the same feedback structure. Each is powered by a strong positive feedback loop early in the simulation; each is subject to a constant rate of decay; and the growth in each system is slowed by the actions of a single negative loop as the system fills up the space available. On the other hand, the feedback structure of the two systems is not absolutely identical. Did you notice the hiring control loop in the sales model? It acts to control new hires to bring the number of employees up to the budgeted number. This is an additional negative feedback loop that is not present in the flowers model. The hiring control loop makes sense for a company with an explicit policy for hiring new employees, but it would not make sense in a flower model since nature does not set explicit targets for the growth in flowers. The two systems differ in this respect, but their similarity in terms of the other three loops is sufficient to guarantee that they both exhibit S-shaped growth over time.

Figure 9.19 shows that a similar collection of feedback loops is responsible for the S-shaped pattern in the epidemic model. The contagion loop is highlighted to draw attention to its role in causing the rapid growth in the infected population at the outbreak of the epidemic. This loop is analogous to the sales growth loop in figure 9.18. The depletion loop provides negative feedback that acts to slow the growth of the epidemic as fewer susceptible people remain. This loop plays a similar role to the saturation loop in the sales model. The final loop in figure 9.19 is the recovery loop, a negative loop similar to the departures loop in figure 9.18.

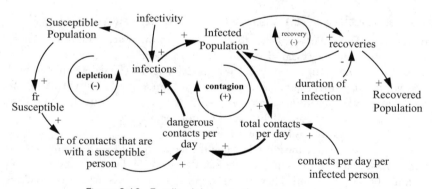

Figure 9.19. Feedback loops in the epidemic model.

Loop Dominance and Density-Dependent Feedback

These examples all show a shift in loop dominance during the course of the simulations. The early years are dominated by a positive feedback loop that gives the system the power of exponential growth. The image in figure 9.20 tells the story—the positive feedback loop outweighs the negative loop in the early stage of the simulations. As the systems fill the space available, dominance shifts to a negative loop such as market saturation or depletion of the susceptible population. The shift in dominance occurs in a smooth, gradual manner, and the system reaches equilibrium.

Figure 9.20. Image of loop dominance in which the positive feedback loop outweighs the negative feedback loop.

You'll be able to see the shifts in loop dominance if you think about the simulated behavior and take the time to draw the causal loop diagram. You'll find that identifying the dominant loops becomes easier after you've seen several systems with the same dynamic behavior (and the same feedback loop structure). Understanding loop dominance is an important theme in system dynamics and in systems thinking (D. H. Meadows 2009, 45). You will learn to identify the dominant loops in most models with the help of the diagramming methods explained in this chapter. (The book's website [the BWeb] describes a variety of mathematical methods to assist in the identification of the dominant loops.)

The term *loop dominance* is not commonly used by modelers in population biology or ecology. The more common term is *density-dependent feedback*, which stands for the growing strength of a feedback effect as the population density increases. With higher density, the negative feedback loop grows in strength and eventually brings a stop to the growth. Such feedback is present in the flowered area model, in the sales model, and in the epidemic model. Indeed, density-dependent feedback has to be present for these systems to exhibit S-shaped growth. For a contrasting example, look at the population model in figure 9.12 with the feedback loops shown in figure 9.15. It has one positive loop and three negative loops, but there is no density-dependent feedback. If you build and simulate this model, you will find that it can exhibit either exponential growth or exponential decay. No matter what parameters you assign, this model will never show S-shaped growth since none of the loops changes strength as the population grows larger over time.

Communicating with Causal Loop Diagrams

Causal loop diagrams help us see the feedback structure in the system, but they are not always needed if the stock-and-flow diagram can communicate the same ideas. For example, the sales model diagram (figure 7.10) and the epidemic model diagram (figure 8.8) highlight the key loops. As you become more familiar with system dynamics, the loops will stand out more clearly in your mind's eye. In these cases, the best way to communicate the loops may be to name them within the stock-and-flow diagram. Short names like "contagion" or "depletion" are best, and they should fit easily inside the loop.

Causal loop diagrams are most useful when the loops are not easy to see in the model diagram (the stock-and-flow diagram). Remember that the causal loop diagrams are for communication, not for simulation. They help us think about the feedbacks in the system. They are used extensively in the system dynamics literature because there is a general consensus that understanding the feedback structure is key to understanding the dynamic behavior.

Although there is a consensus on the fundamental purpose of the diagrams, there is plenty of room for variations in how these loops are communicated. Figure 9.21 shows some of the symbols that might appear in a positive loop. For instance, you have seen the letter *R* in figure 9.16, where it was used to communicate that the positive feedback loop acts as a reinforcing loop. The letter *R* is the first of several symbols in figure 9.21. The next symbol is the snowball rolling downhill. It is sometimes used to remind us that a positive loop can lead to a snowballing effect over time. Some authors might place the exponential growth graph inside the loop to communicate the likely behavior over time. Some wish to communicate the favorable results by placing the smiling face within the loop. If the loop leads to unfavorable results, some authors place the frowning face inside the loop. I do not use these symbols in this book. I prefer the "(+)" sign as a neutral symbol for positive feedback that is likely to serve under a broad range of examples, regardless of the context.

Figure 9.21. Symbols sometimes used for a positive feedback.

Figure 9.22 shows symbols that might appear in a negative feedback loop. The *C* reminds us that a negative loop acts to counteract the effect of outside disturbances. The teeter-totter reminds us of the balancing acts of negative feedback. The exponential decline graph might be placed in a loop to communicate the expected pattern of behavior. Some authors use the frowning face if they are disappointed with the loop's effect; these authors would then use the smiling face if they are happy with the loop's effect. For this book, I rely on the "(–)" sign as a neutral symbol for negative feedback that is likely to serve under a broad range of examples, regardless of the context.

In reading about system dynamics, you should also be prepared to encounter a variety of symbols on the tips of the arrows in causal loop diagrams. Some authors do not place the + sign at the top of the arrow. They find it more descriptive to use an *S* to remind them that the two variables change in

Figure 9.22. Symbols sometimes used for negative feedback.

the same direction. With this convention, they are likely to use an *O* rather than a minus sign when two variables are likely to change in opposite directions. Other authors may follow yet another convention—they will simply leave the arrows without any labels because they believe the accompanying text makes the proper points about the feedback loop structure. Still others will leave the positive arrows unlabeled and will attach a U-turn symbol to the negative arrows to denote that the two variables change in opposite directions.

The variety of labels can be confusing at first. But you now know the fundamental feedback concepts, so you'll be able to understand the meaning of causal loop diagrams that use different conventions. (Think of the different labels as different regional accents that add color to the language of system dynamics.) But you should be cautious when looking at diagrams with similar appearance used within a different field of study. The words may look the same, but the authors may not be "speaking the same language." For example, the words and arrows in the "influence diagrams" used in decision analysis (Kirkwood 1992) may look like the diagrams in this chapter. But the decision analyst's "influence diagram" has an entirely different purpose than a causal loop diagram (BWeb).

Beginners' Difficulties

Learning to use causal loop diagrams is like learning a new language. If your experiences are like those of other students, you will probably find it easier to read a diagram from the literature than to create a diagram of your own. Your first attempts may be cluttered with too many arrows, and the loops may not make sense. A frequent tendency among new students is to create spaghetti diagrams—diagrams with so many arrows that the loops are hidden within a plate of spaghetti. When asked why they included so many arrows, students answer that they want the diagram to be complete. They document each and every interconnection in the model with a separate arrow in the diagram. The end result is a cluttered diagram. These students have forgotten that the purpose is to show the loops in the system, not the clutter. (The better place to document each and every connection is the stock-and-flow diagram.)

Another problem for some beginners is the use of action words in the causal loop diagram. The words should be nouns, not verbs. To illustrate the problem with action words, suppose we are thinking about the supply and demand for cars (see figure 9.7). We might think about the consequences of an increase in the market price and place "market price is going up" at the bottom of the diagram in figure 9.23. The positive arrow to car production makes sense because car production would go up if prices go up, and the negative arrow to sales makes sense as well. Then we explain that inventories will go up because of both higher production and lower sales. So far, so good. But a problem arises with the arrow from inventories to the price. There is no way to determine whether this arrow is positive or negative because of the apparent contradiction between the inventories going up and the market price going up. The problem with this diagram is the initial decision to use action words like "price is going up." Keep these thoughts in your head, but limit the words in the diagram to nouns like "price" and "production." Remember, the "action is in the arrows, not in the words" (Richardson and Pugh 1985, 28).

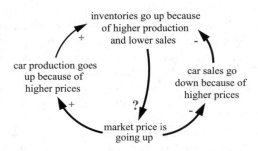

Figure 9.23. Avoid action words in a causal loop diagram.

Causal Loop Diagrams and the Modeling Process

Causal loop diagrams help us see the feedback structure of the system. Information feedback can take the form of positive feedback or negative feedback. Positive feedback loops act to magnify the impact of changes from outside the loop, and they often represent how the system is able to grow in exponential fashion. Positive feedback loops sometimes take the form of "vicious circles" that act to make the situation more and more difficult over time. Negative feedback is entirely different. The negative loops act to negate the impact of the change from outside the loop, and they often show how a system strives to reach a goal. They sometimes cause the system to wear down over time.

I recommend drawing the causal loop diagram after you have constructed the stock-and-flow diagram (see table 1.2). Look for a positive loop if you expect the system to show exponential growth. Look for negative loops if your system is reaching for a goal or wearing out over time. It's the interaction of the different loops that will help us explain the system's dynamic behavior. But you might be wondering, What if the diagram does not show any feedback loops? In this case, your model will not show interesting dynamics on its own. (The dynamics will simply be imposed from the external inputs.) If you find yourself in this situation, return to the earlier steps of modeling. If you expand your perspective to look at the bigger picture, the new model might include the feedback loops needed to understand the dynamic behavior.

Exercises

Exercise 9.1. Joe fills the tank

Draw a causal loop diagram to reveal the feedback loops in the gas tank model (exercise 3.10).

Exercise 9.2. Loops in the two-bottles system

Draw a causal loop diagram for the two-bottles model (exercise 6.10). You should see a negative loop that controls the height of water in the first bottle and a similar loop to control the height in the second bottle.

Exercise 9.3. Study for the grade

Draw a causal loop diagram to reveal feedback in the study habits of a student who invests the time necessary to reach a grade. Assume that the student receives grade feedback from the teacher during the course of the semester and is able to adjust study time up or down as needed to reach the goal.

Exercise 9.4. Study because of growing interest

Draw a causal loop diagram of a student whose study habits are dominated by a growing interest in the subject. More interest leads to more questions, which leads to more study time, which leads to still more questions.

Exercise 9.5. Error

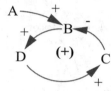

Figure 9.24. Mark the error in this diagram.

Exercise 9.6. Error

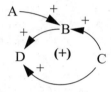

Figure 9.25. Mark the error in this diagram.

Exercise 9.7. Label the loops

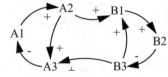

Figure 9.26. Label the three loops as (+) or (−).

Exercise 9.8. Count the loops

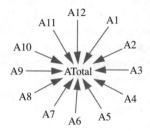

Figure 9.27. How many loops are in this diagram?

Exercise 9.9. Count the loops

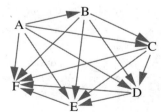

Figure 9.28. How many loops are in this diagram?

Exercise 9.10. Advantage to the advantaged: VCR sales

Figure 9.29 is a start of a causal loop diagram of the competition between two types of videocassette recorders (VCRs). One type is the VHS (Video Home System), which is in common use today. The other is Beta, a solid technology, but one that is not seen in stores today. Let's assume that a larger market share of VHS leads video outlets to stock more prerecorded tapes in VHS format, thereby increasing the attractiveness of VHS recorders. This loop is sometimes called *advantage to the advantaged*, and it is often said to be responsible for the VHS technology's dominating the market. Finish the diagram by labeling each arrow and each loop.

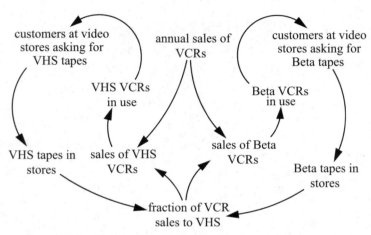

Figure 9.29. Complete this diagram of market share competition.

Exercise 9.11. Advantage to the advantaged: your campaign for president

Imagine you are running for president, and your best consultants are arguing over why you are not making progress in the primary. The first consultant says you needs more voter recognition before your ranking in the polls will improve. The second consultant says you need more campaign contributions so you can buy more television time to get your message out. You suspect that they are both right because of the circular cause and effect in the system. Draw a causal diagram to show the positive feedback loop that influences the success of your political campaign. It should include the factors that both your consultants are talking about. It should also include similar factors for your opponent in the presidential primary.

Exercise 9.12. Loops in the first model of Mono Lake

Figure 9.30 shows the first step in translating the flow diagram of the first model of Mono Lake (figure 5.4). Water in the lake, the only stock variable, is placed in the mid-

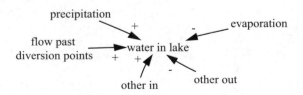

Figure 9.30. Getting started with a causal loop diagram of the Mono Lake model.

dle of the diagram. Each of the flows is included to get you started. Complete this diagram. Do you see any loops? (Remember that the first model did a poor job of explaining the decline in Mono Lake.)

Exercise 9.13. Loops in the final model of Mono Lake

Draw a causal loop diagram of the third model of Mono Lake (figure 5.9). You do not need to include every variable in the diagram, just enough to reveal the loops in the model. You should see one negative loop involving evaporation and a second negative loop involving the specific gravity. There is also a curious positive feedback loop owing to the way precipitation is simulated in the model (BWeb).

Exercise 9.14. Mono Lake export policy

Expand the causal loop diagram from exercise 9.13 to include a policy to control water exports as a function of the lake's elevation (figure 5.14). Assume that higher exports are permitted when the elevation exceeds a safe level. When the elevation declines, exports must decline. Draw any new loops that appear in the diagram because of this policy.

Exercise 9.15. Positive feedback in the land use model

Draw a causal loop diagram of the forest harvesting model (figure 6.6). It should show 6 negative loops and 1 positive loop.

Exercise 9.16. Positive feedback in the circulatory system

Draw a causal loop diagram of the circulatory system model in figure 6.11. It should show 6 negative loops and 1 positive loop.

Exercise 9.17. Positive feedbacks in global hydrologic cycle

Draw a causal loop diagram to show the 3 positive feedback loops in the equilibrium diagram for the global hydrologic cycle in figure 6.2.

Exercise 9.18. Positive feedback in closed systems

Positive feedback loops sometimes have the power to generate exponential growth in a system. Is this the case for the positive feedback loops in exercises 9.15, 9.16, and 9.17?

Further Reading

- Richardson (1986) explains potential problems in drawing and interpreting causal loop diagrams.
- Ford (1978) describes modeling of problems in energy boomtowns; further information is on the BWeb.
- D. H. Meadows (2009) describes the need for courage to look at the big picture and "defy the disciplines."
- Richardson (1991) describes the early history of system dynamics and its links with control engineering.

Acknowledgments

- The image of scales for loop dominance in figure 9.20 is inspired by the cover of *Toward Global Equilibrium* (D. H. Meadows and D. L. Meadows 1973). Several exercises from this book of readings are on the BWeb.

Chapter 10

Homeostasis

Homeostasis refers to our remarkable capability to maintain a relatively stable physiological state even when the outside environment is varying dramatically. This chapter explains homeostasis with examples from Walter Cannon's (1932) classic book *The Wisdom of the Body*. It then turns to general ideas that may be useful beyond physiological systems. The most useful idea may well be the *span of control*. The chapter closes with a discussion of the span of control and its applicability to several of the systems simulated later in the book.

Homeostasis

The term *homeostasis* was invented by Walter Cannon, professor of physiology at the Harvard Medical School. Cannon was intrigued by how the body maintains a stable state. He believed higher organisms had "learned" this ability over eons through gradual evolution. Organisms have had large and varied experience in "testing different devices for preserving stability" in the face of potential dangers. As they have grown to become more complex, Cannon believed that it was imperative that they develop more "efficient stabilizing arrangements."

Cannon opened with the example of blood pressure, a topic of paramount importance because of the need to maintain adequate pressure for the blood to "perform as a common carrier of nutriment and waste and to assure an optimum habitat for living elements" (Cannon 1932, 41). Cannon explains the physiological responses to blood loss in great detail. This chapter combines his descriptions with information from a standard text on medical physiology (Guyton and Hall 1996) to provide a concrete example of homeostasis, feedback, and the span of control.

Blood Pressure Control

Causal loop diagrams are a useful way to summarize some of the physiological responses to a disturbance in blood pressure. Let's start with figure 10.1, which deals with the blood loss from a cut or rupture. The diagram begins with the size of the initial cut or rupture in the upper left. A larger cut leads to a larger wound and greater blood loss. The body quickly senses the trauma to the local vessels and triggers a vascular spasm that acts to reduce the size of the wound.

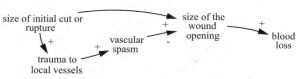

Figure 10.1. External disturbance leads to a wound and blood loss.

The vascular spasm helps reduce the blood loss, but the body's main response is clot formation. A blood clot is composed of a meshwork of fibrin fibers that run in all directions, entrapping blood cells, platelets, and plasma. These fibers adhere to damaged surfaces of blood vessels, so the blood clot adheres to the vascular opening, reduces the size of the wound, and lowers the blood loss.

Figure 10.2 summarizes some of the factors in clot formation. Clot formation begins with trauma to local vessels and the formation of the prothrombin activator, which is said to be the rate-limiting factor in causing blood coagulation. This leads to more thrombin production and greater formation of fibrin fibers, which add to the size of the clot. As the clot grows, there is still greater formation of thrombin and more formation of fibrin fibers and a larger clot. Guyton and Hall (1996, 466) describe the process of rapid clot formation:

> *Once a critical amount of thrombin is formed, a vicious circle develops that causes still more blood clotting and more thrombin to be formed: thus, the blood clot continues to grow until something stops its growth.*

It's the final closure of the wound that eventually stops the growth in the blood clot, as shown by the negative feedback loop in figure 10.2.

Now let's look to the question of how the body responds in the minutes following the loss of blood. It is crucial that the body maintain blood pressure during this dangerous time. Cannon describes several responses that act to prevent loss of blood pressure. Two of the responses act through the muscles

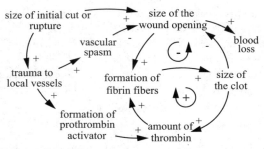

Figure 10.2. Feedback effects for clot formation.

and the spleen, as shown on the right side of figure 10.3.

The loop in the upper-right corner involves the muscles that encircle the vascular system. A decline in blood pressure causes these muscles to contract, and their contraction reduces the capacity of the vascular system. Lower capacity means higher blood pressure, so this loop acts to maintain blood pressure in the face of cumulative blood loss. The loop in the lower-right corner of the diagram works through the spleen. The spleen is a reservoir of blood, and a drop in blood pressure may trigger the spleen to contract. The contraction releases the stored blood into the general vascular system, and the extra blood adds to the total volume of blood in circulation and increases the blood pressure.

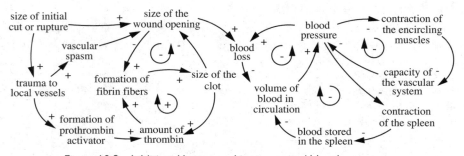

Figure 10.3. Additional loops working to control blood pressure.

Negative Feedbacks and Physiology

These responses illustrate a general point about physiological systems—that is, the majority of the feedback in physiological systems takes the form of negative feedback. Indeed, Guyton and Hall (1996, 8) explain that "essentially all control systems of the body operate by negative feedback." They argue that positive feedbacks generally "lead to instability, and often death." But, fortunately, there are "rare instances" (like clot formation) in which the "body has learned to use positive feedback to its advantage."

Cannon's description of blood pressure control is followed by descriptions of the control of blood sugar, food intake, water intake, and salinity. At first glance, these systems may appear to be entirely different in design and composition. But Cannon saw that they were fundamentally the same when viewed in terms of their ability to maintain homeostatic control within a tolerable range. Physiologists now summarize the effectiveness of the control mechanisms by the normal range (e.g., 98°F to 98.8°F) and the approximate nonlethal limits (e.g., 65°F to 110°F).

The effectiveness of homeostatic systems can also be described by their ability to survive a major disturbance. In the case of blood loss, for example, Guyton and Hall (1996, 8) estimate that the normal body can lose around a liter of blood and still survive. The body's ability to survive a major disturbance is often discussed in terms of the *span of control*, a key concept when we discuss how Cannon's ideas could transfer beyond physiological systems.

Beyond Physiology

In the epilogue to *The Wisdom of the Body*, Cannon (1932, 287) asks if there are general principles of stabilization that could apply beyond physiological systems:

> *Might it not be useful to examine other forms of organization—industrial, domestic or social—in light of the organization of the body?*

His general question could be asked of environmental systems as well. In his text *Fundamentals of Ecology*, Odum (1971, 34) explains that the term *homeostasis* is generally applied to the "tendency for biological systems to resist change and to remain in a state of equilibrium."

It's revealing to turn to *Webster's Dictionary* for definitions. *Webster's* first entry is the physiological definition. Homeostasis is defined as the

> *tendency toward maintenance of a relatively stable internal environment in the bodies of higher animals through a series of interacting physiological processes.*

Webster's then lists an alternative definition that shows how broadly the term has come to be used. Homeostasis is also defined as the

> *tendency toward maintenance of relatively stable social conditions among groups with respect to various factors (such as food supply and population among animals) and to competing tendencies and powers within the body politic, to society, to culture among men.*

When thinking about environmental systems, we should take Cannon's concluding suggestion as a challenge to think about the interacting factors that allow the system to remain stable. His suggestion seems particularly relevant for a system that has managed to survive over a long time period with large variations in the external factors. Cannon's views were predicated on the opportunity for evolutionary change:

> *The perfection of the process of holding a stable state in spite of extensive shifts of outer circumstances is not a special gift bestowed upon the highest organisms but is the consequence of a gradual evolution.*

Odum (1971, 35) stresses this point in his description of homeostasis of ecosystems. His text provides many examples to confirm that

> *really good homeostatic control comes only after a period of evolutionary adjustment. New ecosystems (such as a new type of agriculture) or new host-parasite assemblages tend to oscillate more violently and to be less able to resist outside perturbation as compared with mature systems in which the components have had a chance to make mutual adjustments to each other.*

Another general observation from physiology is that homeostasis of environmental systems will likely arise from a combination of negative feedback loops working in tandem. Once you begin to think in terms of feedback, it will be easy to spot one or two loops that govern system behavior. We should challenge ourselves to look beyond the first few loops that jump immediately to mind.

A final observation that will help in our study of environmental systems is to appreciate the extreme difficulty in verifying theories about homeostatic systems, especially if we focus our measurements on the central controlled variable in the system. By design, homeostatic systems control the central variable to erase the impact of outside disturbances. Imagine, for example, that we wanted to verify the workings of the physiological mechanisms that maintain core temperature at 98.6° F. We change the external conditions and take another temperature reading. What do we get? 98.6° F. In this situation, the body is almost too effective for our experiments to reveal the underlying mechanism, as explained by Riggs (1963, 398):

> *The truth is that in a normal unanesthetized human subject, feedback control of body temperature is so extremely effective that precise quantitative characterization of the mechanisms is well-nigh impossible.*

The temperature example raises questions of what should be measured. Certainly we measure the pivotal variable (e.g., body temperature). But equally important are measurements of shivering and sweating, the actions that the body takes to maintain control. The nature of the threat to the body will be revealed by a combination of the controlled variable and the extraordinary efforts to maintain control.

Look for Positive as well as Negative Feedback

If Cannon's ideas are to prove useful in environmental systems, it is imperative than we expand our thinking beyond negative feedback. Our study of the environment must consider both positive and negative feedback. Cannon did not emphasize positive feedback, but the human body certainly relies on positive feedback to act in a useful manner (e.g., in cell division and clot formation).

Although physiologists are certainly aware of positive feedback, it is negative feedback that dominates the discussion of homeostasis in social and economic systems. Richardson (1991, 48) observed that "to some in the social sciences, the feedback concept became identified, virtually synonymous, with homeostasis," and that "the close association of feedback with homeostasis eliminated completely the consideration of positive feedback loops." The limited perspective of some social scientists should not limit your own thinking about the management of environmental systems. You'll need to consider both positive and negative feedbacks to build your understanding of environmental systems.

Stability and the Span of Control

Figure 10.4 shows the stability testing images from chapter 6. The physiological discussion brings forth the image of the marble resting at the bottom of the cup. If disturbed from the

normal position, the forces of gravity will negate the disturbance and bring the marble back to the normal position at the bottom of the cup. But this story only works if we limit the size of the disturbance. If we push the marble beyond the edges of the cup, the restorative forces are no longer oper-
ative. The marble is now beyond the span of control. Think of the span of control as the width of the cup.

Figure 10.4. Stable, unstable, and neutral equilibrium.

The span of control is depicted in general terms in figure 10.5. The flat portion of the sketch resembles a plateau, so this portion of the sketch is sometimes called the *homeostatic plateau*. (The term *homeostatic plateau* is described by Hardin [1966, 159] and by Odum [1971, 34]. I use the term *span of control* in this book.)

The vertical axis represents the position of an internal variable that is subject to homeo-static forces. The horizontal axis represents an external input that can change over time. (That change is represented as movement to the left or right on the horizontal axis.) As long as the input remains within the span of control, the homeostatic processes maintain control. This portion of the diagram is marked with symbols for negative feedback to remind us that they are responsible for the control. And if you look closely, you'll see that the circular arrows point in opposite directions. This is a visual reminder that the feedback process on the low side of the span of control is probably very different from the feed-back process on the high side.

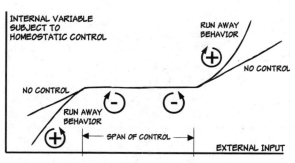

Figure 10.5. The span of control.

Body temperature control provides a concrete example of the meaning of figure 10.5. The vertical axis would represent the body's core temperature, with a normal value around 98.6° F. This value is maintained in the face of cool ambient tem-peratures by a combination of mechanisms including shiver-ing. It is maintained in the face of high ambient temperatures by a different combination of mechanisms, including sweating. The body is capable of "two-sided control," the general example depicted in figure 10.5. The body can maintain the core temperature near the normal value even with the ambient temperature as low as 63° F or as high as 90° F (Hardin 1966, 158). The span of control is from 63 to 90° F.

Now, what happens if the ambient temperature lies outside the span of control? Hardin (1966, 158) describes this situation in detail. The body would experience runaway behavior caused by a vicious circle involving cell metabolism:

> *When the thermostat fails, runaway feedback takes over: higher temperature causes meta-bolic reactions to go faster which produces more body heat which raises the temperature which causes metabolic reactions to go faster . . . and so on. This runaway feedback (if not stopped) leads to death. Below the lower temperature limit, a similar runaway feedback leads to death from stoppage of metabolism.*

The body temperature is listed in table 10.1 as the first of several examples of span of con-trol. Body temperature control is an example of two-sided control, with runaway behavior ex-pected if the external factor is outside the span of control. The next entry is the home heating system in figure 9.3, an example of one-sided control. The home is equipped with a furnace, so it can deal with low outdoor temperatures. The indoor temperature can be maintained at the target value as long as the furnace is large enough to counteract the heat loss on a cold day. Let's imagine a home heating system with a target temperature of 65°F and a span of control from 0°F to 65°F. Now suppose the outdoor temperature were to be −10°F. The indoor temperature would equilibrate at 10°F below the target. If the outdoor temperature were −20°F, the indoor temperature would equilibrate at 20°F below the target. We don't have control, but the situation is definitely different from the runaway behavior of the human body. The label "no control" is used in figure 10.5 to represent this situation.

Table 10.1. Examples of the span of control in this book.

Example (chapter)	External factor (one sided or two sided)	Internal variable (y axis)	Outside the span of control
Body temperature (BWeb)	ambient temperature (two sided)	core temperature	runaway behavior
Home heating (ch 9; figure 9.3)	outdoor temperature (one sided)	temperature inside the house	no control
Blood loss (ch 10)	size of the wound (one sided)	blood pressure	runaway behavior
Daisyworld (ch 11)	solar insolation (two sided)	Daisyworld planet temperature	no control
Salmon population (ch 15)	harvest fraction (one sided)	salmon returns to the Columbia	no control
Climate change (ch 23)	emissions of CO_2 (one sided)	atmospheric CO_2 and temperature	runaway behavior

Runaway behavior can occur when there are vicious circles waiting in the wings if the system is pushed too far. Runaway behavior is what Odum (1971) had in mind when describing ecological systems: "Homeostatic mechanisms have limits beyond which unrestricted positive feedback leads to death." Blood loss demonstrates runaway behavior. It is listed in table 10.1 as a case of one-sided control, with the size of the wound as the external factor. The internal variable that must be controlled is blood pressure. The normal human body holds around 5 liters of blood and has developed extraordinary physiological responses that can limit the size of a wound and maintain blood pressure when there is a major loss of blood. Guyton and Hall (1996, 8) estimate that the natural processes can return the circulatory system to normal operation within 2 hours if the loss is less than 1 liter. But if the body is pushed beyond the span of control, runaway behavior is expected. If there is a loss of around 2 liters, a vicious circle will take over:

> The amount of blood in the body is decreased to such a low level that not enough is available for the heart to pump effectively. As a result, the arterial pressure falls and the flow of blood to the heart muscle through the coronary vessels diminishes. This results in weakening of the heart, further diminished pumping, further decrease in coronary blood flow, and still more weakness of the heart; the cycle repeats itself again and again until death occurs.

Once this vicious circle takes over, death is normally expected within a few hours.

Looking Ahead: Daisies, Salmon, and Climate Change

The final three entries in table 10.1 turn our attention to future chapters. Chapter 11 describes Daisyworld, a make-believe world invented to illustrate how the biota could interact with their physical environment to create a wide span of control. Chapter 15 describes a salmon population with thousands of salmon returning to the mouth of the Columbia every year. The external factor is the fraction of the returning salmon that are harvested. This population will show surprising resilience in the face of unusually large harvest fractions. The final entry is climate change. You'll read in chapter 23 that scientists fear there are vicious circles waiting in the wings if the climate system is pushed too far. The threat of runaway behavior is one of the principal reasons why scientists and policy makers are calling for major reductions in emissions of CO_2 to the atmosphere.

Further Reading

- The book's website, the BWeb, lists system dynamics applications to physiology and provides modeling exercises on body temperature control.
- Many biomedical problems (such as diabetes and hypertension) can be viewed as biological control problems, for which system dynamics is ideally suited (Gallaher 1996).
- Hardin (1966, 159) describes homeostasis with examples of temperature control by different animals and humans. Figure 10.5 is my adaptation of Hardin's sketch of the homeostatic plateau.
- Odum (1971, 34) describes homeostasis in ecological systems, drawing on Hardin's description to depict the homeostatic plateau.
- Richardson (1991) examines Cannon's ideas and their importance to the spread of feedback thinking within the social sciences and the systems sciences.
- Capra (1996) describes the influence of Cannon's ideas in *The Web of Life*, a book for the general reader.

Chapter 11

Temperature Control on Daisyworld

Imagine a planet inhabited by only two plants—white daisies and black daisies. The white daisies have a high albedo. The planet surface covered by white daisies tends to reflect much of the incoming solar luminosity. The black daisies have a low albedo. The surface covered by the black daisies tends to absorb much of the incoming luminosity. Consequently, the mix of daisies on the planet influences the absorbed luminosity and the planet's temperature. Now imagine that the planet's temperature influences the rate of growth of the daisies. If the temperature is close to the optimum value for flower growth, the flowered areas will spread across the planet. But if the temperature is too high or too low, the flowered areas will recede over time.

This imaginary world is called Daisyworld. It was created by Andrew Watson and James Lovelock (1983) to illustrate a world with close coupling between the biota and the global environment. Lovelock used Daisyworld to counter criticism of the Gaia hypothesis.

The Gaia Hypothesis

Lovelock conceived the Gaia hypothesis in the 1960s while working for NASA on the Viking mission to Mars. He joined a team of scientists designing experiments to detect life on Mars. Some experiments were designed to collect samples from the planet's surface, but Lovelock believed that the key insights were to be found in the Martian atmosphere. He reasoned that the atmosphere of a planet without life would be at equilibrium with the rocks. He compared the gas composition of the Martian atmosphere with his own calculations for the Earth's atmosphere under the assumption that Earth was devoid of life. He found a close correspondence and concluded "that there was no life on Mars. It was as dead as a cinder and no one had to spend huge sums of money to go there to find out" (Levine 1993).

Lovelock continued questioning the conditions necessary to sustain life after his NASA assignment. For example, he wondered how the Earth's surface temperature has remained relatively constant over a 3.6-billion-year period in which the heat of the sun has increased by 25% (Capra 1996, 102). He also asked how the Earth manages to maintain atmospheric oxygen concentration at 21%, a value that he feels is just below the safe upper limit for life (Lovelock 1995, 65).

These questions intrigued Lovelock (1990, 102), but they did not capture the attention of the scientific community: "Before the advent of the Gaia hypothesis, such questions were rarely asked, and would have been as pointless as asking an anatomist or a biochemist how human temperature is regulated." Lovelock believed these questions would challenge the scientific community to adopt a systems perspective because "such questions about systems cannot be answered from the separated disciplines of biochemistry or biogeochemistry, nor from neo-Darwinist biology. The answer comes from physiology or control theory." Lovelock was particularly impressed with Walter Cannon's description of the coordinated physiological processes that create homeostasis. Levine (1993, 89) reports that Lovelock's conclusions on the living planet followed from his thoughts about Earth's regulatory features. "They reminded him of Cannon's central principle of physiology: that the living body strives to maintain the constancy of its internal environment." Lovelock believed that the Earth behaves in the same way; its living and non-living parts collaborate to hold temperature and other conditions at reasonably constant levels:

> In effect, the whole Earth follows Cannon's principle—rocks, grass, birds, oceans and at-mosphere all pull together, act like a huge organism to regulate conditions. Lovelock thus construed these observations to mean that Earth itself must also live. (Levine 1993, 89)

The idea of a living organism needed a name, and Lovelock was reluctant to invent a "bar-barous acronym such as Biocybernetic Universal Systems Tendency/Homeostasis." He turned to his neighbor William Golding, who "recommended that this creature be called Gaia, after the Greek Earth goddess." Lovelock added the term *hypothesis* to emphasize the need for a dif-ferent way of thinking if proper experiments are to be designed. He argued that Gaia had al-ready proved her worth in scientific circles:

> The Gaia of this book is a hypothesis but, like other useful hypotheses, she has already proved her theoretical value, if not her existence, by giving rise to experimental questions and an-swers which were profitable in themselves. (Lovelock 1995, 10)

But Lovelock views Gaia as more than a scientific paradigm. He looks to Gaia to provide spiritual support as well:

> The Gaia hypothesis is for those who like to walk or simply stand and stare, to wonder about the Earth and the life it bears, and to speculate about the consequences of our own presence here. (Levine 1993, 11)

Views of Gaia

Levine (1993) reviews Lovelock's ideas, their reception within the scientific community, and their development into a "spiritual elixir." His review closes with a resounding endorsement:

> [Gaia] is an idea with broad powerful appeal. For scientists, Gaia is a launching platform into discovery and cross-disciplinary thinking. For the disenchanted it is a perch, and for those who have strayed from the fold, a source of renewed spiritual energy. Even secular hu-manists may feel her draw as a metaphor for transcendence without God. To be sure, few of us who have seriously dealt with the Gaia hypothesis will ever see Earth and life in the same way again. (Levine 1993, 92)

Further readings on this fascinating topic are listed at the close of the chapter. If you read further, you'll learn that one of the main criticisms stems from Lovelock's early view that Gaia sought to create an optimal physical and chemical environment for life. This sounded too much like teleology to mainstream biologists like Dawkins (1986) and Doolittle (1981). (The word *teleology* is from the Greek *telos*, meaning "purpose".) They argued that "nature does not think ahead or behave in any kind of purposeful manner" (Levine 1993, 90). Capra (1996, 107) labels these critics as "representatives of mechanistic biology." He believes they "attacked the Gaia hypothesis as teleological because they could not imagine how life on Earth could create and regulate the conditions for its own existence without being conscious and purposeful."

Lovelock responded to the teleological criticism with Daisyworld. His assumptions for this simple world were spelled out one by one and converted to a set of differential equations in a paper with Andrew Watson (Watson and Lovelock 1983). The solution to the differential equations give the equilibrium conditions. They show a world with homeostatic properties that do not rely on an explicit teleological assumption.

Watson and Lovelock were not trying to model the Earth, but rather a fictional world that displays a property they believe is important for the Earth. The sheer complexity of the earth's biota and environment argued for a hypothetical world because "the earth's biota and environment are vastly complex and there is hardly a single aspect of their interaction which can as yet be described with any confidence by a mathematical equation." They believed their message would be better understood if they invented "an artificial world, having a very simple biota which is specifically designed to display the characteristic in which we are interested, namely, close coupling of the biota and the global environment."

Daisyworld

The book's website, the BWeb, provides a dynamic version of Daisyworld that shows changing temperatures over time. The world will eventually reach equilibrium, and the equilibrium values match the results reported by Watson and Lovelock (1983). In this chapter we will summarize the wide span of control of Daisyworld and explain the feedback loops that give rise to this homeostatic behavior. Then we'll turn to a special simulator for a different perspective. The exercises ask you to experiment with the simulated world to learn its dynamic response to changes in solar luminosity and to confirm its wide span of control. You will also be asked to design a different world with a wider span of control. But first, you need to know the numerical details of Daisyworld.

The Flowered Areas

The planet surface is comprised of white daisies, black daisies, and bare ground. The flowers are subject to a constant rate of decay of 30%/yr, so you might envision that they live around 3.3 years. The flowers' growth rate depends on temperature, as shown in figure 11.1. The optimum temperature is 22.5°C. If the local temperature were 22.5°C, the intrinsic growth rate would be 100%/yr. Figure 11.1 shows a symmetric relationship. Should the local temperature fall to 5°C (or climb to 40°C), the growth rate would decline to zero, and we expect the flowered area to recede at 30%/yr.

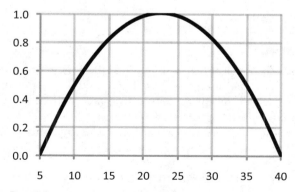

Figure 11.1. Daisies' intrinsic growth rate depends on the local temperature.

The growth rate in figure 11.1 is an intrinsic growth rate. It applies when the flowered areas are close to zero and there is plenty of bare ground. But as the bare ground is used up, the flower growth rate will be reduced to zero. The limitation of bare ground is imposed in a linear manner. For example, if 30% of the bare ground were available, flower growth rates would be 30% of the values shown in figure 11.1.

Watson and Lovelock (1983) focused exclusively on the equilibrium properties of the world. You saw the equilibrium areas in figure 6.5. Figure 11.2 expands the snapshot to include the temperatures. The temperature near the white daisies is 17.46°C; the temperature near the blacks is 27.46°C. The white temperature is located on the cool side of the optimum, and the indicated growth rate for white area is 92%/yr. The black temperature is on the hot side of the optimum; the indicated growth rate for the black area is also 92%/yr. Figure 11.2 shows that 32.6% of the area is available to accommodate new growth. The growth reduction multiplier is the linear assumption, so the multiplier is 0.326. The actual growth rates are 0.92 times 0.326, which is 30%/yr. The decay rates for both black and white daisies is 30%/yr, so both flowered areas are in equilibrium. The white flowers occupy 403 acres, and the black flowers occupy 271 acres.

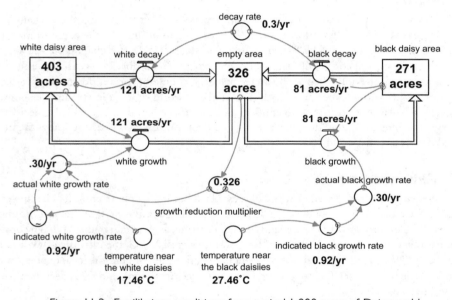

Figure 11.2. Equilibrium conditions for a typical 1,000 acres of Daisyworld.

Temperatures

Now, let's consider the temperature assumptions at the bottom of figure 11.2. Daisyworld's average temperature depends on the solar luminosity and the fraction of the luminosity that is absorbed. In this case, the solar luminosity is constant at 1.0. The fraction absorbed depends on the planet's average albedo, which is a weighted average of the albedos from the three surfaces: white daisy albedo (0.75), bare ground albedo (0.50), and black daisy albedo (0.25). Figure 11.2 shows 40% of the planet covered by whites daisies, 33% by bare ground, and 27% by black daisies.

The planet's average albedo would be 0.533, and we would expect 46.7% of the solar luminosity to be absorbed. The absorbed luminosity in this case is 0.467. The planet's average temperature is a nonlinear function of the absorbed luminosity, as shown in figure 11.3. If the absorbed luminosity is 0.467, the planet's average temperature would be 21.8°C.

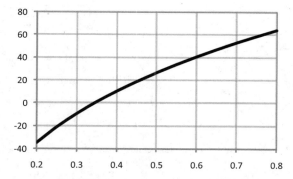

Figure 11.3. Average temperature depends on absorbed luminosity.

An important part of Daisyworld is local cooling in the vicinity of the white daisies and local heating in the vicinity of the black daisies. The seventh equation in Watson and Lovelock's paper causes the temperature near the black flowers to be somewhat warmer and the temperature near the white flowers to be somewhat cooler:

local heating = (20°C) * (planet's average albedo – black albedo) = 5.66°C
local cooling = (20°C) * (white albedo – planet's average albedo) = 4.34°C

These adjustments lead to the temperatures shown in figure 11.2:

temperature near blacks = 21.80 + 5.66 = 27.46°C
temperature near whites = 21.80 – 4.34 = 17.46°C.

Notice that the local temperatures differ by 10°C. This is no coincidence; the 10°C difference will appear year after year because of the 20°C proportionality constant and the fact that the white albedo is 0.5 larger than the black albedo. The 10°C separation is important because it helps create the relative positions of the local temperatures on the growth curve in figure 11.1. For example, the 17.46°C white temperature is located on the cool side of the optimum, while the 27.46°C black temperature is located on the hot side of the optimum. These locations set the stage for the feedback control in the system.

Daisyworld as a Feedback System

There are a total of eight feedback loops in Daisyworld. You've seen three feedback loops in figure 9.17 that control the spread of flowers across the landscape. Daisyworld has two sets of flowers, so there are six feedback loops to give S-shaped growth in the two areas of flowers. The remaining two loops involve the interaction between the flowers and the planet's average temperature. These two loops provide the negative feedback shown in figure 11.4. To confirm the negative feedbacks, let's introduce a disturbance from outside the system. Specifically, let's consider how Daisyworld would react to a sudden increase in the incoming solar luminosity. Figure 11.4 shows that the increased solar luminosity would lead to more absorbed luminosity and an increase in the average temperature. Working around the white daisy loop, we would see an increase in the temperature near the white flowers. Remember that the white flowers are operating on the cool side of the optimum temperature, so an increase in their local temperature will increase their growth rate and lead to a larger white area. This increases the planet's albedo and lowers the fraction of luminosity absorbed, thereby negating part of the disturbance from outside the system.

The black daisy loop also acts to negate part of the disturbance. The temperature near the black flowers is increased, moving the black flowers farther away from the optimum temperature for growth. This leads to less black area, an increase in the average albedo, and a reduction

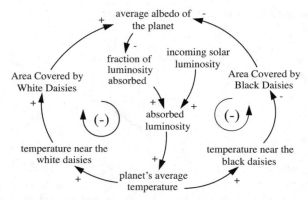

Figure 11.4. Temperature control loops in Daisyworld.

in the fraction of luminosity absorbed. These loops suggest that the planet might be able to survive a change in the incoming luminosity. But they are only suggestive of what might happen. Computer simulation is needed to see if the loops are sufficiently strong to give homeostatic behavior.

Dynamics of Daisyworld

Watson and Lovelock (1983) present Daisyworld as a collection of algebraic equations and differential equations. The equations are nonlinear, so they cannot be solved by analytical methods. But analytical methods can show the equilibrium conditions, and the equilibrium conditions occupied all of their attention in the original article. However, with the Stella version of Daisyworld, we can simulate the nonlinear equations numerically. This means we can show behavior over time. The simulation model allows us to look at both the dynamics and the equilibrium conditions. Let's start with the dynamic response of the world in the years immediately after an increase in solar luminosity.

Figure 11.5 shows the first 12 years of a simulation that starts with 250 acres of white daisies and black daisies and 500 acres of bare ground. These are not the equilibrium conditions from figure 11.2, but we will use the simulation to see if the model finds the equilibrium conditions on its own. The simulation shows the temperatures on a scale from 0°C to 50°C. They are in gradual decline during the first 10 years. This descent is caused by the change in the mix of flowers. It takes only a decade for the flowers to reach the equilibrium with a higher proportion of white daisies. Their larger surface area gives the planet a higher albedo and causes the lower temperatures.

The disturbance occurs in 2010 in the form of an abrupt increase in the solar luminosity. It jumps from 1.00 to 1.25 and remains at the higher value for the rest of the simulation. The jump in solar luminosity causes an immediate increase in all three temperatures. The temperature near the black daisies increases to 43°C, a value off the chart in figure 11.1. The black daisies would not be able to grow, and their decay should shrink their area. The temperature near the white flowers jumps to 33°C, a value above the optimum for growth. However, figure 11.1 shows that the white daisy growth rate would still exceed the rate of decay, so we would expect their area to spread. As it does, the planet's average albedo should increase, and perhaps we will see a reduction in the temperatures. (Indeed, if you look closely at the results from 2010 to 2012, the reduction is already underway.)

Figure 11.6 shows the remainder of the simulation. All three temperatures continue to decline. Within two decades after the disturbance, the world has found a new equilibrium, with

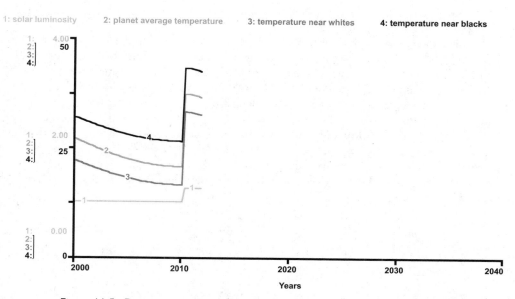

Figure 11.5. Daisyworld's immediate response to an increase in solar luminosity.

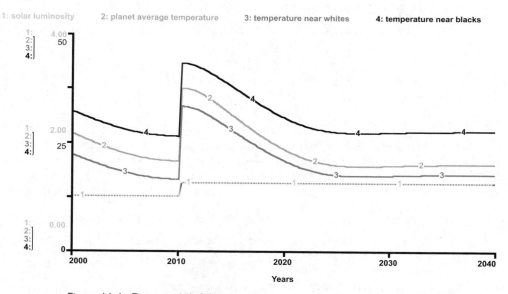

Figure 11.6. Daisyworld's full response to an increase in solar luminosity.

temperatures nearly the same as before. Indeed, if you look closely, you'll see that the planet's average temperature is slightly lower than the temperature prior to the disturbance. This test shows that the planet can deal with a 25% increase in solar luminosity, and it can do so within a few decades. The rapid response is made possible by the relatively short lifetime for the daisies.

Span of Control

This simulation shows that Daisyworld can maintain homeostasis with the luminosity raised to 1.25. But what if it were raised higher? This question leads us to investigate the span of control. This can be done by repeated simulations similar to figure 11.6. We simply try different values for the solar luminosity and observe the equilibrium conditions. Figure 11.7 shows the planet's average temperature at the end of 12 such tests. The span of control is from 0.7 to 1.4, and the temperature is controlled relatively close to 20°C. However, if the luminosity were to reach 1.5, not even the white daisies could survive. When they disappear, the world loses their reflective power, and the temperature would be nearly 60°C. At the opposite extreme, not even the black flowers would survive if the luminosity were lowered to 0.6 or below. When the black flowers disappear, the world loses their absorptive power, and the temperature would be over 9°C below zero.

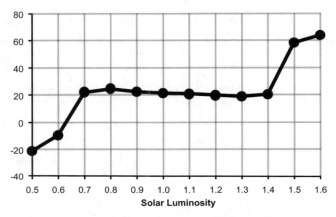

Figure 11.7. Equilibrium value of average temperature in 12 tests with different solar luminosities.

The equilibrium values of the areas are displayed by a stacked graph in figure 11.8, with the sum of the areas fixed at 1,000 acres. The horizontal axis is the solar luminosity from the 12 simulations. The boldfaced values draw our attention to the range of luminosities that will permit flowers to survive. Let's look at the case described in figure 11.2. It would be positioned in the middle of figure 11.8 since the solar luminosity is 1.0. There are 271 acres of black daisies, 403 acres of white daisies, and 326 acres of bare ground. The model does not show where the flowers are located in space. For display purposes, figure 11.8 provides a stacked graph of areas. The stack begins with 25% of the bare ground at the bottom. This is followed by the black area, another 50% of the bare ground, and the white area. The stack is completed by placing the remaining 25% of the bare ground at the top of the display

Now let's focus our attention on the extreme edges of the span of control. If the

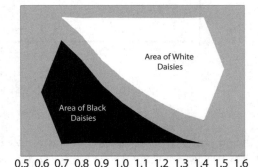

Figure 11.8. Equilibrium areas in 12 tests with different solar luminosities.

luminosity is 0.7, we have around 700 acres of black daisies and only 3 acres of white daisies. The black flowers dominate this world, and their absorptive power helps maintain a temperature that allows them to survive. But if the luminosity is lowered one more step, even the black flowers could no longer survive. Daisyworld would be entirely bare ground, and the world has lost control of temperature.

At the other extreme, the white flowers can survive if the luminosity were equal to 1.4. In this case, the world would end up with around 700 acres of white flowers and only 1 acre of black flowers. The white flowers dominate this world, and their reflective power helps maintain a temperature that allows them to survive. But if the luminosity were raised one more step, even the white flowers would no longer survive Daisyworld would be entirely bare ground and no longer in control of temperature.

Learning from Daisyworld

Daisyworld illustrates how the living world can interact with the physical environment to create a wide span of control. The world achieves homeostatic control of temperature, which makes life possible. The model represents a fictional world that was defined in the simplest possible manner to illustrate homeostatic control without invoking an explicit teleological assumption.

The simplicity and fictional nature of Daisyworld makes it an excellent platform for modeling exercises. Since we are dealing with a fictional world, we are free to invent our own theories about a new world. We can then examine the dynamics of the new world through computer simulation. Expanding Daisyworld is a fun and effective way to practice modeling. As we do so, it's important to keep the span of control in mind. With each new variation, we should ask whether the new world will achieve a wider span of control.

Exercises with the BWeb Model

This chapter includes two sets of exercises. The first ten exercises make use of the BWeb downloadable model. You'll work on the interface layer to simulate the existing model under a wide variety of conditions. The second set of exercises call for changes in the underlying structure of the model. You will make the changes on the map and model layers of the software, then simulate the new model to learn if it has a wider span of control.

Exercise 11.1. Download and verification

Download the Daisyworld simulator (BWeb) and click on the information buttons to learn about the model. Select the heat shock scenario, the second of four scenarios for the solar luminosity. Select variety #5 of the flowers and run the model. You should see the results in figure 11.9.

Exercise 11.2. Checking the scorekeeper

The simulator includes a point system for keeping score. This is to promote some friendly competition as you experiment with the model. The points are explained in the information button and are depicted by the red graphical input device. Run the model with the heat shock scenario and variety #5 and watch the score device. You should be earning 7.1 points per year prior to the heat shock. You will be losing 10 points per year in 2011. By the end of the simulation, your cumulative score should be 168 points.

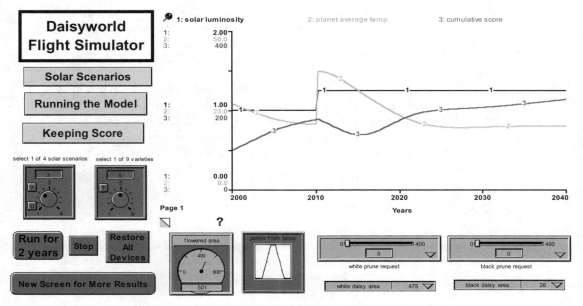

Figure 11.9. Simulation with the heat shock scenario and variety #5.

Exercise 11.3. Is the ninth variety better?

Watson and Lovelock's world has only one variety of daisies, but you will be given the opportunity to experiment with the nine varieties in table 11.1. Variety #5 represents the flowers in the original paper by Watson and Lovelock (1983). They grow best at 22.5° C. But perhaps we would do better with the ninth variety in a heat shock scenario because the flowers grow best when the temperature is 25.0° C. Select the ninth variety and simulate the heat shock scenario. You should see around 45% of the world covered

Table 11.1. The nine varieties of flowers available in the simulator.

Labels	Variety	Black albedo	White albedo	Optimum temperature (°C)
	1	0.10	0.90	20.0
White & black	2	0.10	0.90	22.5
	3	0.10	0.90	25.0
	4	0.25	0.75	20.0
Light & dark	5	0.25	0.75	22.5
	6	0.25	0.75	25.0
	7	0.40	0.60	20.0
Salt & pepper	8	0.40	0.60	22.5
	9	0.40	0.60	25.0

by the white daisies at the end of the simulation. But the world is uncomfortably hot by the way we keep score. (Your cumulative score will be –95 points.) Why did variety #9 perform so poorly compared to variety #5?

Exercise 11.4. Which variety might be best in a heat shock scenario?

Put the model aside and ask yourself which variety would score best in the heat shock scenario. Write down your reasons for the best variety. Then pick another variety that might come in second and explain its advantages.

Exercise 11.5. Find the best variety in a heat shock scenario

Experiment with the model to find the best variety (using the scorekeeper). Did the best variety win for the reasons given in the previous exercise?

Exercise 11.6. Which variety might be best in a cold shock scenario

Think about which variety would score best in the scenario with the luminosity dropping to 0.75. Pick another variety that might come in second and explain its advantages.

Exercise 11.7. Find the best variety in a cold shock scenario

Experiment with the model to find the best variety. Did it win for the reasons given in the previous exercise?

Exercise 11.8. Causal loop diagramming

Figure 11.4 shows a causal loop diagram of the temperature control loops in Daisyworld. How would this diagram be changed if we are describing variety #2 rather than variety #5?

Exercise 11.9. Pruning and the heat shock scenario

Select the winning variety in the heat shock scenario and select the heat shock scenario. Now use the simulator to learn if you can improve the score with pruning. If you achieve a higher score, explain why your interventions lead to a better world. If you cannot achieve a higher score, explain the reasons for your failure.

Exercise 11.10. Pruning and the cold shock scenario

Select the winning variety in the cold shock scenario and select the cold shock scenario. Now use the simulator to learn if you can improve the score with pruning. If you achieve a higher score, explain why your interventions lead to a better world. If you cannot achieve a higher score, explain the reasons for your failure.

Modeling Exercises

Exercise 11.11. Verify the high end of the span of control

Use the BWeb model to test a heat shock scenario. Be sure to select variety #5; then experiment with heat shocks that raise the solar luminosity to 1.4 and to 1.5. The temperatures at the end of the simulation should match figure 11.5. The areas should match figure 11.6.

Exercise 11.12. Verify low end of the span of control

Use the BWeb model to test a cold shock scenario. Don't forget to select variety #5; then experiment with cold shocks that lower the solar luminosity to 0.7 and to 0.6. The temperatures at the end of the simulation should match figure 11.5. The areas should match figure 11.6.

Exercise 11.13. Adding black ponds to Daisyworld

Expand the BWeb model to include the effect of ponds. For every 1,000 acres of landscape, there is a pond that typically holds 1,000 acre-feet of water and covers a surface area of 200 acres. (Lower the initial value of the empty area by 200 acres to make room for the pond.) The volume of water in the pond would be a new stock in the model. It would be fed by runoff that is constant at 400 acre-feet/yr. The volume is reduced by evaporation, which is the surface area multiplied by the evaporation rate. Set the surface area to a nonlinear function of the volume of water. (Create a nonlinear graphical function and make sure you have 200 acres when the pond holds 1,000 acre-feet.) The normal evaporation rate is 2 feet/yr, so we expect evaporation to be 400 acre-feet/yr, which keeps the pond in dynamic equilibrium. But the actual evaporation rate depends on the temperature near the ponds. The ponds are located next to the black daisies, so the temperature near the black daisies influences the evaporation rate. Assume that a 5°C increase in the black daisies temperature would cause the evaporation rate to double. And finally, set the albedo of the black pond surface to 0.25, and include the pond surface area in the calculation of the planet's average albedo. Simulate the new model with the solar luminosity at 1.0. Can the ponds maintain their normal size (200 acres) under these conditions?

Exercise 11.14. New feedback loops with black ponds

Expand the causal loop diagram in figure 11.4 to show any new feedback introduced by the ponds. Based on the new feedback, how would the ponds change the span of control?

Exercise 11.15. Heat shock scenario with black ponds

Simulate a heat shock scenario with the solar luminosity at 1.5 (just beyond the upper end of the normal span of control). Does the new world provide homeostatic control?

Exercise 11.16. Kirchner's pathological daisies

Kirchner (1989) provides a detailed and informative critique of the Gaia hypothesis and the Daisyworld model. Review his example of a "pathological Daisyworld" in which the optimum temperature for the black daisies is 15°C higher than the optimum temperature for the white daisies. Alter the Daisyworld model to replace the original daisies with Kirchner's daisies. Simulate the new model to test its performance over time. Does the model behave as Kirchner suggests?

Exercise 11.17. Competition between Kirchner's daisies and Lovelock's daisies

Expand Daisyworld to be home for two groups of daisies. The first group is Lovelock's daisies; the second group is Kirchner's "pathological" daisies. You will now have four stocks of flowers. Set the initial values to 200 acres for all four types of flowers, and set the initial value of empty area to 200 acres. Simulate the model for 40 years with the solar luminosity constant at 1.0. What kind of world emerges at the end of the simulation? Do you see a world with homeostatic temperature control dominated by Lovelock's daisies? Or do you see a world dominated by Kirchner's daisies?

Exercise 11.18. Competition among daisies and homeostasis

Repeat the previous exercise with an increase in the solar luminosity from 1.0 to 1.25 in the 20th year of the simulation. How does the world respond? Do you see a barren, lifeless planet with temperatures that vary entirely with the amount of incoming solar luminosity? Or do you see a response like the one shown in figure 11.6?

Exercise 11.19. Permafrost on Daisyworld

One of Kirchner's criticisms of Daisyworld is that the structure is slanted in favor of the stabilizing loops. Let's add permafrost to Daisyworld to see the effect of a destabilizing loop. Begin by assigning a separate stock with 100 acres initially covered by permafrost. (Reduce the initial value of the empty area by 100 acres.) Assume that the permafrost albedo is 0.85 (typical for fresh snow), and include the permafrost area in the calculation of the planet's average albedo. Then assume that the permafrost expands when the local temperature falls below $0°C$ and that it shrinks if the local temperature rises above $0°C$. Assume that the permafrost temperature is $20°C$ below the planet's average temperature. So, imagine that the average temperature rises above $20°C$ and that the permafrost temperature rises above freezing. This will trigger shrinkage in the permafrost and a reduction in its reflective power. The world would then become warmer causing further shrinkage in the permafrost. This is the destabilizing effect we are looking for. Build this new model and simulate it with the solar luminosity constant at 1.0. Does the new world find equilibrium conditions on its own?

Exercise 11.20. Permafrost and the span of control

The span of control in the original model is from 0.7 to 1.4 (as shown in figure 11.7). Test the model with permafrost with a jump in the solar luminosity to 1.4. Does the new world maintain temperature close to $20°C$?

Exercise 11.21. Lovelock's variations

Lovelock (1991, 70–72) has designed several versions of Daisyworld. Instead of just two, there are many species of daises with varying pigments in the new models. Also, there are models in which the daisies evolve and change color. There are models in which rabbits eat the daisies and foxes eat the rabbits. And there are examples with major catastrophes that wipe out 30% of the daisies at regular intervals. According to Capra (1996, 110), Lovelock "finds that Daisyworld is remarkably resilient under these severe disturbances." Review these variations in light of Kirchner's criticism that Lovelock's point of view is slanted in favor of the stabilizing feedbacks in nature. Do you believe Lovelock's many variations are slanted in favor of negative feedback?

Further Reading

- To form your own view of Gaia, read Lovelock's work (1991, 1995, 1998) and his work with Lynn Margulis (Lovelock and Margulis 1974; Margulis and Lovelock 1989).

- It is also useful to read Lovelock's critics (Doolittle 1981; Dawkins 1986; Kirchner 1989) and the various interpretations of the Gaia controversy (Joseph 1990; Levine 1993; Capra 1996).
- Lovelock and Kump (1994) describe a model of climate regulation with terrestrial plants and ocean algae influencing the average albedo. The model is used to show failure of climate regulation under simple but "qualitatively realistic" conditions.

Chapter 12

Hitting the Bull's-Eye

The structure of system dynamics models may be explained with a combination of stock-and-flow diagrams, equilibrium diagrams, and causal loop diagrams. These diagrams are effective if you are communicating with a group that is familiar with stocks, flows, and feedback. But these concepts are not familiar to everyone, and sometimes you simply don't have time to explain the model in detail. This chapter explains the bull's-eye diagram, a compact and effective way to describe the model. The diagram will remind you of the dartboard (where the inner ring is called the bull's-eye). If you have played darts, you'll know that the successful players are those who can hit the bull's-eye. System dynamics modeling is similar; the successful modelers are those who can create models with the key variables in the center of the bull's-eye diagram.

Example of the Sales Company

Let's begin with the sales company model whose feedback structure is depicted in figure 12.1. The diagram draws our attention to the sales growth loop, the positive feedback loop that gives the company the power of exponential growth. There are six variables in this loop, and these are called *endogenous*, meaning proceeding from or derived from within. An *endogenous variable* is any variable in a feedback loop, or any variable affected by a variable in a feedback loop. In figure 12.1, the saturation loop includes the same variables plus the effectiveness, so the effectiveness is also endogenous. The remaining endogenous variable is the departures. It

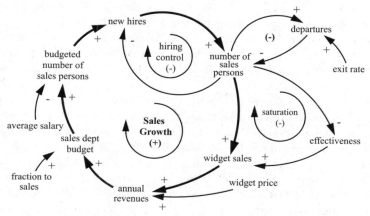

Figure 12.1. Causal loop diagram of the sales company model.

appears in a negative feedback controlled by the exit rate. The exit rate is an *exogenous variable* (*exogenous* meaning originating from outside or derived externally), as it originates from outside the system. The exit rate is one of four exogenous variables in the model; the others are the widget price, fraction to sales, and the average salary. These are sometimes called the *inputs* to the model.

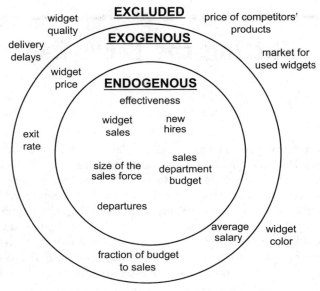

Figure 12.2. Bull's-eye diagram of the sales model.

Figure 12.2 is a bull's-eye diagram for the sales model. The endogenous variables are placed in the inner ring, and the exogenous variables in the outer ring. The goal is to communicate the scope and focus of the model, so you do not need to include every variable in the diagram. In figure 12.2, for example, the budgeted number of salespersons does not appear in the inner ring because its inclusion would not change our impression of the scope and focus of the model.

Figure 12.2 also shows a few excluded variables that are placed outside the outer circle. In this case, the diagram is used to emphasize that the model does not deal with widget quality, delivery delays, or the color of the widgets. We also ignore the price of competing products and the market for used widgets. At this point you are probably thinking that we could list hundreds of other variables that are excluded from the model. So, how many variables should be placed in the diagram? The custom is to list a few variables that might have been included because they are closely related to the model purpose. These might be candidates for inclusion in a future version of the model.

Bull's-eye diagrams are a compact way to portray the system boundary, and they are particularly useful for describing a model that has grown too complex to be easily displayed with flow diagrams. The diagram is also a convenient way to show the relative balance between required inputs and endogenous variables. A quick glance at figure 12.2 confirms that the model has more variables in the center of the diagram. This is a good sign. Models with a large number of endogenous variables are more likely to generate interesting behavior from "inside the system." This is especially true when the exogenous inputs are constant, as was the case in the sales force simulations in chapter 7.

An Alternative Model of a Sales Company

Let's turn now to a somewhat different view of a sales company, as shown in figure 12.3. This model could be used to simulate the change in the company's cash balance over time. The revenues from widget sales adds to the cash balance, and the balance is drained by salary payments and a variety of other payments. Salary payments are based on four categories of salespeople, each earning a different salary. Other payments include payments for rent, furniture, utilities, and the like. The model shows that widget sales are broken down into black, green, blue, and red, and a separate price is assigned to each color.

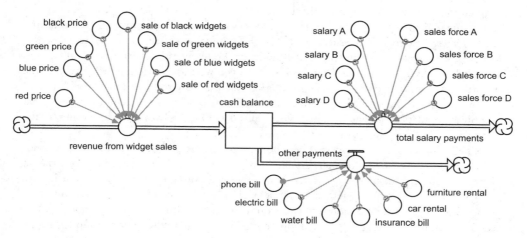

Figure 12.3. Model to simulate the cash balance at the sales company.

Now suppose you were to draw a bull's-eye diagram for the new model. Which variables would you place in the inner ring? Some students view the cash balance as an "output" of the model and interpret it as an endogenous variable that should appear in the center of the bull's-eye diagram. But the endogenous variables are not simply outputs of a model; they are variables that appear somewhere in a feedback loop (or they are influenced by another variable that is in a feedback loop). If you look closely at figure 12.3, you will see that there is absolutely no feedback in this model. All the variables are exogenous, so nothing would appear in the center of the bull's-eye diagram.

At first glance, the new model appears to be more complex than the previous sales model, and perhaps a more complex model will deliver more useful insights about the company. The new model includes categorical complexity, as we now include many categories of workers, products, and expenses. But is the model structurally complex? Apparently not, because it does not contain a single feedback loop. Since it has no feedback structure, it is not likely to teach us anything about the dynamic behavior of the company. Any changes in the cash balance over time will simply be imposed on the system by the values assigned to the 24 converters in figure 12.3. Models of this sort are sometimes called *bookkeeping models*. Their job is to perform a series of well-defined calculations based on an extensive set of inputs. The interrelationships are often clearly understood or are tautological. If you see a need for such a model, you could certainly use one of the stock-and-flow programs to construct a model like figure 12.3. But why would you bother to do so? System dynamics models should be reserved for the study of systems with structural complexity, not categorical complexity.

The Mono Lake Case: Hitting the Bull's-Eye

System dynamics models are used to study systems with feedback, so the bull's-eye diagram should show the key variables appearing in the center of the diagram. If the important variables are missing from the center of the diagram, the model won't convey useful insights about the dynamic behavior of the system. The Mono Lake case in chapter 5 illustrates the approach. Recall the first model in figure 5.4; it contained no feedback, so all the variables were exogenous. The lack of feedback was the reason for the model's poor behavior. The second model (figure 5.7) performed much better because of the inclusion of feedback through the surface area of the lake. The model gave the general pattern of behavior, but it overstated the lake evaporation. We then turned to the third model (figure 5.9) to represent the change in evaporation due to changes in the specific gravity of the water. This model proved useful in testing the lake's response to a changes in export. The third model was the final model in chapter 5, but the exercises continued the process of model development. With each new addition, you included new feedback relationships. For example, you were asked in exercise 5.5 to base the export on the elevation of the lake. This change moved export to the center of the bull's-eye diagram. And finally, figure 5.13 encouraged the development of an interdisciplinary model that combines population biology with hydrology. This new model moves the brine shrimp to the center of the bull's-eye diagram.

The Mono Lake case is reviewed here because the pattern of model improvement is not the direction that many newcomers are inclined to take. When asked about the best way to improve a model, many students first respond with suggestions on how to improve the inputs. They sometimes express the view that a model is simply a device to "convert inputs into outputs," and that their primary job is to come up with a more accurate or a more extensive collection of inputs. But this line of development adds no new feedback to the model, so the changes do little to improve our understanding of dynamic behavior. The more productive approach is the one followed in the Mono Lake case. Rather than adding more inputs, ask yourself whether some of the current inputs are better represented as endogenous variables. Changing exogenous inputs to endogenous variables may seem strange if you are new to modeling. But as you become more familiar with system dynamics, you'll discover that the prospects for insight are much greater when the key variables are in the center of the bull's-eye diagram.

Diagrams for New Models

The previous examples are familiar to you from earlier chapters in the book. But what if you see a diagram for a model that is entirely new to you. What impression can you gain from a single diagram? Bull's-eye diagrams will be useful for a variety of models, regardless of their size or their methodology. Their usefulness was demonstrated in *The Electronic Oracle* by Meadows and Robinson (1985). They described nine different computer models dealing with policies on population growth, agricultural production, and economic development. The models employ a variety of methodologies, including system dynamics, econometrics, linear programming, and input-output analysis. Many of the models were quite complex, both in size and in the eclectic combination of methodologies. Nevertheless, the bull's-eye diagrams provided excellent summaries that fit on a single page.

Figure 12.4 shows one of the diagrams. This is a boundary diagram of a model of the people, livestock, and rangelands of the Sudano-Sahel area in Africa. It was constructed by Tony Picardi (1975) as part of a project for the U.S. Agency for International Development (USAID). Picardi used the model to simulate the overgrazing and soil degradation that is widespread in the Sahel, and to simulate various USAID proposals. The simulations revealed that some of USAID's proposals would only intensify the system problem. The model contains

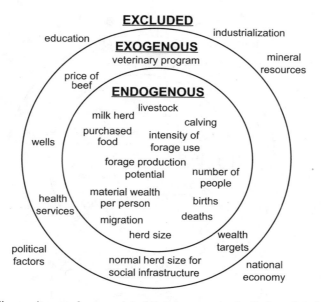

Figure 12.4. Bull's-eye diagram for a model of development in the Sudano-Sahel region of Africa.

almost 200 equations. Nevertheless, you are able to appreciate the main thrust of Picardi's work from the bull's-eye diagram in figure 12.4.

The next two bull's-eye diagrams convey the main thrust of models to appear later in the book. Figure 12.5 shows the key variables in a model to simulate the cycles in real estate construction. Figure 12.6 shows selected variables from a model to simulate the overshoot of the deer population on the Kaibab Plateau in northern Arizona. The diagrams convey a sense of the focus of these case studies, and they reinforce the point made earlier—that the bull's-eye diagram is a compact and effective way to describe a model.

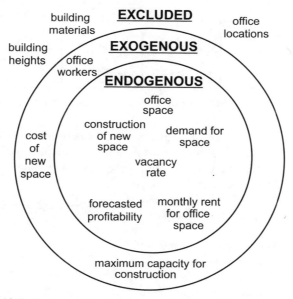

Figure 12.5. Bull's-eye diagram for a model of real estate construction.

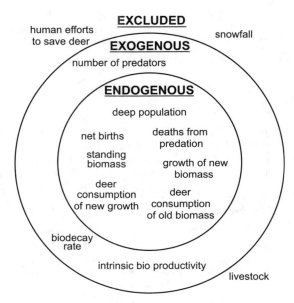

EXCLUDED
human efforts
to save deer   snowfall
EXOGENOUS
number of predators

ENDOGENOUS
deep population

net births   deaths from
predation
standing
biomass   growth of new
biomass
deer
consumption   deer
of new growth   consumption
of old biomass

biodecay
rate

intrinsic bio productivity
livestock

Figure 12.6. Bull's-eye diagram for a model of the deer herd on the Kaibab Plateau.

Review of Dynamic Patterns

This chapter concludes part I of the book, so this is a good place to review the dynamic patterns that have been simulated so far. Figure 12.7 reminds you of the six shapes for the dynamic patterns that are simulated in the book. At this point, you have seen four of the dynamic patterns. You saw exponential growth and exponential decay when first experimenting with Stella and Vensim software in chapter 2. The exponential approach was characteristic of the decline in Mono Lake. And we have simulated S-shaped growth in flowered areas, a sales company, and an epidemic.

Exponential   Exponential   Exponential
Growth   Decay   Approach
S-Shaped
Growth   Overshoot   Oscillations

Figure 12.7. Six shapes for dynamic patterns simulated in the book.

It's useful to keep these previous examples in mind the next time you face one of these dynamic patterns. It's also useful to think of the minimum combination of stocks, flows, and feedbacks that are required to generate each of the shapes. The minimum structures for the four shapes simulated in part I are shown in figures 12.8–12.11. Figure 12.8 starts with exponential growth. The minimum structure is a single stock and flow with a positive feedback to make the growth larger and larger over time. The minimum structure for exponential decay is shown in figure 12.9. There is a single stock reduced by decay that is proportional to the remaining value of the stock.

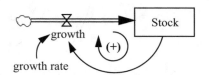

Stock
growth
(+)
growth rate

Figure 12.8. Minimum structure
to generate exponential growth.

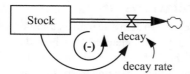

Figure 12.9. Minimum structure to generate exponential decay.

Figure 12.10 shows the minimum combination to see the exponential approach to equilibrium. This diagram will remind you of the Mono Lake model, where the stock corresponds to the volume of water in the lake. The inflow corresponds to an exogenous variable (e.g., runoff), and the outflow is controlled by negative feedback to bring the system into equilibrium.

Figure 12.10 is a minimum structure; its value is to alert us to look for similar relationships when we are simulating the same pattern in a new system. In the case of Mono Lake, we are alerted to the need for a negative feedback to control the size of the outflow and to see the lake approach equilibrium. The target pattern appeared when we made the lake's area depend on the volume of water in the lake. However, notice that the final Mono Lake model (figure 5.9) looks much different from the minimum structure. The final model illustrates appropriate modeling for clarity and commutation. Even though the minimum structure requires only two flows, the larger model (figure 5.9) does a much better job of showing the flows in the basin.

S-shaped growth has received the most attention so far. You have seen models to simulate the spread of flowers across the landscape, the growth in a sales force, and the growth in the number of people affected by an infectious disease. Each model takes a somewhat different form, but you might wonder about the minimum combination of stocks and flows to generate an S-shaped pattern.

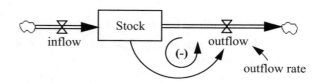

Figure 12.10. Minimum structure to generate the exponential approach to equilibrium.

Figure 12.11 shows an example with a single stock and two flows. The growth of the stock is controlled by an intrinsic growth rate when the stock occupies only a small part of the available space. There is a limited capacity to serve the needs of the stock. As the stock fills up the space, the density-dependent feedback becomes sufficiently strong to bring the growth and decay into balance. In this diagram, the density-dependent feedback acts upon the growth. An alternative minimum structure would leave the growth rate fixed, and the density-dependent feedback would act upon the decay.

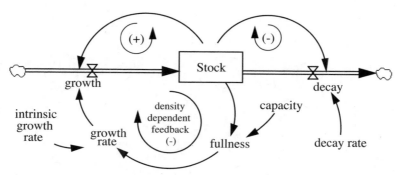

Figure 12.11. Minimum structure to generate S-shaped growth.

Looking Ahead to Overshoot and Oscillations

Understanding the overshoot pattern is crucial if we are to achieve sustainable management of human and natural systems. We'll see an example of the overshoot in chapter 21, which describes the overshoot in a deer population in the early part of the 20th century. Oscillations are the sixth of the six shapes in figure 12.7. They appear in many systems, ranging from the beating of the human heart to oscillations in animal populations. Chapters 18–20 explain oscillations with examples from business systems and wildlife populations.

Exercises

Exercise 12.1. Number of exogenous variables

How many exogenous variables do you see in figure 9.26? How many in figure 9.27?

Exercise 12.2. Boomtown problem

There are 9 variables in figure 9.11, only 3 of which are exogenous. Which are the exogenous variables?

Exercise 12.3. Blood loss model

There are 15 variables in figure 10.3, only 4 of which are exogenous. Which are the exogenous variables?

Exercise 12.4. Forest land use model

Figure 6.6 shows a stock-and-flow diagram, so the feedback loops do not stand out as clearly as in a causal loop diagram. Nevertheless, you should be able to distinguish between the exogenous and the endogenous variables. Which are the exogenous variables in this model?

PART II

INTERMEDIATE MODELING

Chapter 13

The Modeling Process

A model is usually built up in steps of increasing complexity until it is capable of replicating the problematic behavior of the system. It is then used to determine if a change in policy variables could lead to improved behavior. You've seen examples of good modeling in the previous cases, and you've become familiar with the steps of modeling. Table 13.1 lists the eight steps in sequence with an A, B, C mnemonic to help you remember the steps. The most important thing to remember, however, is that the modeling process is seldom executed in a sequential manner. You should not strive for perfection on each step before proceeding to the next step. Experienced modelers know the value of iteration, and they make sure to reserve time to go through all the steps several times before exhausting the project resources.

Table 13.1. The steps of modeling.

Qualitative Modeling

Step 1.	**A** is for get Acquainted with the problem.
Step 2.	**B** is for Be specific about the dynamic problem.
Step 3.	**C** is for Construct the stock-and-flow diagram.
Step 4.	**D** is for Draw the causal loop diagram.

Quantitative Modeling

Step 5.	**E** is for Estimate the parameters.
Step 6.	**R** is for Run the model to get the reference mode.
Step 7.	**S** is for Sensitivity analysis.
Step 8.	**T** is for Testing the impact of policies.

This chapter explains each step in the modeling process. The first four steps are qualitative and conceptual. They help us organize our thinking about the system and the underlying causes of the problematic behavior. The remaining four steps are quantitative in nature. We estimate parameters and use computer simulation to test our theory of the problematic behavior. I believe computer simulation is essential if we are to check the theory that grows out of the qualitative discussion. Computer simulation provides concrete examples of dynamics, and it adds clarity to the concepts discussed in the qualitative stages. Finally, dynamics are complicated, and we can't expect to simulate them in our head.

Environmental problems are increasingly the subject of group modeling exercises involving a combination of experts and stakeholders. System dynamics has been put to good use in recent years to help groups gain a shared understanding of their environmental challenges. This chapter describes the forms of iterative modeling that have emerged from these group exercises. The chapter concludes with a discussion of model validation.

Problem Familiarization

Step 1 in the modeling process is to get acquainted with the problem. Your first task is to meet the people and learn their views of the problem. You should also learn about the larger organization in which the people work. The organization could be a business, a public agency, or a broad community of concerned citizens. Learn how the organization measures performance and why the system is performing poorly. Become familiar with the policies that have been proposed to "fix" the problem, and learn the pros and cons of each proposal. If the organization has a history of quantitative analysis, learn about any models that have been developed. Ask yourself whether a system dynamics model would add to the previous modeling results.

Take the time to identify the client for the modeling study. (A *client* is a person or group that uses the services of a professional. The professional could be an accountant, an architect, or perhaps an expert on modeling.) The client may be an individual in a leadership position, or it may be a team assigned to the problem. Sometimes the client is a broad community of stakeholders with different views of the problem and conflicting claims on the benefits of the system. And, finally, remember that the client may be you. Sometimes the purpose of a model is to test your own idea about how a system works and how it might work better. When you have identified the client, be sure to ask the most pragmatic question of all—how much time can be devoted to the modeling study?

Problem Definition

Step 2 is to be specific about the dynamic problem. This is probably the most important step in the process and the easiest step to skip over in a rush to get to work on the computer. As you become acquainted with the organization, you should ask yourself if the organization really has a dynamic problem. If it does, you should be able to draw a graph of an important variable that changes over time. The graph should summarize why the people are dissatisfied with the system's performance. The graph should not be an abstract representation that you create to make the modeling easy. Quite the contrary. The people in the organization should immediately recognize the graph as a summary of their problem.

This graph is called the *reference mode*, and it serves as a target pattern of behavior. The reference mode will always have time on the horizontal axis and an important system variable on the vertical axis. The length of time is called the *time horizon*, and your selection of the time horizon will guide decisions on what to include or to exclude in the model. The time horizon should be sufficiently long to allow the dynamic behavior to be seen. In the Mono Lake case, for example, around 100 years was required for the lake to reach equilibrium. In the epidemic model, on the other hand, we only needed to simulate 20 days to see the epidemic run its course. If your problem seems to involve dramatically different time horizons, the best course is to think of two models, each with their own reference mode.

The reference mode will probably correspond to one of the six dynamic patterns in figure 13.1. These patterns were selected because they cover most of the dynamic patterns you are likely to encounter in environmental systems (see the book's website, the BWeb). In some cases, the reference mode will be known from historical performance, or it will be a relatively simple extension of the historical trend, as was the case with Mono Lake. And in some cases, your team may be concerned about a possible pattern of behavior that has not yet materialized. In this situation, think of the reference mode as a hypothesis about the possible pattern, as in the overshoot reference mode used in *The Limits to Growth* (D. H. Meadows et al. 1972).

Figure 13.1. The reference mode corresponds to one of the six dynamic patterns.

The reference mode is the best way to be specific about the dynamic problem. It's also helpful to specify a key policy that could lead to an improvement in system behavior. You should be clear about what you mean by improved behavior. Drawing a second time graph can help make your goals clear. The second graph will show the same variable seen in the reference mode, and it will have the same time horizon. The new graph should show a qualitatively different pattern of behavior. The new pattern should be clearly recognized by people in your organization as a more desirable pattern of behavior.

Drawing a reference mode is easy advice to give, but it is hard advice to follow. Do not be discouraged if you don't immediately arrive at a clear reference mode. However, if you discover that you cannot draw this time graph, you should reconsider why you are building a model. Perhaps you have discovered that the organization does not really have a dynamic problem. In this case, there is no point in building a dynamic model.

Model Formulation: Stocks, Flows, and Feedback

Steps 3 and 4 deal with stocks and flows and the feedback relationships that control the flows. *Step 3* is to construct the flow diagram. You know the drill—start with the stocks, then add the flows, and then add the converters to explain the flows. Be sure to specify the units for each variable and to check for unit consistency as you add each new variable. The flow diagram should contain the variable shown in the reference mode, and it should also contain a variable to represent the policy. If you are having trouble identifying the stocks and flows in the system, you probably haven't spent enough time getting acquainted with the system.

Drawing the causal loop diagram is *step 4*. You've learned the diagramming conventions in chapter 9, and you know that the diagrams can be constructed in a step-by-step manner by translating from the stock-and-flow diagram. The purpose of the diagram is to help us see the key loops in the model. In some systems, we must be prepared for the loop structure to become rather complicated. When this happens, you may find it helpful to draw several diagrams. Remember, causal loop diagrams are a communication tool, not an analytical tool. You don't need to show each and every relationship in the diagram. Your goal is to show the loops, not the clutter.

At this stage, you should have an opinion on the minimum combination of feedback loops that must appear in your model. If your reference mode shows exponential growth, for example, you need to see at least one positive feedback loop in the diagram. And if you're simulating S-shaped growth, look for loops like contagion, depletion, and recovery—the three loops that give S-shaped growth in the epidemic model.

Now, what if the diagram does not show any feedback loops? This is a sign that you should return to step 3 and revisit the stock-and-flow diagram.

Is Simulation Needed?

Figure 13.2 lists the eight steps of modeling with connecting arrows to represent moving from one step to the next. The first four steps are conceptual and qualitative in nature. We can learn from the discussions of problem definition and the stocks, flows, and feedbacks without ever assigning a number to any of the variables. Think of some of the diagrams from previous chapters:

- Control of shaft speed through a centrifugal governor (figure 9.9)
- Control of blood pressure through physiological responses (figure 10.3)
- The bigger picture of the boomtown housing problem (figure 9.11).

These diagrams give important insights into system behavior without discussing the numerical values. The insights should be viewed as starting points for further exploration. The exploration can take the form of engineering experimentation (such as trying different designs of the centrifugal governor). Or there may be biomedical experiments (such as those described by Walter Cannon in regard to animal loss of blood pressure). But such experimentation is not possible in many social and environmental systems. In the boomtown situation, for example, it makes better sense to test the ideas about productivity problem through computer simulation (BWeb).

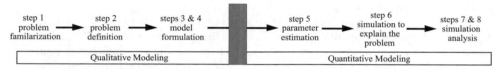

Figure 13.2. The steps of modeling portrayed in a linear, sequential fashion.

The diagram in figure 13.2 depicts the qualitative and quantitative parts of the modeling process as if they were separated by a wall. The diagram conveys an impression held by some—that a major barrier exists between the qualitative and quantitative parts of the modeling process. We all have our own predilections: Some prefer to discuss the key concepts and interrelationships in the system. Others prefer to dig into the numbers and make sense of how they add up. Those with a preference for conceptual modeling sometimes argue that parameter estimation and computer simulation are too difficult and too time consuming. They argue that crossing the wall is likely to swallow up valuable resources that are better concentrated on the conceptual discussions.

Others take the view expressed in this book—that the ideas emerging from the qualitative discussions need to be tested through computer simulation. The major environmental problems of our day involve complicated dynamics that cannot be simulated in our head. They deserve our best efforts at both qualitative and quantitative modeling. This means we need to cross the wall and go to work on parameter estimation.

Parameter Estimation

Step 5 is to estimate the parameters. They are best estimated in a one-at-a-time fashion using independent information best suited for each parameter. In the Mono Lake case, for example, the effect of specific gravity on the evaporation rate came from evaporation pan studies, but the estimate of Sierra gauged runoff was based on 50 years of data on the gauged streams. Be prepared for a wide range of uncertainties in environmental systems. Some parameters may be known with perfect accuracy; others may be known to within 10% accuracy. But what about parameters whose values are so difficult that our estimates may be within 50%? How much effort should be invested to pin down the highly uncertain parameters? Should we aim to have all parameters estimated to within 10%? Can we proceed if the two-thirds of the parameters are known with great certainty, or do we need to pin down values for 90% of the parameters?

These are some of the questions addressed by Richardson and Pugh (1981, 230) in a discussion of policy-directed modeling in business systems. They reasoned that if the policy implications of a model do not change when its parameters are varied, then the modeler does not need to spend more time estimating their value:

> To decide how much effort to put into estimating a given parameter value, one ought to know how sensitive the behavior of the model is to the value of that parameter. Yet to know that, one must run the model, and that requires parameter values. To resolve the dilemma, the modeler picks some values rather quickly and simulates the model. Initial estimates are made carefully, to be sure, with as much concern for accuracy as can be easily mustered, but keeping in mind that one can always go back and estimate more carefully if it makes a difference.

This is sound reasoning for models of environmental systems as well. Don't let uncertain parameters derail you from the modeling process. The only way you will learn if the uncertainty is important is to move forward with your best estimate.

Sources of Information

In estimating parameters, you should take advantage of all sources of information at your disposal. Useful information may be stored in the organization's numerical data sets and in their annual reports and memos. And don't forget the vast body of information stored in the minds of the participants. Forrester (1980) refers to these three categories of information as the *numerical database*, the *written database*, and the *mental database*. He emphasizes that the numerical database may be minuscule when stacked up against the written database and the mental database. We will make more progress if we draw on all three bases of information when structuring a model and estimating the parameters.

The range of information sources at your disposal are portrayed along the information spectrum in figure 13.3. Sources of information are sometimes labeled as *hard* and *soft*. The hard sources are on the left, and the soft sources are on the right of the figure. The results of physical laws and experiments are often available for portions of environmental systems, so many of the physical parameters might be assigned values with a narrow range of uncertainty. Environmental modeling is best performed in an interdisciplinary fashion where the people and their organizations are simulated along with the hydrology, biology, and other physical aspects of the system. Models of social and economic systems are often estimated by drawing on the sources on the softer side of the spectrum.

physical laws	controlled experiments	uncontrolled experiments	statistical information	case studies	expert judgment	stakeholder knowledge	personal intuition

Figure 13.3. The information spectrum.

The most common information source discussed in social and economic studies is statistical information, typically in the form of time series data or cross-sectional data. Various statistical techniques (e.g., econometrics) may be used to take advantage of time series and cross-sectional data. However, these methods are not one-at-a-time methods. Rather, they provide estimates of several behavioral parameters to give the best fit for the main output to time series data. Most system dynamics practitioners are weary of conventional statistical techniques such as ordinary least squares or generalized least squares for parameter estimation in feedback-intensive systems (Senge 1977; Richardson and Pugh 1981). A more suitable approach to estimating parameters in environmental systems would be statistical methods based on the Kalman filter (Kalman 1960; Schweppe 1973; Mass and Senge 1980).

Social system case studies are positioned one step farther on the soft side of the information spectrum. Case studies are frequently used when time series data are not available. A case study sometimes takes the appearance of a good story about the institution and the problems (Wilber and Harrison 1978). A well-told story will frequently provide insights that could not be obtained by inference from the limited numerical data. Expert judgment is used extensively in system dynamics projects. Frequently, the experts are part of the modeling building team. In some cases, it may be useful to call on a panel of outside experts. They might be polled in an iterative fashion following the popular Delphi method (Turoff 1970). The end result of a Delphi survey could be better estimates of highly uncertain parameters as well as new ideas for additional variables to be included in the model.

Stakeholder knowledge is an important source of information even though the views of stakeholders are sometimes discarded as unscientific. All too often, a stakeholder is viewed as a person with a stake in the system but not necessarily with knowledge about how the system operates or its past performance. Thankfully, this view is gradually being replaced by an awareness that stakeholders often possess useful insights (Stave 2002; Tidwell et al. 2004; van den Belt 2004; Cockerill, Passell, and Tidwell 2006; Beall 2007). Local knowledge is especially important in natural resource management. It can take the form of "street science" accumulated in a community (Corburn 2005), or it can be experiential knowledge acquired by simply living in the system. And finally, "local experiential knowledge will be essential to implementing specific solutions" in ecological systems in different locations (Costanza and Ruth 1998, 72).

Personal intuition is listed at the right (soft) side of figure 15.2. Intuition should be used when the other sources of information are not helpful. At this point, you might be wondering about a straight-out guess. Maybe it would be better to eliminate the highly uncertain parameter from the model. Kitching (1983, 41) provides good advice in his description of difficult parameters in systems ecology:

> The measurement of real values may be difficult or even impossible. Under these circumstances the value used in the model, legitimately, may be a straight-out guess. Of course, such a value cannot be regarded and treated in the same way as other parameters in the model. It must be treated tentatively and its role can be evaluated using techniques of sensitivity analysis.

Kitching's advice applies to a wide range of areas, not just to systems ecology. A straight-out guess is a legitimate part of useful models, so we should resist the temptation to eliminate the highly uncertain parameters. Remember that excluding a parameter is equivalent to adopting an implicit estimate of the numerical value. Specifically, you would be setting the parameter value to zero. If you are confident that you can do better than zero, take advantage of your personal intuition, your best guess, and proceed with the modeling process. You won't learn until later whether your guess needs to be replaced with a better estimate.

What If the Scientific Foundation Is Missing?

Environmental systems are sometimes poorly understood owing to a lack of information on the fundamental processes. If the scientific foundation is lacking, it may make sense to put the modeling process on hold. On the other hand, many poorly understood systems present serious problems, and decisions are being made today, one way or another. System dynamics can contribute in these difficult situations if the model includes the highly uncertain processes in an explicit fashion. The model may then be used to simulate the impact of current policy proposals in the face of the scientific uncertainty. This approach is illustrated in the case of the threatened salmon population in the Pacific Northwest (chapter 15) and the case of the threat of global climate change from CO_2 accumulation in the atmosphere (chapter 23).

However, in some instances, it makes sense to put the modeling process on hold while new research is conducted. You'll recognize such a situation when you find that so little is known about a process that you cannot assign a name to the variables, let alone estimate the parameter values. An example might be the modeling of reclusive animal populations whose birth processes or habitat associations are simply not known.

In such cases, environmental scientists design a research plan, one that might involve years of data collection followed by statistical interpretation of the data. Experimental research is a complex process, with results unfolding over time. Some projects conclude with ambiguous results and regrets over the experimental design. System dynamics models may be used to anticipate the problems that could arise in experimental research. System dynamics has been used to support evaluation studies in business and social systems (BWeb) through the use of *synthetic data* analysis. A similar approach could be useful in environmental systems. The approach starts with one of the theories on the uncertain process. The theory is inserted into the model using the familiar tools of stocks, flows, and feedbacks. The model is then expanded to include random conditions in the environment and perhaps some measurement error in the experiment. The model would then be simulated over the time period proposed for data collection. The simulation results for the measured variables provide a proxy for the actual data from a costly experiment. This simulation output is synthetic data that can be used to pretest the statistical methods proposed for the project. The end result may be an improved experimental design and more realistic expectations for the conclusions from the research.

Should We Include Stakeholder Values?

So far, we have said little about values and the modeling process. We have yet to mention the values and priorities of the stakeholders, nor have we talked about the difficult trade-offs in environmental systems. Participants in environmental systems often disagree on the fundamental goals for the system, and the debate over policy making would seem to boil down to a debate about values. What is the contribution of system dynamics in these situations?

The Mono Lake case is instructive because it presents a clear example of a value trade-off. Some argued that a high export of water should be allowed because the Los Angeles depends on a low-cost supply of clean water from the Mono Basin. Others argued that the lake is a scenic and ecological treasure, and therefore it should be allowed to return to higher levels. The Mono Lake models in chapter 5 helped us appreciate the dynamics of the lake's response to a change in export, but they did not touch on the value questions. There were no statements about right or wrong, and there were no declarations on whether one set of values is more worthy than the other. The Mono Lake case illustrates the approach taken in most system dynamics applications. The focus is on dynamic behavior, and values are set aside. The goal is improved understanding of a dynamic problem. The applications are successful if the participants reach general agreement on what constitutes "improved behavior" when discussing policies.

But general agreement may be missing in some social and environmental systems. The boomtown problem from figure 9.11 is an example. The housing shortage is shown as part of a vicious circle that leads to productivity problems for the construction manager. Housing shortages are a problem when viewed through the eyes of the construction worker or the construction manager. From their point of view, the goal of policy changes is to avoid housing shortages and the ensuing productivity problems. But others see the situation differently. Some current residents may view housing shortages as a beneficial result, especially if they are hoping to sell their homes for a high price during the boom.

So, how do we make progress reaching a consensus on policy when participants have fundamental differences in values? The most common response from system dynamics practitioners is to steer clear of the value debate. This makes sense because the key concepts (i.e., stocks, flows, and feedbacks) do little to shed light on value disagreements. The common approach is to focus on dynamics and hope that the insights on dynamic behavior will be useful, regardless of value differences.

An alternative approach is to pay attention to the participants' values by borrowing from the field of behavioral decision analysis (Keeney and Raiffa 1976; Edwards 1977). Formal evaluation tools allow for a quantitative expression of the values held by different groups of stakeholders. Formal evaluation can be especially useful in environmental management because there are often numerous indicators of interest (Holling 1978). One of the more useful evaluation methods is multiattribute utility measurement (MAUM). It provides a step-by-step process of selecting performance attributes and assigning weights to reflect the importance of each attribute. Performance along each of many attributes can then be scored based on the preferences of the participants. The key to applying formal scoring procedures to social and environmental systems is to ensure that participants select their own individual weights and preferences. The end result can be the discovery that groups often agree more often then previously thought (Gardiner and Edwards 1975). Combining MAUM with system dynamics modeling can be useful when participants have dramatically different goals for the system. The combination is demonstrated by Gardiner and Ford (1980) in the boomtown situation described in figure 9.11. We found that disagreements over values do not necessarily stand in the way of reaching consensus on policy (Ford 1978).

Evaluation methods are not discussed further in this book, but you can experiment further with the Idagon River simulator (BWeb). The model simulates a large-scale river system typical of the western United States (figure 13.4). The system is hydrologically complex, as there are interactions between rivers, reservoirs, aquifers, and irrigated areas. The system is also com-

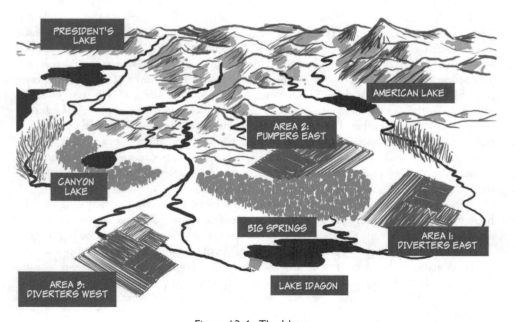

Figure 13.4. The Idagon.

plicated by debates over water policy that sometimes harden into arguments over "my water versus your water." You may experiment with the river management policies, with the simulated results scored by the weights you assign to each of the goals in table 13.2. There is no requirement that you and your classmates agree on how to keep score. Indeed, you will probably find great diversity of opinion on which of the goals deserve the highest weights in keeping score. You may search for the best way to manage the river given your own priorities for the system; your classmates may do the same. Perhaps you will discover common ground for managing the river.

Table 13.2. Six goals for keeping score on the Idagon (BWeb).

1. Increase the Idagon Economic Product, the annual value of agricultural and electricity production.

2. Maintain agricultural land use in areas 1 and 2 close to historical values.

3. Achieve high flows below Canyon Lake in April to aid the salmon smolts migrating downriver.

4. Meet minimum flow requirement below Canyon Lake to protect the environmentally sensitive area.

5. Maintain high discharge at Big Springs to protect water quality below the springs.

6. Meet minimum flow requirement below American Lake to protect the environmentally sensitive area.

Simulation to Explain the Problem

Step 6 is to run the model and compare it to the reference mode. This is the first opportunity to test the model. Robinson (1980, 262) describes it as "the intellectual highpoint" of the modeling process:

> *In a sense, formal models are built to allow testing. Were mathematical models not amenable to a diverse spectrum of testing procedures, they would have little advantage over verbal models. Procedurally, too, testing is a climactic activity. Suspense tends to build from the time the model is conceptualized to the time it is ready to be tested, as the modeler wonders how it will work.*

The important question at this stage is whether the simulation result matches the reference mode. If it does, you have reached a major milestone in the modeling process, one that often requires several iterations. But what if the simulation does not match the target pattern? It's important to study the results carefully at this stage, looking at many of the intermediate variables as well as the main variable that does not match the reference mode. You may well find that your stocks and flows are poorly formulated or that the model lacks the necessary feedback to generate the target pattern. Then it's back to the model formulation steps.

On the other hand, we should not be too quick to blame the computer model when there are unexpected simulation results. Sometimes it's our mental model, not the computer model, that is poorly formulated. The Mono Lake case tells the story of a change in one student's thinking when the hydrologic model did not show downward momentum following a reduction in water export. The moments when the model results make us stop and think are special moments in the modeling process. These are the moments of insight that make modeling worthwhile.

Simulation Analysis

Step 7 in the modeling process is sensitivity analysis; *step 8* is policy analysis. Both steps involve running multiple simulations to see if changes in parameter values lead to important changes in the simulations. When the parameter is an uncertain input beyond our control, it is sensitivity analysis. We run the model several times with variations in the parameter values to learn if the results are "sensitive" to changes in the parameter. Pay particular attention to the uncertain parameters and be sure to allow them to vary across their full range of uncertainty. Check to see if you get the reference mode after each test. If you do, you have reached another important milestone in the process—you have discovered that you have a robust model.

A model is called *robust* when it generates the same general pattern despite the great uncertainty in parameter values. Sensitivity analysis may be conducted in an informal, one-at-a-time fashion based on our instincts about which parameters are likely to be important. (Comprehensive analysis with simultaneous changes in many parameters is explained in appendix D.) Experienced practitioners have learned that a well-structured model will frequently turn out to be robust. They expect the key to the behavioral tendencies to be the underlying feedback loop structure, not the exact numerical values assigned to the parameters. If, on the other hand, you find a new pattern of behavior with changes in parameter estimates, it's time to return to step 5 (estimating the parameters). As you invest more time on the sensitive parameter, you will do so with the knowledge that you are working on one of the truly important parameters in the model.

Step 8 in the model process is policy analysis. This final step is perhaps the most rewarding step of all. We run the model several times with variations in the values assigned to the policy variables. These simulations will reveal whether the policies lead to the desired changes in the simulated behavior. If you learn that the policy test is promising, you should return to sensitivity analysis to learn if the policy performs well under a wide range of values assigned to the uncertain parameters. Also, if you have ignored random variations up to this point in the analysis, you should introduce randomness to produce a more realistic setting for the policy analysis. If the policy results are encouraging, you will probably wish to define the policy variables in a more detailed manner. At this stage, you would return to step 3 (constructing a stock-and-flow diagram) to describe the details of policy implementation.

Why We Iterate

Experienced modelers know that modeling is a highly iterative process. The iterative nature of model building is illustrated in the Mono Lake case. Three iterations were required before we were ready to test the impact of changes in the water export policy. You might wonder why it is necessary to iterate. You might be asking yourself

Why don't we get it right the first time around?

If you pose this question to experienced modelers, they will probably tell you that the iterative process was crucial to their success. They often observe that communication among the team members was vastly improved by starting with a simple model and gradually building more realism over time. Voinov (2008, 21) describes the process well:

You don't build a model going down a straight path. You build a model going in circles.

Many experienced modelers will observe that the modeling process turned out to be more important than the final model (it's the journey that counts, not the destination).

A useful rule of thumb is to complete the initial iteration within the first 25% of the time interval available for the modeling project. If you have four years, make sure you complete the first iteration within a year. If you only have four months, complete all the steps within the first month of the project. This will provide ample opportunity for reflection on the initial model. With the initial results completed, the team members will be in a better position to provide concrete suggestions for improvement. Also, a short time interval between model construction and simulation results is important. It will minimize the loss of client interest and understanding that occurs while the team is engaged in model construction (Robinson 1980, 262).

I try to follow the 25% rule in my own work because I have seen a dramatic increase in the contributions from members of a project team once a "demonstration model" is available. At times, the reaction of team members resembled the reaction when a seed crystal is dropped into a supersaturated solution. Suddenly, there are useful ideas popping up everywhere! After several of these experiences, I came to realize that the useful suggestions don't rise to the surface until team members see a concrete example of the system dynamics approach applied to their organization. I remind students to complete all eight steps with plenty of time for iteration. Sometimes a jingle is helpful:

Two, four, six, eight; don't forget to iterate.

How We Iterate

System dynamics modeling has been put to use in a wide variety of resource and environmental issues. These studies are exemplary for their commitment to stakeholder involvement (Vennix 1996; Rouwette, Vennix, and van Mullekom 2002; Stave 2002; Tidwell et al. 2004; van den Belt 2004; Cockerill, Passell, and Tidwell 2006; Beall 2007). They are summarized briefly here to draw lessons on the iterative process that is actually followed in environmental modeling.

The sage-grouse study by Beall and Zeoli (2008) is a good example. The project was conducted for a conservation district in central Washington. The district is home to a threatened sage-grouse population (photo 13.1), and the district was responsible for developing a multiple-species habitat conservation plan. A system dynamics modeling project was conducted over a three-month interval with participation by nine groups. The project was successful in gaining full participation by all groups and in helping to shape the habitat conservation plan. Beall (2007) attributed the success to the iterative nature of the modeling process. "Simulate early and often" was the team's guiding principle.

Photo 13.1. Greater sage-grouse.

The sage-grouse study is listed alongside other group modeling projects in table 13.3. The study leaders shared a commitment to iterative modeling with clarity and transparency. Most of all, the teams were committed to stakeholder participation in the formulation and use of the models. Several cases involved a large number of stakeholders. Thirty groups were represented in the water resource modeling of the Okanagan Basin in British Columbia and in the Baixo Guadiana Basin in Portugal (and Spain). The largest and most extensive project dealt with management issues in the Ria Formosa National Park (in the Algarve region in the south of Portugal). Sixty groups participated in the Ria Formosa study, which is described in full detail by Videira (2005). The project duration was 18 months, and the team eventually arrived at a model with 40 stocks which proved useful in creating guidelines for the development of a management plan for the protected area.

Table 13.3. Recent environmental modeling studies with group participation.

Topic and study area (issues or purpose)	Number of groups	Project duration (months)	Model size (stocks)
Sage-grouse in central Washington (endangered species)	9	3	10
Okanagan River in British Columbia (future water supply)	30	12	9
Problem bears in New York (human-bear interactions)	1	18	16
Fishery in Gloucester, MA (sustainable fishery)	3	18	28
Rio Grande in New Mexico (water supply management)	5	24	20
Upper Fox River in Wisconsin (scoping study)	10	4	6
Ria Formosa Estuary, Portugal (scoping study)	10	4	24
Ria Formosa National Park (management)	60	18	40
Baixo Guadiana Basin, Portugal (group learning)	30	9	15

The studies in table 13.3 were reviewed by Beall (2007). She began with the hypothesis that all groups would follow the "simulate early and often" strategy that had worked well in her sage-grouse study. She selected nine studies with much in common. All nine groups were committed to stakeholder involvement in model formulation and discussion. All the groups used system dynamics software so that participants could see the stocks, flows, and feedbacks at work in the system. All the groups knew the value of the conceptual and qualitative stages of modeling, and they were committed to the quantitative stages as well. Each of the groups developed models that could be tested through computer simulation. Taken as a whole, the nine groups delivered a common lesson about the power of system dynamics to engage stakeholders in the modeling process.

But the studies revealed differences in how the groups iterated through the steps. The first five studies followed the "simulate early and often" strategy. Each of these teams completed the eight steps of modeling several times during the study period. This was intentional, as the team leaders believed in the benefits that would follow from the participants seeing simulation results early in the process. Figure 13.5 portrays the iterative modeling process by listing the steps in a circular fashion with the dark arrows representing advancement to the next stage of modeling. The gray arrows represent the many possible return paths to repeat previous steps. The gray wall was shown previously in figure 13.2. It appears in the new diagram to separate

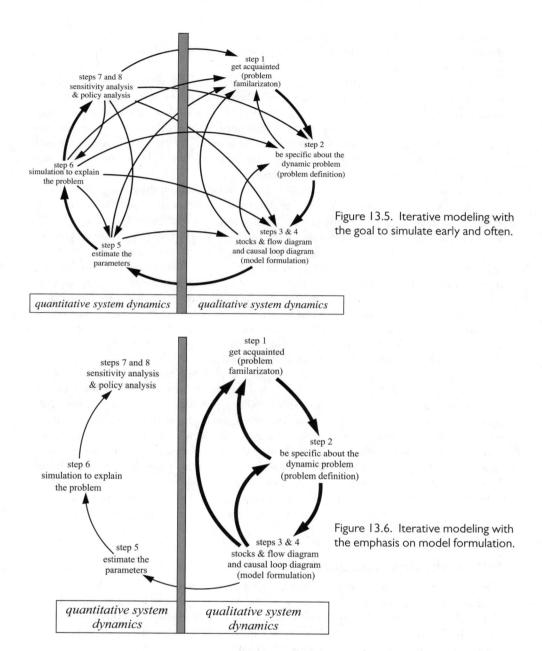

Figure 13.5. Iterative modeling with the goal to simulate early and often.

Figure 13.6. Iterative modeling with the emphasis on model formulation.

the qualitative and quantitative stages of modeling. The many arrows make it clear that the participants "crossed the wall" many times during the course of the projects.

The four studies in the lower portion of table 13.3 took a different approach to iteration. Their efforts were focused much more strongly on the qualitative stages of problem definition and model formulation. There was extensive stakeholder involvement in the discussion of the system, the problem, and the stakeholders' goals, as depicted in figure 13.6. The stakeholders were heavily involved in model formulation, including learning to use the software and constructing the stock-and-flow diagrams alongside the team leaders. Each of the teams crossed over to the quantitative stages and produced a simulation model that was used for both sensitivity analysis and policy analysis. These steps were typically performed near the end of the project, and the project concluded with presentation of the simulations and their implications for policy making.

The review of the modeling studies raises the question of how the "simulate early and often" groups were able to iterate so many times. Perhaps a longer time interval made multiple iterations possible? The 24-month study of the Rio Grande in New Mexico was the longest-running project, and it was able to complete multiple iterations. The two scoping studies in the lower half of table 13.3 were only 4 months in duration, so perhaps there was not sufficient time for an iterative approach. On the other hand, the multiple iterations in the sage-grouse study were completed in a 3-month project.

Time constraints play a role, but I believe the dominant factor is the willingness of the participants to allow the modeling specialists to do a good portion of the model formulation back at the office. For example, the nine groups in the sage-grouse study were quite happy to allow Beall and Zeoli (2008) to formulate the stocks and flows for the bird population and the land uses and to present the working model at district meetings. The participants were comfortable with both the model structure and the model results, even though they did not engage in hands-on construction. The Ria Formosa National Park study took a quite different approach. Their goal was to engage a large group of participants in hands-on modeling so the stakeholders would have more ownership in the model. This approach created more stakeholder familiarity with the step-by-step workings of the model, and a good understanding of the many alternative models that were not adopted. The approach led to a simulation model with useful results that were presented toward the end of the project.

So, which approach is better? I have had good experiences with the "simulate early and often" strategy, so I automatically head in that direction in my own studies. But which approach makes sense when you sit down with a new group of stakeholders? Van den Belt (2004) addresses this question in *Mediated Modeling*. She reports that the transition from qualitative to quantitative modeling can go smoothly in some situations, especially when specialists are able to work on the quantitative aspects "behind the scenes." And she reports that the transition tends to be easier in groups that "were less involved in the quantification of the model." But she concludes that the literature on environmental modeling does not allow one to anticipate in advance the extent to which the "participants should be involved in quantification of a model." Her view is that this is a "wide-open area for research and experimentation" (Van den Belt 2004, 255).

What about Validation?

We now turn to the question often asked at the end of a modeling project: What has been done to validate the model? Up to now, validation has not even been mentioned in any of the eight steps of modeling. Perhaps there should there be a ninth step for validation? I believe validation is too important to be left to the end. It makes more sense to think about validation throughout the modeling process. This chapter explains how this might be done. But first, let's set the stage with a concrete example of the confusion that can arise in discussing model validation.

A Validation Scene

Imagine that you and your colleagues have just constructed a computer model to improve your understanding of a dynamic problem within your organization. You have followed the steps of modeling, and you are pleased that the model generates the dynamic behavior specified as the reference mode. You feel that the stock-and-flow structure is reasonable, and that the parameter values are reasonable as well. You have drawn causal loop diagrams, and the combination of feedback loops helps you appreciate the reasons for the dynamic behavior. Finally, and most important, you have used the model to simulate policies that might be implemented by your organization. The simulations suggest that the policies could lead to improved behavior, so you are eager to spread this encouraging news within the organization.

Fortunately, your timing couldn't be better. The organization has just convened an executive committee to examine the very dynamic problem you have been studying. The committee is comprised of top leaders from across the organization, and it has recruited an excellent staff with expertise in computers and modeling. The committee is open to ideas from throughout the organization, so they give you an opportunity to present your results. You and your colleagues show them everything—stocks and flows, equations, simulation results, causal loop diagrams. You conclude the presentation with the policy simulations that you feel deserve further investigation.

The committee is impressed by your use of computers and models. They assure you that bits and pieces of your model agree with their own views. They are especially impressed by the way the bits and pieces are assembled to provide a holistic perspective on the organization. They congratulate you for an illuminating presentation. Then they turn the next stage of the meeting over to the chief of staff. The chief of staff is an expert in computers with a long history of success implementing computerized systems within the organization. He opens his portion of the meeting with a question:

Can you prove the model is valid?

You are not sure how to respond because you told the committee everything in your opening presentation. But you muster a response that highlights the reasonableness of the stock-and-flow structure and the relative accuracy of the parameter values. You conclude by emphasizing the close correspondence between the simulation results and the organization's dynamic problem. As you sit down, you are left with the unsettled feeling that your description does not constitute "proof" that the model is valid.

The chief of staff seems to agree. He tells the committee that your work is certainly interesting, but that you have failed to prove the model is valid. He advises the committee that it would be dangerous to base important policy changes on an unproven model. The committee is inclined to agree with the chief of staff, given his expertise in computers. They adjourn the meeting by inviting you to return when you are able to prove the model is valid.

Asking the Right Question

The preceding scene is all too common in the real world of computer modeling. The use of formal mathematical methods, sophisticated software, and high-powered computers seems to leave an impression that computer models will be constructed as perfect replicas of the system under study. But computer models are the same as any model. By design, models are simplifications of the system under study. To better understand how the system works, we build a simplified picture of the system. The key to the model's usefulness is leaving out the unimportant factors and capturing the interactions among the important factors. Once a factor is left out, the model is automatically subject to the criticism that it is invalid.

Such criticism is pointless. It reveals that the critic does not understand the nature of modeling. No model, however how carefully designed, can be validated. All we can do is invalidate it by exploring the implications of its assumptions and testing how far the results diverge from reality (Holling 1978). The question is not "Is the model valid?" but "Is the model useful?" This is a much more pragmatic question that may be addressed in a concrete manner. We would begin by specifying the purpose of the model and the alternative methods of achieving the same purpose. Then we ask ourselves about procedures for building confidence in the model.

Greenberger, Crenson, and Crissey's (1976) description of the use of models in the policy process is particularly instructive here. They reviewed several modeling methodologies used to influence public policy, including system dynamics, econometrics, and linear programming. The cases ranged from highly visible battles between Congress and the White House over budgets to discussions of fire station locations in New York City. They concluded that

> *there is no uniform procedure for validation. No model has ever been or ever will be thoroughly validated. Since, by design, models are all simplifications of the reference system, they are never entirely valid in the sense of being fully supported by objective truth. Useful, illuminating convincing, or inspiring confidence are more apt descriptors applying to models than valid.* (Greenberger, Crenson, and Crissey 1976, 70)

Greenberger argues that one can bolster confidence in a model by having it reproduce past behavior of the reference system, exploring its response to perturbations, critically examining the premises and theories on which it is based, and finally, putting it to use. Along the way, we should remember that "such tests are aimed more at invalidating than validating the model." Each new test may "only reveal the presence (not the total absence) of errors. However convincing a model, there is always a chance that its next test or use will turn up a serious shortcoming." Their review focused on public sector models, but their line of reasoning applies to ecological and biological models as well. Rykiel (1996, 229) explains that the validation of ecological models is not a procedure for testing scientific theory, but a means to learn if the model is acceptable for its intended use. Levins (1966, 430) outlines a general strategy for model building in population biology and concludes with the observation that

> *all models leave out a lot and are in that sense false, incomplete, inadequate. The validation of a model is not that it is "true" but that it generates good testable hypotheses relevant to important problems. A model may be discarded in favor of a more powerful one, but it usually is simply outgrown when the live issues are not any longer those for which it was designed.*

Although several authors may agree on the meaning of the term *validation*, there is plenty of room for confusion. *Webster's Dictionary* defines *valid* as "sound, just, well founded," and the verb *validate* means to "substantiate." These definitions are well aligned with Greenberger and Levins's views. But *Webster's* also lists definitions with a legal or procedural meaning, as in whether you have a valid contract or a valid passport. With these definitions, *validate* refers to the act of proving that a contract is legally binding or verifying that a passport was issued properly.

Greenberger encourages us to think of validation differently. He argues that "validation is not a general seal of approval" but a more general "indication of a level of confidence in the model's behavior under limited conditions and for a specific purpose." He suggests that "data provide a tangible link between a model and its reference system, and a means for gaining confidence in the model and its results." Greenberger believes that a "model that closely reproduces data on observed past behavior of the reference system gains credibility and wins the acceptance and trust of potential users."

Concrete Tests to Build Confidence

Matching historical behavior is one of several tests that can be used to build confidence in a model. Others have described a wide assortment of useful tests (Forrester and Senge 1980; Richardson and Pugh 1981; Kitching 1983; Rykiel 1996). From my own experiences, five tests stand out as especially useful in environmental systems:

1.   Historical behavior
2.   Verification
3.   Face validity
4.   Extreme behavior
5.   Detailed model check

Historical Behavior

The most common test is to set the inputs at their historical values and see if the outputs match history. Indeed, this test is usually what people automatically think of when discussing model validation. (Some people use the term validate as synonymous with conducting the historical behavior test.) Kitching (1983, 43) describes several variations in the historical behavior test for ecological systems. Model parameters might be based on data from one ecological site, and the model could be simulated to learn if the population projections match data from a different site. Or a model might be estimated from data over one time period, and the simulated population compared with data from a later time period.

You will see several examples of the historical behavior test in this book. For example, the Mono Lake model (chapter 5) was checked to learn if it showed the gradual decline in lake elevation if exports were maintained at the historical values. In the Kaibab deer case (chapter 21), you will see that the model reproduced the rapid growth and overshoot that is believed to have occurred in the 1920s. The historical behavior test is especially informative if the model is designed with a large number of endogenous variables and only a limited number of exogenous inputs. However, not all models adopt the system dynamics emphasis on feedback. Many models are constructed with little attention to the feedback loop structure of the system, and, in some cases, the models may not contain a single feedback loop. In this extreme situation, each and every variable is exogenous, and the historical behavior test is meaningless. If you arrange for all the inputs to follow historical trends, it should be obvious that the outputs will match history as well.

Verification

A model may be verified when it is run in an independent manner (typically by a different group on a different computer) to learn if the results match the published results. The goal is simply to learn if the computer model "runs as intended" (House and McLeod 1977, 66) and with correct program logic (Rykiel 1996, 232). Greenberger, Crenson, and Crissey (1976, 70) describe verification as a "test of whether the model has been synthesized exactly as intended. Verification of a model indicates that it has been faithful to its conception, irrespective of whether or not it and its conception are valid." Kitching (1983, 42) warns that verification may sound tautological, but it is "nevertheless a necessary check that the mechanisms of the model are in fact doing what the modeler thinks they are doing."

You have verified many of the models in this book as part of the student exercises. In each case, your goal was to demonstrate that the model behaves as published, regardless of whether you agreed with the model. Verification is essential if the field of modeling is to build from past models. Think of model verification in the same way as a scientist thinks of experimental verification. The first step in a serious scientific endeavor is to verify the previously published results!

Although verification seems like an obvious test, verifying models in a real organization can often be exceedingly difficult. The problem may arise from the sheer size of the model. In some models, for example, the challenge of checking the model may increase "in relation to some power of the number of computer instructions" (House and McLeod 1977, 66). But size is not the only obstacle. You may have trouble verifying a model because of inadequate documentation or the commercial interests of the developers. If you encounter these problems, you may be better off discarding the model from further consideration.

Face Validity

The "face validity" test is a common sense test. You simply ask if the model structure and parameters make sense. This test relies on your understanding of the system to judge the structure (the combination of stocks and flows). You would also rely on your common sense to check the parameter values. If you see flows pointed in the wrong direction, or if you see negative values of a parameter that must be positive, you know there is a fundamental problem with the model. You have performed the face validity test with most of the models in the book. It was the natural thing to do, so you might be wondering why we need to invent a title for something that happened automatically in each new chapter.

A face validity test is important to remember as you move from the classroom into a large organization. An unfortunate aspect of models used in large organizations is the difficulty in checking their face validity. Many models grow to become so complex that one cannot see inside to perform this simple test. Such models are sometimes called *black boxes*. If you encounter a black-box model, you may be better off discarding the model from further consideration.

Extreme Behavior

One of the most revealing tests is to make a major change in model parameters and see if the model's response is plausible. In chapter 21, for example, we will subject the Kaibab deer model to an extreme combination of parameter changes just to see if the model's response is plausible. Also, a model is automatically placed under extreme conditions if you subject it to the comprehensive sensitivity analysis described in appendix D. When two or three parameters take on unusually low values, for example, the model may be placed in an unexpected situation. If the model is structurally flawed, the flaws will be revealed by a simulation with clearly erroneous behavior.

You should not be surprised if the extreme behavior test is not conducted by modeling teams in large organizations. Their reluctance may seem reasonable—why subject a model to extreme conditions when it was probably designed to simulate most likely conditions? Also, specialists within different disciplines may resist the need for their portion of an interdisciplinary model to deal with extreme conditions (Holling 1978). The reluctance for extreme behavior testing might also stem from the fact that many models are likely to exhibit erroneous behavior when subjected to major changes. These problems typically arise from a variety of features that improve the cosmetic appearance of the model under most likely conditions. If the cosmetic features are not structurally sound, the model is likely to exhibit erroneous behavior under extreme conditions.

Detailed Model Check

If you are working on an important topic within a large organization, it is quite possible that there are other models that provide a more detailed and accurate representation of some aspect of the system. If this is true, you should take advantage of these models to provide benchmark simulations that may be used to check your own model. An illustrative example is given in the model of nitrogen accumulation in a catchment in appendix G. The simple model is checked against a detailed model that accounts for the spread of nitrogen through the spatial dimensions. And you will read in chapter 24 about models that were used in the electric power industry. Detailed model checks were extremely useful in bolstering confidence in these models (Ford 1994, 1997).

Implementation

This chapter closes with comments on the implementation of a model or the implementation of policies suggested in a modeling study. Let's start with model implementation.

Some models are designed to be used once by their developer, and the model is discarded after the findings have been reported. They are like paper plates: use them once and throw them away. But other models are constructed to serve a longer-lasting need within an organization. This situation often occurs when the organization faces similar questions and reporting requirements year after year. In these circumstances, a large-scale modeling effort may be justified to address the reoccurring questions in a sustained manner. When organizing a sustained effort, validation is best viewed as a continuous, sometimes arduous process of building confidence in the model. Richardson and Pugh's (1981, 311) observations are particularly relevant to large-scale efforts. They advise us not to think of validation as a "one-time process that takes place after a model is built and before it is used for policy analysis" and they warn us that "there is a tendency to think of validation as a process similar to warding off measles: a model, susceptible to contagious criticism, gets validated and becomes immune to further attack." Several system dynamics groups have achieved implementation of models in large organizations. Their success comes, in part, from the continuous attention to model testing to build confidence within the organization.

Policy implementation is the primary goal of pragmatic modeling. At the end of the day, the final test of model usefulness is whether the modeling process leads to better understanding of policies to improve system behavior. Forrester (1961, 115) explains that "the ultimate test is whether or not better systems result from investigations based on model experimentation." Richardson and Pugh (1981, 313) echo his view. They argue that "the ultimate test of a policy-oriented model would be whether policies implemented in the real system consistently produce the results predicted by the model." If your organization is focused on scientific research, models may provide the greatest benefit in helping researchers to design useful experiments. Watt (1968, 349) reports that a common characteristic of successful application of large-scale simulation studies in biology is "an enormous amount of feedback between the experimental work and the computer activity." If you are designing a model to test the impact of a policy, you should not expect that policy implementation will be automatic. The key to actual implementation is sustained client interest and close communication between the client and the modeling team throughout the process.

My own experiences with implementation involve energy and the environmental problems. Lessons from the actual use of models to shape energy policy are described in the concluding chapter of this book.

Chapter 14

Software: Further Progress with Stella and Vensim

This chapter describes additional features of Stella and Vensim that amplify the power of system dynamics modeling beyond the main tools of stocks, flows, and feedbacks. I have selected a collection of features that are particularly useful in environmental modeling. Both Stella and Vensim come with excellent help pages, so there is plenty of online documentation to supplement what you read here.

Read and Verify

This chapter is similar to chapter 2; it is written in workbook style. Each model is described in a step-by-step manner as if you are following along, executing each step with your copy of the software. (As you do so, remember that these results are from version 9.0.3 of Stella and version 5.3 of Vensim PLE, the Personal Learning Edition.) This chapter is written for readers who wish to learn how to build and test models. If you are reading this book for general ideas and concepts, you can skip over the step-by-step instructions. (The figures will give you a general impression of what can be accomplished with the software.)

Control Panels

Both Vensim and Stella facilitate the development of control panels. A control panel for a computer model is similar to the control panel on a television—it provides the key controls and a display of the results. Models with control panels are sometimes called *flight simulators*. This term reminds us of a pilot flight simulator, an electromechanical model of an airplane. Pilot flight simulators are complicated and expensive representations of a real airplane. They provide a sophisticated interface resembling cockpit controls and the view out the window. Pilots practice in the simulator to improve their flying instincts. Computer-based flight simulators have been developed to serve a similar purpose in business systems; they help students and managers develop their business instincts (Morecroft 1988; Senge 1990; Sterman 1992). Flight simulators can be put to good use in environmental systems as well.

The first control panel in this book is the Daisyworld simulator in figure 11.9. It uses many of Stella's interface tools to make it easier to experiment with different scenarios and different varieties of flowers. You can learn about the interface tools by downloading the Daisyworld simulator (from the book's website, the BWeb). Experiment with each object, and read

Stella's documentation. You can also learn by deleting and re-creating each of the objects. With a little practice, you will soon be able to construct your own control panels.

Control panels can be constructed with Vensim, as illustrated in figure 14.1. This is a control panel to experiment with the sales model from chapter 7. Six sliders allow one to change the value of six inputs up or down within the assigned range. Vensim calls these *input-output devices*. You select the inputs to appear on the control panel and their range of values. Figure 14.1 shows a custom graph of the sales force, new hires, and departures. (A custom graph is also an input-output device in the Vensim software.) Vensim's SyntheSim tool makes experimentation easy, and we see the effect of changes instantly. Four of the inputs were readily available in the sales model shown in figure 7.10. Two new inputs are included in figure 14.2 to facilitate more extensive testing. The initial sales force is now specified as a named variable so it can be explored through the SyntheSim tool. Another input is the saturation size, which is set at 1,000 sales persons. The modified model calculates a saturation ratio as the size of the sales force divided by the saturation size. The ratio is then used to find the effectiveness multiplier. The actual effectiveness is the product of the maximum effectiveness and the multiplier. A lookup for the effectiveness is used to create the relationship shown previously in table 7.1. The modified model is now ready for interactive experimentation from the control panel (BWeb).

Control panels make models easy for others to use, so they can promote more experimentation and discussion. Control panels also help the model developer. I recommend building a control panel whenever the model has reached the stage where you have explained the reference mode. The next steps are sensitivity testing and policy testing. Developing a control panel

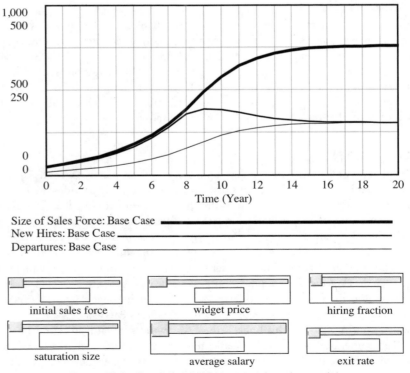

Figure 14.1. Control panel for testing the sales model.

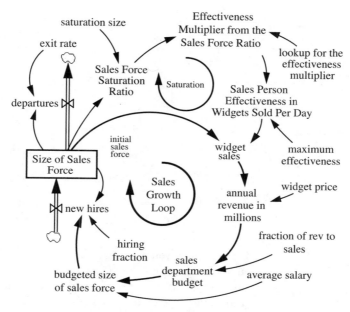

Figure 14.2. The modified sales model.

is a good way to think about the main results that should be in view when performing these tests. It also helps one think about the main inputs that are worthy of experimentation. And, finally, control panels promote experimentation by providing rapid feedback.

But there is a downside to control panels. Simulation analysis can sometimes be too easy, and we fall into the trap of playing with the model as if it were a video game. We should guard against the video game syndrome by taking the time to think about each new simulation. It's helpful to guess the result—take the time to draw the pattern you expect to see when you click the "run" button. Then pay particular attention to results that do not match your expectations. This will slow down the process of experimentation, but it is thoughtful experimentation that will lead to more learning in the long run.

Randomness and Stochastic Simulation

Up to now, all simulations in the book show a dynamic pattern similar to one of the six dynamic patterns (figure 13.1). The trajectories show a simple, smooth shape over time. These are called *deterministic simulations* because the results are determined by the rules built into the model. If we run the model a second time with the same inputs, we will get the identical pattern of behavior. But environmental systems are exposed to randomness that can deflect the system from the smooth trajectories. Randomness sometimes creates irregularities that appear to be superimposed on the underlying trend. Randomness may also excite a dormant pattern. Simulations with randomness included are called *stochastic simulations.*

The flowers model in figure 14.3 illustrates stochastic simulation. The flowers' growth rate is influenced by temperature, and the temperature varies from a low of 20°C to a high of 30°C in a random manner. Let's assume that the temperatures are equally distributed across this range and turn to Stella's RANDOM function. (This function is found in the alphabetical list of built-ins located next to the numeric pad when you open the equation for temperature.) The temperature is defined as RANDOM (low temp, high temp, seed), where the seed

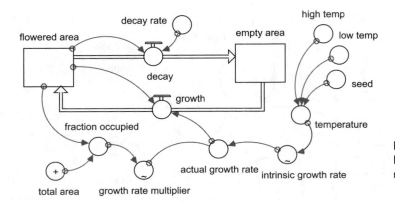

Figure 14.3.
Flowers model with
randomness.

Table 14.1. Growth rates.

Degrees C	Growth rate
20	0.00
21	0.20
22	0.70
23	1.00
24	1.20
25	1.20
26	1.20
27	1.00
28	0.70
29	0.20
30	0.00

is any number you wish. Its purpose is to guarantee that the statistical function delivers a replicable sequence of random numbers. Stella will draw a new number from the uniform distribution in each time step of the simulation. For this illustration, DT is 0.25 year, so the temperature will change four times a year.

The changes in temperature lead to changes in the intrinsic growth rate, as noted by the ~ on the converter. Table 14.1 shows some of the growth rates: 1.20/yr if the temperature is between 24°C and 26°C; zero if the temperature is at the extreme values of 20°C or 30°C. The upper graph in figure 14.4 shows the temperature varying between 20°C and 30°C. The intrinsic growth rate is shown in the lower graph. Notice that it is around 1.20/yr whenever the temperature is near the optimum value of 25°C.

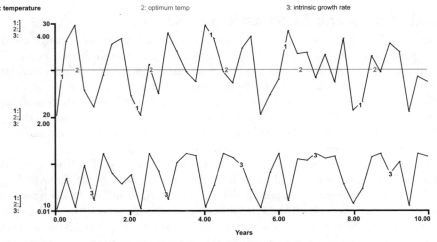

Figure 14.4. Random behavior of temperature (upper graph) and
intrinsic growth rate (lower graph).

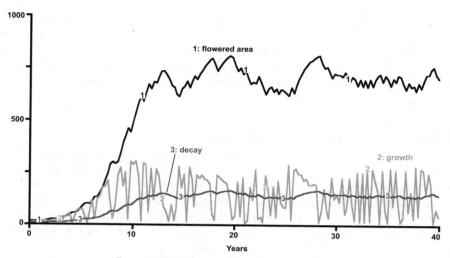

Figure 14.5. Results of a stochastic simulation.

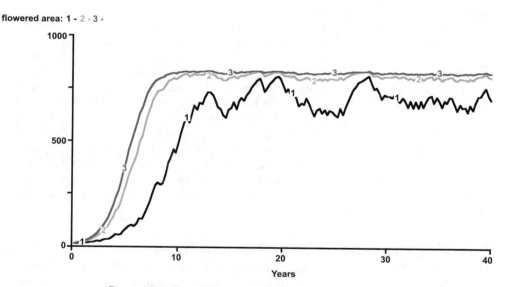

Figure 14.6. Flowered area in three stochastic simulations.

Figure 14.5 shows the results of the stochastic simulation. The growth in the flowered area exhibits the greatest variation, ranging from zero to almost 300 acres/yr. The flowered area climbs to around 700 acres by the 15th year of the simulation. The general pattern of S-shaped growth is discernable, but there are considerable variations around the equilibrium value. The average value of the flowered area is somewhat less than in the deterministic simulation shown in figure 7.5. The lower equilibrium arises because the temperature varies across a wide range of values, and the average intrinsic growth rate is not as large as before.

Now suppose the temperature randomness is over a smaller range. We can repeat the stochastic simulations and compare results. Figure 14.6 is a comparative time graph of the flowered area in three simulations. The first case was shown in figure 14.5. Table 14.1 tell us that the average value of the intrinsic growth rate is 0.67/yr. The second case allows the temperature to vary randomly between 22°C to 28°C, and the average of the relevant entries in table

14.1 is 1.0/yr, the same value as in figure 7.5 the deterministic simulation with the flowered area reaching an equilibrium at 800 acres. The second case in figure 14.6 shows essentially the same result. The third case limits the temperature changes to between 23°C and 27°C. These values are much closer to the optimum, and the average intrinsic growth rate would be 1.20/yr. The flowered area would grow somewhat more rapidly in the third case, and its equilibrium would be just over 800 acres.

Advice on Stochastic Simulation

Environmental systems are constantly exposed to external inputs that vary in an unpredictable fashion. Such variations are often called *random* (which is the word used when we do not know how to explain the variation in a deterministic manner). The standard approach is to draw values from the appropriate statistical distribution and then use stochastic simulation to learn how the changes in the external input can alter the simulated behavior. The external variations will sometimes create only minor variations superimposed on a familiar pattern (such as S-shaped growth). In other cases, the randomness may give the system the opportunity to escape a local equilibrium point and emerge with dramatically different behavior. And in cases with oscillatory behavior, the external disturbances may change a dampened oscillation into an oscillation that continues indefinitely.

Stochastic simulations are a good test of model behavior. But they are not the first test that should be performed. Indeed, they are seldom the second or third test. The best way to learn about dynamic behavior is to concentrate on a deterministic explanation on the initial iterations through the steps of modeling. Randomness should be excluded until we are coming close to explaining the reference mode. At that point, it is useful to test the model with simple disturbances. You saw an example in the stability disturbance test in chapter 6. We disturbed models that were in equilibrium and watched their response to learn if the equilibriums were stable. This simple test is like tapping a tuning fork—tap the fork once, and it responds with its characteristic frequency of vibration. Most system dynamics models are tested in this manner, and both Stella and Vensim provide a standard set of test functions (e.g., STEP, PULSE, and RAMP). Stochastic simulations are useful after we have learned the model's response to a simple change in an external input. External variables that appear to be random can then be characterized statistically, as explained below.

Selecting the Properties of Random Inputs

The flowers model uses Stella's RANDOM function. It makes sense if the random input is equally likely to be found across a range from low to high. The uniform distribution is one of several distributions available with Stella and Vensim. Stella provides the uniform, normal, log-normal, exponential, and Poisson distributions. Vensim offers the same choices plus others like the beta and Weibull distributions. Both programs give you the option of setting a seed (to generate a replicable sequence) or to ignore the seed and see a different sequence with each simulation. And both programs sample from the distribution every DT (delta time, the step size of the numerical simulation).

The DT sampling is important because DT is a very small number. (It must be sufficiently small to guarantee numerical accuracy.) But a small value of DT means that we sample from the statistical distributions very frequently. The end result is randomness that resembles high-frequency noise. (*Noise* is an engineering term often used to describe randomness, as in the background hiss on your radio.) *High-frequency noise* refers to random changes that occur extremely quickly compared with the rest of the system. As an example, think of a high-frequency sound and its effect on our eardrums. The frequency of ultrasonic sounds is too high for our ears to hear the sound. This type of disturbance would have no effect on the system.

The random disturbances in environmental systems are much different from high-frequency noise. In watersheds, for example, we might see unusually high runoff in one year and low runoff in the next year. In this case, it would be best to sample every year rather than every DT. Figure 14.7 illustrates the difference in values over a 10-year simulation. The "uniform noise every DT" is drawn from the uniform distribution from 20 to 30 with DT = 0.25 year, but the "Sampled Noise" changes values every year.

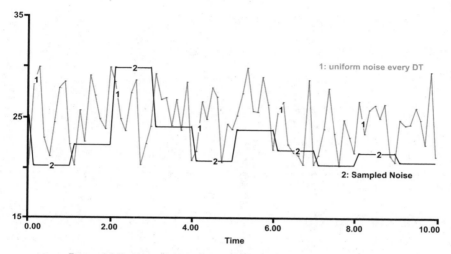

Figure 14.7. Example of sampled noise that changes every year.

Figure 14.8 shows a simple stock-and-flow structure to sample a random variable and hold its value for a specified length of time (called the *noise interval*). The "uniform noise every DT" uses the RANDOM UNI-FORM function. DT is set to 0.125 in this test, so we will see eight values per year. The "time to select?" is a binary variable whose value is either 1 or 0, with the value 1 meaning yes and the value 0 meaning no. When "time to select?" is 1, we know a year has passed, and it is time to pick a new value for the "Sampled Noise." The "time to select?" will be 1 at the end of each year, but it will only re-

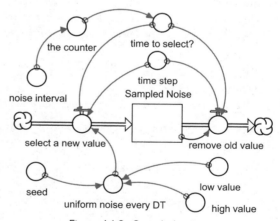

Figure 14.8. Sampled noise.

main at 1 for one time step. The selected value is divided by DT to ensure that the flow delivers the new value from the uniform distribution. (The flow has to be eight times larger to deliver its value into the stock in just 1/8 year.) As the new value is added to the stock, the other flow discards the previous value. The equations are listed in table 14.2. They use some logical functions explained below.

Table 14.2. Equations to sample and hold for a noise interval.

Sampled_Noise(t) = Sampled_Noise(t – dt) + (select_a_new_value – remove_old_value) * dt

INIT Sampled_Noise = 25

INFLOWS:
select_a_new_value = time_to_select?*uniform_noise_every_DT/time_step

OUTFLOWS:
remove_old_value = time_to_select?*Sampled_Noise/time_step

high_value = 30

low_value = 20

noise_interval = 1

seed = 888

the_counter = MOD(time,noise_interval)

time_step = .125

time_to_select? = IF(the_counter=0) THEN 1 ELSE 0

uniform_noise_every_DT = RANDOM(low_value,high_value,seed)

The Logical Functions

Stella provides a variety of logical functions such as AND and OR. These functions work with expressions that can be evaluated as true or false. The numerical value 1 means true; the value 0 is for false. There are only two values, so we call these binary variables. To illustrate their use, imagine that your home furnace is wired up to thermostats in three different locations, and the three targets are 61° F, 62° F, and 63° F. Now, suppose the furnace is designed to come on if the room temperature in any of the locations is below target. We could define a binary variable "furnace on?" with the OR function. And suppose the furnace operates at full capacity (say 4,000 BTU/hr) whenever it is on. The furnace operation could be simulated with three equations:

> Furnace on? = (T1<61) OR (T2<62) OR (T3<63)
> Capacity = 4000
> Furnace Energy Production = Furnace on?*Capacity

Another logical function is the IF THEN ELSE function. It looks at a logical expression that can be evaluated as true or false. In the furnace example, we could say:

> Furnace Energy Production = IF (Furnace on? = 1) THEN Capacity ELSE 0

Or, to illustrate with a different example, suppose an air conditioner is scheduled to come on between 10 a.m. and 6 p.m. each day. Let's assume we are simulating time in hours, and we create an *hour of day* to run from 0 to 24. Let's define a binary variable to represent whether the air conditioner is operating:

> AC Operating? = IF ((hour of day>10) AND (hour of day<18)) THEN 1 ELSE 0

The expression next to the IF will only be true if the hour of day is greater than 10 and less than 18. This interval corresponds to 10 a.m. to 6 p.m. for each day. During these hours, the binary variable is set to 1; otherwise it is set to zero.

Now we need to calculate the hour of day. Suppose our model is designed to simulate 10 days, so time will run from 0 to 240 hours. Think about the value of time when the air conditioner turns on:

time = 10, it turns on for the first time
time = 34, it turns on for the second time,
time = 58, it turns on for the third time

These times all have the same thing in common: they leave a remainder of 10 when you divide time by 24. We use the MOD function to check for the remainder. (MOD is short for modulo, the remainder left over when you divide two numbers.)

The MOD Function and the Calendar

The MOD function will be quite useful in environmental models since it often helps to pay attention to daily or seasonal changes. When keeping track of the calendar, the syntax is normally MOD(time,interval). If we are looking at the hour of a day, the interval is 24. Let's put the MOD function to use in the model of flowers that was simulated in figure 7.5 in years. The model made no distinctions between the seasons in the year. If we change time to months, we can represent the seasonal effects on growth and decay, as shown in figure 14.9. The month of the year is MOD(time,12). The decay rate is 0.2/yr; the average annual intrinsic growth rate is 1.0/yr. The growth rate multiplier is from figure 7.4. The seasonal effect on growth rate is a nonlinear graphical function. It concentrates the growth in months 3, 4, and 5. The seasonal effect on decay rate concentrates the decay in months 9,10, and 11. So we expect to see the flowered area expand in the spring and retreat in the fall. Since time is in months, the flows are simulated in acres/month. To make the units clear, we add a named variable for the 12 months per year.

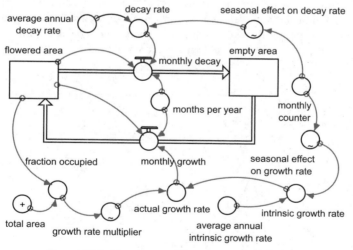

Figure 14.9. Flowers model with seasonal changes.

Figure 14.10 confirms that the monthly model works as expected. It shows the same general pattern of S-shaped growth from chapter 7. (The flowered area reaches an average of around 800 acres, the same as in figure 7.6.) The seasonal variations are superimposed on the smooth trajectory. The flowered area expands each spring and retreats each fall.

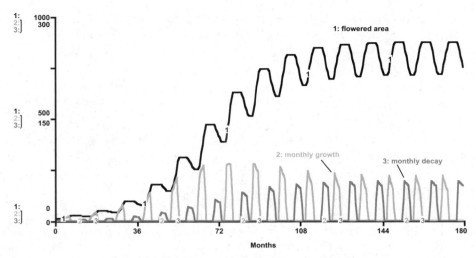

Figure 14.10. Flowered area with seasonal changes.

The SMOOTH Function

The SMOOTH function is the most important of the built-in functions because it can be used to represent the delayed effect of information in the system. Information delays are present in all systems, so the software developers have provided several ways to characterize these lags. To illustrate, let's suppose that there is a sudden increase in price, and that the SMOOTH function represents delay in the perception of the price increase. Figure 14.11 shows the price jumping from $20 to $30 in day 4 of a 16-day simulation. The product price is displayed alongside the smoothed values using the first-order SMOOTH function. The quickest response occurs with a 1-day lag. Think of this curve as representing the weighted average price perceived by the population. Some people get the word right away; others are not paying as much attention. If 60% are aware of the new price, we could say the perceived price is $26. Figure 14.11 shows a

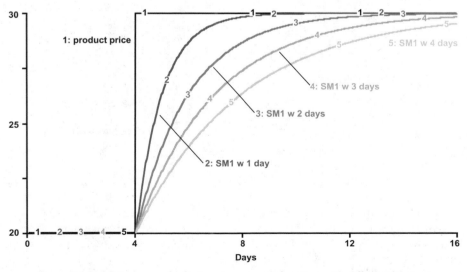

Figure 14.11. Response of a first-order smooth function with different lag times.

curve for SM1 with 1 day at around $26.40 after one day. It's as if 64% of the people are aware of the new price. After 3 days, the delayed price is $29.50. We could say that 3 days is sufficient for 95% of the price increase to be perceived. The remaining three curves show the smoothed price with longer lag times. The 4-day lag curve is the slowest to respond. It follows the same pattern as the 1-day lag, but it takes four times longer to see the same results (i.e., 95% of the people are aware of the price increase after 12 days).

The first-order SMOOTH is useful when an important fraction of the population is paying attention. If this is true, it makes sense for the smoothed curve to respond immediately to the step increase. But we need a different delay process if none of the population get the word for a day or two. The higher-order SMOOTH functions meet this need, as illustrated in figure 14.12. The product price increases from $20 to $30 in day 4, and we show 1st-order smoothing with a 4-day lag, the same as before. The other three curves use 3rd-order, 6th-order, and 12th-order smoothing, all with a lag time of 4 days. The 12th-order SMOOTH is the slowest to respond initially. It would be used if we thought almost nobody gets the word about the price increase until day 6. Once the word gets out to some, however, the 12th-order SMOOTH delivers a rapid response (i.e., almost everybody is aware of the new price by day 9).

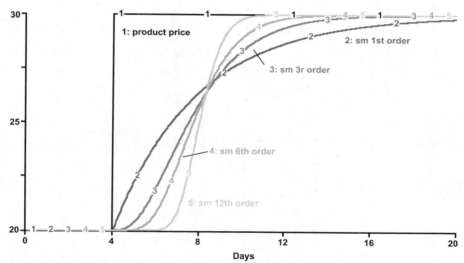

Figure 14.12. Response of first-order and higher-order smooth functions.

You can learn more about the SMOOTH function from the software documentation, and we'll talk about the importance of information delays in chapter 18. The equations used in these illustrations are shown in table 14.3.

Table 14.3. Equations for testing the smooth functions.

product_price = 20 + STEP(10,4)	*product_price = 20 + STEP(10,4)*
SM1_w_1_day = SMTH1(product_price,1,20)	sm_1st_order = SMTH1(product_price,4,20)
SM1_w_2_days = SMTH1(product_price,2,20)	sm_3r_order = SMTH3(product_price,4,20)
SM1_w_3_days = SMTH1(product_price,3,20)	sm_6th_order = SMTHN(product_price,4,6,20)
SM1_w_4_days = SMTH1(product_price,4,20)	sm_12th_order = SMTHN(product_price,4,12,20)

Advice on Using the SMOOTH Function

The SMOOTH function is used extensively in system dynamics models because delayed effects are pervasive in all systems. Our job is to pick the pattern (first order, third order, etc.) and the length of the lag time. The lag time should be long enough to represent the total delay in the simulated effect. In figures 14.11 and 14.12, the lag time represented the delay in changing perceptions of a price increase. But suppose we are keeping track of the perceived price to calculate the change in demand. In this case, the lag time would be increased to represent the additional delays in responding to the perceived price.

Figure 14.13 illustrates the new situation with a combination of converters to obtain the effective price that will determine the demand. This example uses a third-order SMOOTH with an 8-day lag for the price response. The equations are listed in Table 14.4. Figure 14.14 shows the results with a step increase in the price. It jumps from $20 to $30 in day 8, and the

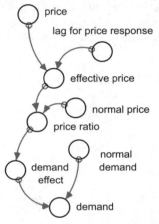

Figure 14.13. Demand depends on the lagged price.

Table 14.4. Equations for the converters in figure 14.13.

demand = demand_effect*normal_demand

demand_effect = 1/price_ratio

effective_price = SMTH3(price,lag_for_price_response,20)

lag_for_price_response = 8

normal_demand = 150

normal_price = 20

price = 20+STEP(10,8)

price_ratio = effective_price/normal_price

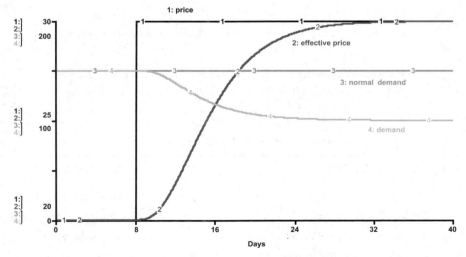

Figure 14.14. Demand responds to a step increase in the price.

effective price follows along with the third-order pattern. We expect to see the effective price at 95% of the way to $30 after three lag times. This occurs by day 32. The effective price is now $30, and the price ratio is 1.5. The effect on demand is defined as the reciprocal of the price ratio. The initial demand is 150 products/day, so the demand eventually turns out to be 150/1.5 or 100 products/day. The model is 95% of the way to the new demand by day 32.

The lines in figure 14.13 represent the flow of information. The delay in the flow is represented by the link from the price to the effective price. We will sometimes draw attention to the delay by the name. (Perhaps *lagged price* would be a better name.) The software documentation will explain the stocks and flows for the SMOOTH function. You'll read that a single stock appears in a first-order SMOOTH, and that three stocks appear in the third-order SMOOTH. These stocks do not appear in figure 14.13, but they are there, hidden away in the SMOOTH function. You have read in chapter 9 that there is least one stock in every feedback loop (otherwise, you'll won't be able to proceed—*sorry, but that would create a circular connection*). The stock can be explicit and visible, or it can be an implicit stock hidden inside the SMOOTH function. If you have trouble closing a loop, remember that the missing stock could take the form of a SMOOTH (BWeb).

Advice on the Other Built-in Functions

The SMOOTH function is an important function that can appear anywhere in a model as needed to represent the delays in information. But the SMOOTH is the only function that should be used throughout the model. My advice is to limit the other functions to the exogenous inputs. Think of the RANDOM function used to shape the exogenous temperature in figure 14.3. It makes sense because we are not trying to explain why the temperature varies up and down. We are simply assuming a random pattern for temperature to see the consequences for the flowered area. On the other hand, suppose we were tempted to use the RANDOM function to shape the planet temperature in Daisyworld. This would not make sense because temperature is at the heart of the Daisyworld model. The endogenous variables should be explained by the main tools of system dynamics—stocks, flows, and the feedbacks to explain the flows.

The same advice applies to the mathematical functions (e.g., MAX, MIN, ABS) and the logical functions (AND, OR, IF THEN ELSE). If you are tempted to use one of these functions to change an endogenous variable, think again. Ask how nature would shape the endogenous variable. This advice is easy to give, but many students find it hard to follow. The difficulty may be that the functions are more familiar than stocks, flows, and feedbacks. In the Mono Lake model in figure 5.4, for example, some students are tempted to define the actual value of water in lake as MAX(water in lake,0). The MAX is a clearly understood function, and they are eager to put it to use. In this case, it will certainly prevent the actual value of water in the lake from going negative. But you know that imposing this function does not help us understand how nature controls the volume of water in Mono Lake.

A second word of advice on functions is to test them immediately after building them. The software developers have worked hard to explain the syntax, but there is plenty of opportunity for error. I recommend testing each new function to verify that it gives the expected pattern of behavior. To illustrate the importance of testing, let's suppose that the population in figure 14.15 is an external input to a model, and we calculate its value using the EXP, the exponential function.

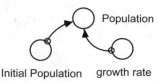

Figure 14.15. Population uses the exponential function.

The equations would be:
Population = Initial Population*EXP(growth rate*time)
Initial Population = 100
growth rate = 0.07

and we test the population with a 40-year simulation that pauses every 10 years to allow for a change in the growth rate. Figure 14.16 shows the population doubling from 100 to 200 during the first 10 years. So far, so good. Then we cut the growth rate to zero and expect the population to remain at 200. But the test shows the population dropping abruptly to 100 and remaining at 100 for the next 10 years. The growth rate is returned to 0.07 in the 20th year, and we expect the population to resume growing. Instead, we see an abrupt jump to 400. The problem with this formulation is that population should be represented as a stock if we want it to stay at its recent value when we set the growth rate to zero. The population equation illustrates the importance of testing each new function. Their syntax may seem obvious, but the functions do not necessarily perform as we intended. We should immediately test each new function under a wide variety of conditions. It's easy to make mistakes; the key is finding and correcting mistakes quickly. The last thing we need is an undetected mistake buried inside the model.

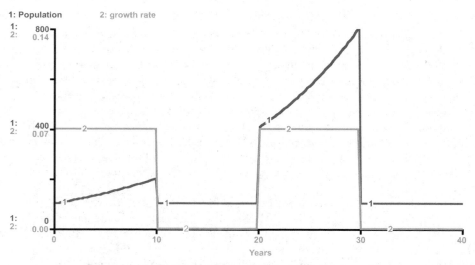

Figure 14.16. Testing if the population growth is reasonable.

Arrays

You've seen arrays in the epidemic modeling in chapter 8. They are useful when we find ourselves using the same stocks, flows, and equations many times with only changes in the parameters for each element of the array. The Stella model in figure 8.13 is an example of a one-dimensional array. We use Stella's array editor to define a *dimension* G with *elements* G1, G2, G3, G4, and G5. Figure 8.16 shows a similar model in Vensim (in the DSS version). We use the *subscript control* to specify a *subscript* G with *elements* G1, G2, G3, G4, and G5. Stella automatically assigns a 3-D icon to array variables, as shown in figure 8.13. Vensim does not change the icon, but we can make changes if we wish to remind ourselves of the subscripts. For example, we could use plural names (as in figure 8.16), or we could assign a thicker border to the stocks. Both programs come with a collection of functions designed to work with arrays.

The epidemic models use one-dimensional arrays. Stella allows two dimensions, so we could have a dimension G for groups of people and a second dimension C for cities. Two-dimensional arrays are also helpful in spatial models, as illustrated in the model of spatial nitrogen flows in appendix G. Vensim allows up to eight dimensions (or subscripts). We use four subscripts in a model of the electric power system in the western United States. One subscript stands for the seven areas of the West; another represents eight types of technologies, and a third represents each of the 10 transmission lines. The final subscript is "hour," to represent the 24 hours in a typical day. The subscripts for areas, technologies, and tie lines are typical for arrays. They allow for more compact equations (we avoid writing the same equation over and over). The subscript *hour* is quite different. This is an innovative approach to allow hourly operation to appear in a model that runs for 240 months. Appendix F describes this unique approach to simulating hourly operations within a long-term model of the electric power industry.

Import and Export

Both Vensim and Stella allow for data to be imported from a spreadsheet. The data can be moved into position with Edit/Copy, then Edit/Paste, or by dynamic data exchange, as explained in the online documentation. And both programs allow for model results to be exported to spreadsheets. Spreadsheets are useful when we wish to display the results in a graphical form not provided by the standard graphs in Stella or Vensim. A simple spreadsheet is also a useful form of display of spatial information, as illustrated in appendix G. Spatial information is also nicely displayed by dynamic connection to a spreadsheet, where it can then be linked to Stella's spatial map tool, as explained in appendix G.

The Conveyor Stock: Simulating Schooling Behavior

This chapter concludes with the conveyor stock, a special version of a stock to help us simulate material moving through a system in a tightly controlled pattern. The name makes us think of a conveyor belt, and the stock stands for the amount of material stored on the belt. The flow off the belt will be identical to the flow onto the belt, except shifted in time. The time interval is called the *transit time* for the conveyor.

Conveyors will be useful in environmental modeling because of the propensity toward *schooling* behavior, as when a school of fish starts and ends their migration in a tight pattern. Similar behavior is seen in the life cycle of animals, and in the seasonal pattern of plants. And, of course, there is schooling behavior in the way students are advanced and graduated from school. We'll use a 12-year schooling process to illustrate the conveyor, starting with figure 14.17. The stock is initialized at zero, and transit time is set to 12 years. The nonlinear graph (~) creates an inflow of 100 students in the 10th year to test the model. Figure 14.18 show the results; it confirms that the entire group of 100 students would graduate exactly 12 years later.

The conveyor stock is quite convenient when the students move through the system in a tightly controlled pattern. The alternative is to string together a collection of conventional stocks and set the output from each stock as the size of the stock divided by the time interval

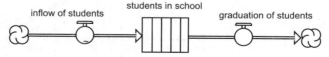

Figure 14.17. Conveyor stock for students in school.

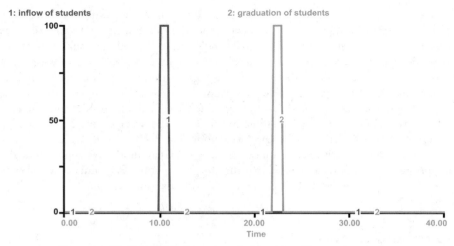

Figure 14.18. Testing the conveyor stock with a 12-year transit time.

for each stock. A 1st-order approach would have only one stock with a 12-year interval. A 3rd-order approach would string three stocks together, with 4 years assigned to each stock. A 6th-order approach would require six stocks, with a 2-year interval for each stock. And, finally, we could turn to a 12th-order approach, with a 1-year interval for each stock. You might suspect that the 12th-order approach would provide a good representation since it sounds most familiar when we recall the 12 years spent in school. Figure 14.19 shows the graduation patterns from each of these approaches. The results bear little resemblance to the pattern of graduates that appears in a school system that aims to advance and graduate the students in a tightly controlled pattern. When systems deliver tightly controlled patterns, we need to make use of the conveyor.

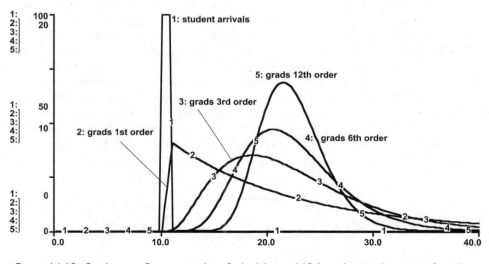

Figure 14.19. Student outflows using 1st-, 3rd-, 6th-, and 12th-order combination of stocks.

Conveyor stocks are designed to receive an input and then deliver an identical outflow that is delayed in time by the *transit time*. We simply specify the transit time and leave the job of controlling the outflow to the software. Stella permits a second outflow, which it assumes is leakage from the conveyor. Figure 14.20 illustrates with a second outflow to represent student dropouts. Stella recognizes the second outflow as leakage, and it assigns the downward arrow to distinguish the leakage from the primary outflow. Our job is to specify the leakage fraction for the entire interval for the time in school. For example, if 20% of incoming students will drop out sometime between 1st grade and 12th grade, we set the leakage fraction to 0.20.

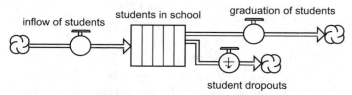

Figure 14.20. Addition of a leakage outflow to represent student dropouts.

Advice on Material Flows

The conveyor will be useful whenever material moves in a tightly controlled pattern. The material could be people advancing through the grades of school or salmon migrating to the ocean. A school of fish is similar to a cohort of students in school. Both the salmon and the students tend to stick together for a variety of social and survival reasons. Conveyor stocks are also useful in simulating the life cycle of animal populations, such as the brine shrimp population depicted in figure 5.13.

These examples are in Stella, but Vensim provides an equivalent capability. You would select a conventional stock to receive the inflow and then set the outflow as a DELAY FIXED version of the inflow. The main difference between the two programs appears when we assign an initial value to the stock, as you will learn in the exercises.

Chapter 15 uses conveyors to simulate the salmon life cycle. Conveyor stocks are just what we need in this case study. But what should we do when the material flows through the system in a more diffuse pattern? In these cases, we string together a combination of conventional stocks to re-create the pattern of material flow. Figure 14.19 shows the patterns from stringing together 1, 3, 6, or 12 stocks. You can see the trend developing, so you would pick the number of stocks to get a good match for your system. If you are unsure of the pattern of material flow, you could use the first-order approach to get started. It's the easiest to implement and to explain. You could then test the importance of the first-order approach by trying a higher-order approach later in the modeling process.

Exercises

Exercise 14.1. Initial value of overwintering eggs

Build the model in figure 14.21 with a 6-month incubation period. There is no inflow, so we are only able to have eggs hatching if there are some eggs initially assigned to the stock. Let's assume that there are 120 eggs initially. If we write the number 120, Stella will assume that the 120 eggs are equally distributed across the conveyor. Confirm that the eggs hatching will be 20 eggs per month for the first 6 months.

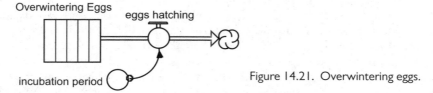

Figure 14.21. Overwintering eggs.

Exercise 14.2. Control the month for eggs to hatch

Experiment with different ways to set the initial values. Try 0,0,0,0,0,120 to see all 120 eggs hatch in the sixth month. Try 120,0,0,0,0,0 to see all the eggs hatch in the first month.

Exercise 14.3. Model of an insect population

Build the model in figure 14.22 with the initial value of Overwintering Eggs set to 0,0,0,120,0,0. Set the fraction female to 0.5 and the eggs per female to 2. The insects have a 12-month life cycle with a 6-month incubation period, a 5-month maturation period, and a 1-month adult period.

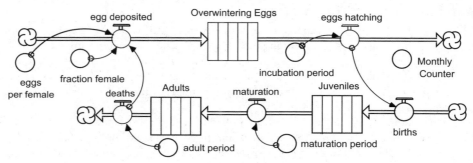

Figure 14.22. Insect model with no losses.

The model assumes that the insects deposit eggs immediately before their death. Simulate with time running from 0 to 48 months. The simulation starts in January, so the juveniles and adult stocks are initialized at zero. Document your results with a time graph of the eggs hatching during the 48-month simulation. You should see the same number hatch each year. And you should see the hatching appear in the same month of each year. (The monthly counter is included so you can check its value when the hatching occurs.)

Exercise 14.4. Expand the insect model for losses

Expand the previous model (figure 14.22) to include a leakage flow for egg loss. Set the leakage fraction to 0.50, and set the eggs per female to 4. Simulate the model for 48 months and verify that you see the same-size insect population year after year.

Exercise 14.5. Simulate a growing insect population

Change the model in exercise 14.4 by increasing the eggs per female to 8. You should see the population growing over time. Simulate the model for 480 months. How many adults do you see at the end of the simulation? Do you see any density-dependent feedback in this model?

Exercise 14.6. Create Vensim equivalent of a conveyor

Create the Vensim model in figure 14.23. The initial value of the stock is zero, and time will run from 0 to 40 years. Set the lookup to give the inflow of students shown previously in figure 14.18. The time interval is 12 years, and the outflow is controlled by the DELAY FIXED function:

Outflow = DELAY FIXED(inflow, time interval for school,0).

Run the model and verify that the outflow matches the results in figure 14.18.

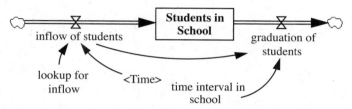

Figure 14.23. Vensim version of conveyor stock for students in school.

Exercise 14.7. Reformulate the population model

Figure 14.15 shows a population that is supposed to grow in exponential fashion. But if we change the growth rate to zero midway through the simulation, we get erroneous results. Formulate a different way of simulating the population and verify that your formulation is working properly with a time graph to correspond to figure 14.16.

Chapter 15

The Salmon of the Pacific Northwest

The declining salmon runs on the Snake and Columbia river system are one of the most serious environmental problems in the Pacific Northwest. The salmon have existed in the region for over 12 million years, and their historical migration at the mouth of the Columbia has been estimated at between 11 million and 16 million per year (NPCC 2005). By the early 20th century, salmon runs were depleted through a combination of overharvest and habitat degradation. Later in the 20th century, hydroelectric development on the Snake and Columbia rivers aggravated the problems. Some salmon habitats were no longer accessible; other habitats were more difficult to reach owing to reservoirs and dams. By the 1990s, less than 1 million adult salmon and steelhead were migrating up the Columbia each year.

These declines have occurred despite a public and private investment of more than $1 billion by the early 1990s. By 2002, cumulative spending on all fish and wildlife programs by the Bonneville Power Administration reached almost $7 billion. Salmon counts on many rivers improved in 2001–2004 (Fisher and Hinrichsen 2005), with the total counts on the Columbia reaching 2 million. The reasons for the recent improvement are difficult to explain, but many attribute the higher runs to cooler, more favorable ocean conditions (NPCC 2003). Other contributing factors include changes in dam operations and improvement in salmon habitats. As of 2008, agencies and tribes in the Northwest have committed to hatchery improvements, stream restoration, and changes in the screens and spillways at the dams. These and other actions are described in the *2008 Biological Opinion* issued by the National Oceanic and Atmospheric Administration (NOAA 2008). The cost of implementing the plan is difficult to gauge, but one estimate (BPA 2008) puts the cost at over $900 million for a 10-year period.

This chapter uses system dynamics to help us understand the dynamics of the salmon population in one of the rivers in the Northwest. The model simulates the long-term trends in the population over successive generations. The model is then used to simulate the impact of land use development, hydroelectric development, and harvesting. The chapter concludes with exercises to let you play the role of a fishery manager. A second set of exercises challenges you to expand and improve the model.

The Spring Chinook of the Tucannon River

This chapter focuses on the spring Chinook that spawn in the Tucannon River. The Tucannon rises in the Blue Mountains of Oregon and flows 50 miles to join the Snake River in southeastern Washington. The Tucannon provides 50 miles of habitat suitable for the salmon to spawn. The spawning is the concluding act of the salmon's complex life cycle. The adults return to the Tucannon to create nests—called redds—in the gravel beds. Each redd contains thousands of eggs that hatch in the spring. The hatchlings live for a month or more on nutrients stored in their yolk sacs. Once the sac is absorbed, the young fish—called fry—must find and capture food. The juveniles spend 1 year in the Tucannon competing for food. Those that survive undergo a biochemical change called smoltification that triggers the migration urge. The smolts migrate 50 miles to the Snake and then another 400 miles down the Snake and Columbia rivers. The migration is difficult, and only a small fraction may survive to reach the ocean.

The ocean provides the larger and more abundant food that the salmon require to grow to maturity. The ocean conditions vary over decadal cycles of warm-water and cold-water conditions. The salmon experience good feeding conditions during the cold-water years. Some remain in the ocean for 3, 4, or maybe 5 years. And some adult males—called Jacks—return after only 1 year in the ocean. We'll simplify the life cycle by assuming that adults return after 2 years in the ocean. They arrive each spring and migrate upriver to the spawning grounds. The redds are created in the fall, and the eggs are fertilized as the final stage in the life cycle.

The Tucannon was studied by Bjornn (1987) in research for the U.S. Soil Conservation Service. He developed estimates of various population parameters to project changes in the population from one generation to the next. He used a spreadsheet to find the harvest that would be possible if the population were managed for a sustainable yield. With predevelopment assumptions, for example, around 20,000 adults could return to the mouth of the Columbia each year. Bjornn estimated that this population could support a sustainable harvest of around 13,000 per year.

By Northwest standards, the Tucannon is a small river. The entire basin covers only 210,000 acres. So you might be skeptical about 20,000 fish returning to such a small river. The entire Columbia River basin is around 800 times larger, so think of the number of returning adults if we amplify the Tucannon estimate 800-fold. The amplified number staggers the mind: around 16 million salmon would return to the mouth of the Columbia every year! Such a migration is hard to comprehend with current conditions on the river. Nevertheless, it matches the best information from the era prior to development.

Impact of Development

Bjornn (1987) studied the impact of development along the Tucannon and on the Snake and Columbia. He represented development by changes in the habitat parameters due to land use changes in the Tucannon Valley. He also changed migration parameters due to hydroelectric development on the Snake and Columbia. His new calculations revealed that around 2,400 adults could return, and around 600 could be harvested in a sustainable manner. The general conclusion is that development has reduced the overall size of the fishery by around 10- to 20-fold.

This large reduction is similar to trends in the Columbia River basin as a whole. The Columbia River salmon and steelhead runs have been impacted by a combination of factors, including harvesting, dams, irrigation, mining, and livestock grazing. Before any of these impacts, up to 16 million wild salmon and steelhead returned to spawn in the streams where they were born. By the end of the 1980s, the total was "around 2.5 million, including known fish harvested in the ocean, with about 0.5 million of these as wild fish" (Army Corps of Engineers et al. 1992, 1–6).

Reference Mode

The purpose of the model is to simulate the trends in the salmon population over several decades. We will initialize the model with a small number of fish and the pristine conditions that a small number of fish probably encountered when they first found their way to the Tu-cannon. They would have discovered suitable conditions for growth, and would have grown from one generation to the next until their numbers reached the limit of the Tucannon. Our initial objective is to develop a model that simulates growth in the salmon population to reach a population of around 20,000 returning adults. Our reference mode is S-shaped growth since we expect the population to grow in exponential fashion before accommodating to the space available in the Tucannon. But the S-shaped pattern will be more complicated than the examples in previous chapters because of the salmon life cycle. (There may be no fish in the river one month and thousands of fish the very next month.) We'll use conveyor stocks to represent the life cycle, following the approach in the insect exercises from chapter 14.

The S-shaped pattern is a useful target, but it is only the starting part of the reference mode. We will then introduce changes in the habitat and migration parameters to represent the large decline in migrating adults. We will also experiment with the model to test the effect of policies. The main policy is the harvest fraction. The model can also be used to test the impact of habitat restoration projects, as you will see in the exercises.

Model Design

The model is designed with conveyor stocks to keep track of the population (in thousands) in seven phases of the life cycle, as shown in figure 15.1. The salmon move through the phases

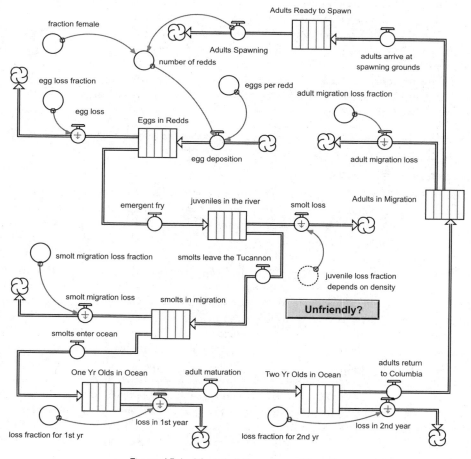

Figure 15.1. Model of the salmon life cycle.

in tightly controlled patterns, so it makes good sense to use conveyors. The life cycle begins in the fall when spawners build the redds. The eggs hatch 6 months later.

The juveniles need 12 months to grow into smolts, and the smolts need 1 month to reach the ocean. After 2 years in the ocean, they return to the Columbia, where their upstream migration requires 4 months. Table 15.1 lists the time spent in each stage. These are the transit times for the conveyor stocks in the model. The total life cycle is 48 months, or 4 years.

The model uses seven population parameters (table 15.2) that are assumed to remain constant over time. These are sometimes called *density-independent* parameters because their values do not change with changes in the size of the population. The first three parameters match Bjornn's (1987) assumptions for the Tucannon under predevelopment conditions. The 50% female is typical of Chinook salmon. Each female would deposit around 3,900 eggs, and 50% of the eggs would survive under pristine conditions. The 25% adult migration loss fraction is taken from Bjornn's estimate of conditions prior to the hydroelectric development on the Snake and Columbia. The loss fraction during the smolt migration and the two loss fractions in the ocean were not available from Bjornn's analysis. These loss fractions are discussed on the book's website (BWeb). Their combined effect is similar to the losses expected by Bjornn.

Table 15.1. Length of each phase of the salmon life cycle.

Adults about to spawn	1 month
Eggs in redds	6 months
Juveniles in Tucannon	12 months
Smolts in migration	1 month
1-year-olds in ocean	12 months
2-year-olds in ocean	12 months
Adults in migration	4 months

Table 15.2. Density-independent parameters in the salmon model.

Fraction female	0.50
Eggs per redd	3,900
Egg loss fraction	0.50
Smolt migration loss fraction	0.90
Loss fraction for 1st year	0.35
Loss fraction for 2nd year	0.10
Adult migration loss fraction	0.25

Checking the Numbers for Exponential Growth

If we are to see S-shaped growth, the salmon population needs the power to grow when it occupies a small portion of the Tucannon habitat. Let's work through the assumptions to see if growth is possible. A good place to start is the top of the model diagram (figure 15.1)—suppose there are 2,000 adults about to spawn. With 50% female, there would be 1,000 redds formed in the Tucannon. (You would see 20 redds if you walked a mile of the river.) With 3,900 eggs per redd, there would be 3.9 million eggs in the gravel nests. Half of these would survive to emerge as fry the following spring. Now we have 1.95 million fry in the river. You'll learn shortly that only around 280,000 of these juveniles will survive the first year in the river. These are the smolts that migrate to the ocean the following spring. With 90% migration losses, 28,000 reach the ocean. And with 35% losses in the first year and 10% in the second year, around 16,000 adults will return to the mouth of the Columbia 2 years later. The adult migration loss is expected to be 25%, so around 12,000 adults will reach the spawning grounds. This calculation starts with 2,000 spawners. Four years later, there are 12,000 spawners. The conditions are certainly suitable for rapid growth in the population.

You know that no system grows forever, so it is logical to ask ourselves about the limits to the growth in the Tucannon population. There are only so many redds that can be constructed in the river. We'll leave this limit as an exercise at the end of the chapter. The other limit involves the room for the juveniles to grow and survive in the Tucannon. Bjornn used the term *carrying capacity* to describe juvenile survival.

The Tucannon Carrying Capacity

There are limits to the number of juveniles that can survive their first year in the Tucannon. During the summer, the juveniles must compete for a limited number of feeding sites. Later in the fall, they must compete for a limited amount of cover. Bjornn believed that summer conditions constrain the size of the juvenile population, and he argued that juvenile survival is heavily dependent on density, as shown in figure 15.2. The horizontal axis represents the millions of fry that emerge in the spring. The vertical axis is the number of juveniles that will still be in the river 1 year later. These are the smolts that will migrate to the ocean.

Figure 15.2 shows a steep slope at the origin, but the curve is almost flat when the number

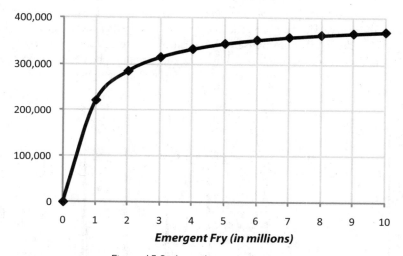

Figure 15.2. Juvenile survival curve.

of fry reaches 10 million. No matter how many fry emerge, there can never be more than 400,000 smolts 1 year later. The steep slope at the origin suggests that 1 million emergent fry would lead to over 200,000 smolts at the end of the year. We can obtain half of the river's carrying capacity with only 1 million fry. Further increases do little to increase the number of smolts. If the fry were to increase from 5 million to 10 million, for example, the surviving smolt population would only increase from around 340,000 to 370,000.

The shape of the nonlinear survival curve requires an equation that goes beyond friendly algebra (see the unfriendly button in figure 15.1). Surviving smolts are calculated with the Beverton-Holt equation (a popular recruitment curve in fishery studies):

$$\text{surviving smolts} = \frac{\text{Fry}}{(\text{Fry}/CC) + (1/S)}$$

The carrying capacity (CC) was estimated at 400,000 under pristine conditions. The slope (S) was estimated at 0.5, which means that half the fry will be able to survive the juvenile stage when there are only a few emergent fry. We will use the Beverton-Holt equation directly so the carrying capacity will be one of the named inputs in the model. (By naming the variable, we make it available for experimentation, as you will see in the exercises. The BWeb describes the details of converting the juvenile survival curve into a juvenile loss fraction suitable for a model that simulates time continuously in months.)

Do We See S-Shaped Growth?

The first simulation is shown in figure 15.3. The population is initialized at a low level, and we simulate for 120 months to see if it will grow over time. The time graph shows the number of smolts (scaled from 0 to 400 thousand) and the adults in migration (scaled from 0 to 40 thousand). The smolt migration lasts only 1 month, so we see a series of 10 spikes, one for each year of the simulation. The smolts begin the simulation at around 100,000, a value that results from the initial value assigned to the stock of juveniles in the Tucannon. The number of smolts dips in the second year but increases strongly in the third year. Eventually, the number of smolts reaches around 380,000, and this number appears year after year. To check for S-shaped growth, you should pencil in a smooth curve connecting the top of each spike in the simulation. Your curve will show the general pattern of S-shaped growth. But it will also reveal a 4-year cycle in the number of smolts. Such oscillations are to be expected given the particular initial conditions (think of the baby boom or the baby boom echo in a human population). The first simulation shows that these small oscillations will dampen out over time.

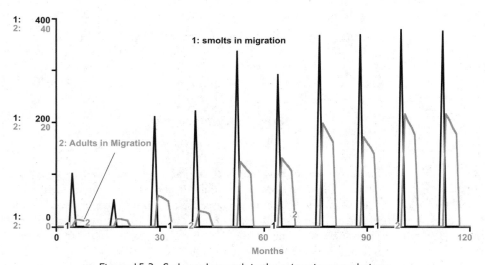

Figure 15.3. S-shaped growth in the migrating populations.

The second variable in figure 15.3 is the number of adults migrating upstream. The simulation begins with around 1,000. The adult migration lasts 4 months, and 25% of the adults will fail to survive the migration. These losses show up as a downward slope in the adult number. Penciling in a smooth curve connecting the peak number of adults shows S-shaped growth and a small cycle as the model reacts to the initial conditions. These small oscillations dampen out rather quickly, and the number of adults levels off at around 22,000, which matches the estimate by Bjornn (1987).

Checking the Numbers for Equilibrium

It's useful to check the numbers whenever a situation reaches equilibrium. This is normally done in an equilibrium diagram, but the salmon's complex life cycle makes it difficult to prepare a traditional equilibrium diagram. Nevertheless, we should be able to check the numbers to understand why the same number of adults return to the Columbia year after year. Let's start with the 22,000 adults, which appears near the end of simulation. With 25% migration losses, around 16,500 will reach the spawning grounds, so there will be around 8,000 redds and a total of around 30 million eggs. With 50% egg loss, around 15 million fry would emerge in the following spring. Their numbers would be off the chart in figure 15.2. The number of juveniles to survive the year in the river would be around 380,000, just slightly less than the river's total carrying capacity. About 38,000 would survive the migration to the ocean. Ocean losses are 35% in the first year, followed by 10% in the second year, so we would expect around 22,000 to return to the Columbia after 2 years at sea. The 22,000 is the same number used to start the numerical exercise—it shows we have dynamic equilibrium.

Disturbance Test

Another useful test is to disturb the system and see if the equilibrium is stable. The salmon have been in the Columbia and Snake for over 12 million years, so we would expect a realistic model to show a stable equilibrium. Stability tests in chapter 6 were performed by adding a new flow to remove from or add to one of the stocks in the model. We can also test the model with a temporary change in one of the external inputs, as shown in figure 15.4.

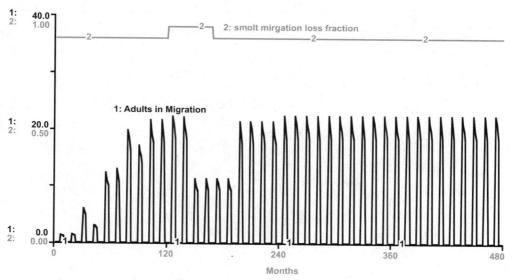

Figure 15.4. Test with four difficult years for the smolts' migration.

The smolt migration loss fraction represents one of the highest sources of mortality in the life cycle. The regular value is 90%, but we will increase the value to 95% to represent very difficult migration conditions. For example, higher losses could appear in years with unusually low runoff (BWeb). This test imposes the disturbance not just for 1 year, but for 4 years in a row. This will guarantee that none of the cohorts in the 4-year life cycle escape the difficult migration conditions.

This disturbance is imposed in the 120th month, and figure 15.4 shows that the number of adults would decline 2 years later. Then the number of migrating adults is cut in half. The number remains at around 11,000 for 4 years in a row. This result makes sense since only half as many smolts survive the migration to the ocean, and the ocean losses are constant. Then we see something rather unexpected. The number of adults returns to the same equilibrium value as before. This test shows an amazing ability of the simulated population to recover from a disturbance. We'll look to the feedback structure to explain this resilience.

Feedback Structure

The salmon model exhibits S-shaped growth over time, so we should expect to see a feedback loop structure similar to previous examples. Specifically, we should expect to see at least one positive loop that gives the population the power to grow over time. And here should be at least one negative feedback loop that acts to slow the rate of growth as the salmon fill the river. And you will also see many small loops involving fixed rates of mortality at different stages of the life cycle.

Figure 15.5 shows the main positive feedback loop highlighted by the darker arrows. We can follow their course around the perimeter of the causal loop diagram. If we start at the top with an increase in eggs, we will expect more fry, more juveniles, more smolts in migration, more salmon in the ocean, more adults entering the Columbia, more adults spawning, and a subsequent increase in eggs. The control loop generates negative feedback that becomes stronger as more and more fry emerge in the spring. More fry mean higher juvenile losses, which lead to fewer smolts in migration to the ocean. This leads to fewer salmon in the ocean, fewer adults returning to spawn, fewer eggs, and fewer fry emerging in the future. This loop gradually applies the brakes to the growth.

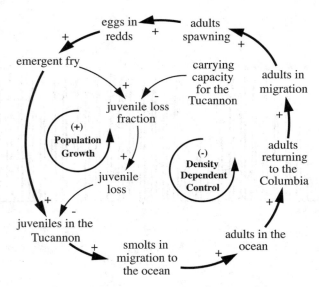

Figure 15.5. Key loops in the salmon model.

Stochastic Simulation

The previous disturbance test imposed a simple change in migration losses to test the model. Now that we understand the response to this simple test, it is useful to experiment with a stochastic simulation with randomness in the smolt migration loss fraction. The normal value of 90% applies for the first 120 months. Then the loss fraction varies in a uniform fashion from a low of 85% to a high of 95%. All other parameters are maintained at the same values used in the disturbance simulation. The results are shown in figure 15.6. Notice the frequent variations in the smolt migration loss fraction. The DT is 1 month, so the loss fraction is changing every month. These rapid changes may remind you of the high-frequency noise discussed in chapter 14. But there is no need to sample and hold the random numbers, as the smolts only spend 1 month in migration. The simulation reveals low runs of around 10,000 and high runs over 30,000. We have a three-fold variation in the number of returning adults owing to random changes in just one of the model parameters. The large variations point to the need for patience in judging the size or abundance of population. The variations also suggest the possibility of a cyclical pattern that may be caused by the 4-year life cycle.

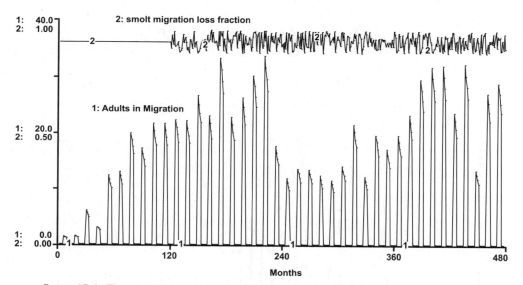

Figure 15.6. Test simulation with random variations in the smolt migration loss fraction.

Harvesting Policy

Let's now expand the model to simulate harvesting of the adults as they enter the Columbia. Figure 15.7 shows a user-specified harvest fraction to calculate the adult harvest. The escaping adults then enter the conveyor for adults in migration, where they continue upriver as in the previous model (figure15.1). We should set the harvest fraction to zero for the first 10 years to allow the population to grow to equilibrium. Then we can experiment with a wide range of values for the harvest fraction.

The first test is in figure 15.8. It shows 95% harvesting after the 120th month. This is an extremely aggressive policy (with around 20,000 returning adults, only 1,000 would escape the harvest). Look closely at the spikes immediately after the 120th month in figure 15.8, and you will see no change. The next four runs remain at over 20,000. The impact of the harvesting does not show up until 4 years later, when the number of returning adults drops by almost 50%. We then see four successive runs of around 11,000, and then the runs fall again. The population follows this downward pattern for the remainder of the simulation. The conclusion from this first test is that 95% harvesting is not sustainable.

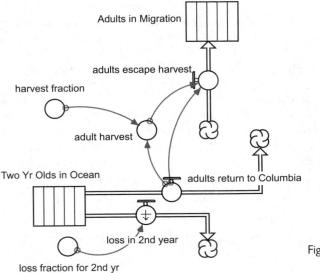

Figure 15.7. Add harvesting.

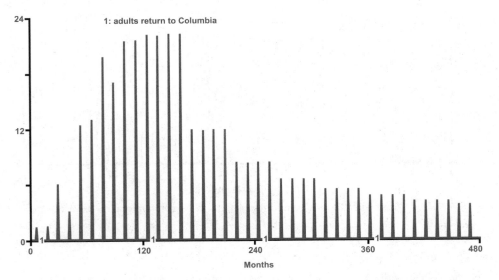

Figure 15.8. Adults returning to the Columbia in a test with 95% harvesting.

The next test implements 50% harvesting after the 120th month. The results are shown in figure 15.9, but you will have to look closely to see the impact of harvesting. There is no impact in the four runs after the 120th month, but you are familiar with this pattern. What may surprise you is the barely discernible impact on the number of adults for the remainder of the simulation. Their numbers continue at nearly 22,000 year after year, even though 11,000 adults are harvested at the mouth of the Columbia. How can this possibly be?

These surprising results arise from the nonlinearities in the juvenile stage of the life cycle. Recall the equilibrium conditions around the 120th month—there were around 16,000 spawners; 8,000 redds; 30 million eggs; and 15 million emergent fry. With 15 million fry

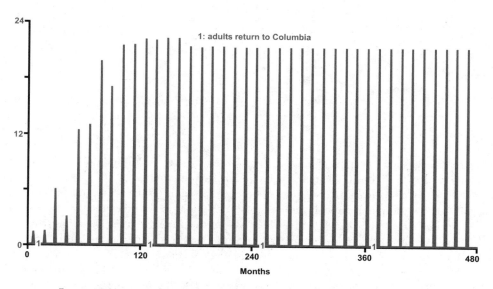

Figure 15.9. Adults returning to the Columbia in a test with 50% harvesting.

emerging each year, the Tucannon is at its limit—around 380,000 juveniles would survive their first year in the river. Now ask yourself what would happen if 50% of the adults were harvested as they entered the Columbia. Half as many would escape, so we would expect to see around 7.5 million emergent fry. According to figure 15.2, the number of surviving juveniles is almost the same as before. This simulation suggests that half of the adult population could be harvested with only a minor impact on the number of adults that would return to the Columbia each year.

This incredible result arises from the competition for space among the juveniles during their first year in the Tucannon. Harvesting cuts the number of escaping adults in half. But the progeny from the escaping adults will have a much greater chance of survival during their first year in the Tucannon. The population's resilience under harvesting is similar to the resilience following the four years of bad migration conditions in figure 15.4. These simulated responses may seem "too good to be true" and perhaps call into question the structure of the model. But remember that the salmon have existed in Northwest rivers for over 12 million years; resilience is what we should expect to see in a realistic model.

Harvesting Policy and the Maximum Sustainable Yield

It is common practice in the study of fisheries to calculate the MSY, the maximum sustainable yield. MSY is defined by Botkin and Keller (1998, G-10) as "the maximum usable production of a biological resource that can be obtained in a specified time period." Watt (1968, 404) explains the search for MSY, starting with minimal harvesting: "If we fish too little, the fish population left in the water after fishing will build up to densities at which intraspecific competition for food stunts fish growth and diminishes the probability of survival." However, "if we fish too hard, too few adults are left behind to spawn, and the stock goes into a decline." You can infer from the previous simulation that the 50% harvest fraction corresponds to "too little" fishing and the 95% corresponds to fishing "too hard." Somewhere in between, we expect to find the maximum sustainable yield.

Figure 15.10 summarizes the results of five experiments with a constant harvest fraction imposed after the 120th month of the simulation. The 50% results are highlighted at the left

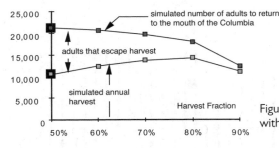

Figure 15.10. Results of five experiments with a constant harvest fraction.

edge of the diagram. The number of returning adults is just over 21,200 and the annual harvest is just over 10,600. With 60% harvesting, the annual harvest is around 12,400. With 70% harvesting, we see a somewhat smaller number of returning adults. But since we are harvesting a larger fraction, the annual harvest turns out to be higher than in the previous experiment. The highest annual harvest is 14,300 (with 80% harvesting), so the MSY is 14,300 salmon/yr.

However, the MSY is not the important conclusion to be drawn from these experiments. The important observation is the wide range of harvest fractions that could deliver a sustainable harvest. The population is so resilient under pristine conditions that annual harvests of at least 11,000 are possible with the harvest fraction anywhere between 50% and 90%. The resilience reminds us of the discussion of homeostasis and span of control in chapter 10. With pristine conditions, the salmon system's span of control is from 0 to 90% for the harvest fraction.

Graphical Analysis

It's useful to show graphical analysis to illustrate the type of analyses that might be conducted when managers and regulators do not have a dynamic model. The graphical approach begins with the juvenile survival curve in figure 15.11. Juvenile loss is the only loss in the model that depends on density. All other losses may be combined in a single line whose slope varies with changes in the harvesting fraction. Figure 15.11 demonstrates with four lines corresponding to harvest fractions of 0%, 50%, 75%, and 90%. The intersection of the lines with the survival curve tells us that the harvest fraction is sustainable (BWeb). Now, think of where you would draw the line for 95% harvesting. It would climb quite steeply and we would not see an intersection point. This tells us that 95% harvesting is not sustainable.

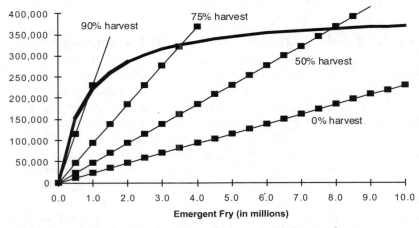

Figure 15.11. Graphical analysis to find sustainable harvest fractions.

The results in table 15.3 confirm the findings from the simulation experiments. That is, the maximum sustainable yield is around 14,000. But, more important, it is possible to obtain sustainable yields of over 11,000 with harvest fractions ranging anywhere from 50% to 90%. It is encouraging to see that the graphical results confirm the findings from the simulation model. However, you might be wondering at this point why we should bother with a simulation model.

Table 15.3. Some results from the graphical analysis.

Harvest fraction	Number of smolts	Annual harvest
50%	380,000	11,000
75%	330,000	14,000
90%	220,000	11,500

Fishery Management: Simulation or the MSY?

You should be aware of the highly restrictive assumptions that are required for the graphical approach to arrive at the intersection points. First, we must be willing to assume that the system is predominantly linear. The graphical analysis allows for a nonlinear relationship for the smolt survival curve, but all other relationships must be linear. However, in simulation modeling, we are free to consider each relationship separately. If a nonlinear representation seems more realistic, we can simply employ a graph function (~) and repeat the simulation.

A second restriction in the graphical approach is the assumption that the system is operating in a sustainable manner. For the graphical analysis to make sense, we must assume that the same number of fish appear in the system from one generation to another. This assumption is suitable in searching for the MSY, but it certainly restricts our ability to analyze a broad range of situations (such as the oscillating situation shown in the stochastic simulation in figure 15.6).

Despite these problems, the MSY appears frequently in the literature. Indeed, the idea was considered "virtually sacred to many wildlife managers until quite recently" (Botkin and Keller 1998, 232). But you should regard the MSY with great skepticism since it is an artificially determined number. To appreciate the artificiality, think of the unusual conditions that were created in the simulation experiments to find the MSY. Each experiment held the values of all parameters constant over time. There were no random disturbances, and the harvest fraction was implemented once the population reached equilibrium. We then held the harvest fraction constant for 30 years and recorded the simulated response. This was done not once, but five times. The results indicate that the MSY is 14,300 and it could be obtained with a harvest fraction of around 80%.

You shouldn't pay much attention to these two numbers. The important conclusion is not the 80% or the 14,300. What we really learn from the previous simulations is that the pre-development conditions allow for major harvests that could be sustained year after year if we set the harvest fraction anywhere from 50% to 90%. The key is to maintain the habitat and the migration corridors in good condition.

The Impact of Development

Bjornn was primarily interested in land use development that has lowered the quality of the salmon habitat. An example is the heavy erosion from dry cropland and cattle grazing that adds sediments to the Tucannon (Harrison 1992). Bjornn suggested that the impact of land use may be simulated by changing the egg loss fraction and the carrying capacity, as shown in table 15.4. Egg loss increases as sediment degrades the quality of the gravel beds. Land use has also degraded the habitat that juveniles will experience after hatching. These changes could lower the carrying capacity from 400,000 to 170,000 smolts. Development has also brought important changes on the Snake and Columbia. Hydroelectric development has placed six dams and reservoirs between the smolts and the ocean. This makes migration more difficult, especially for the smolts.

Table 15.4. Parameter changes to reflect development in the Tucannon habitat and along the Snake and Columbia.

Model parameters	Pristine	Developed
Egg loss fraction	0.50	0.75
Smolt carrying capacity	400,000	170,000
Smolt migration loss fraction	0.90	0.95
Adult migration loss fraction	0.250	0.375

Figure 15.12 shows the number of returning adults if all four of these changes are implemented after the 120th month of a simulation. The first 10 years are the same as previous simulations. The population is allowed to grow to the equilibrium value of 22,000. The impact of development is abrupt and dramatic. The number of returning adults declines within a few generations to less than 20% of the predevelopment population. The population is simulated to find a new equilibrium and remain stable at slightly below 4,000.

At this point, one may experiment with different harvesting policies to learn if the post-development salmon population could be harvested in a sustainable manner. For example, 25% harvesting begun after the salmon reach the postdevelopment equilibrium could deliver a sustainable harvest of around 900 fish per year. You may try alternative harvest fractions to find the MSY under developed conditions as an exercise at the end of the chapter. You'll learn that the harvest is nowhere close to the harvests that were possible with predevelopment conditions.

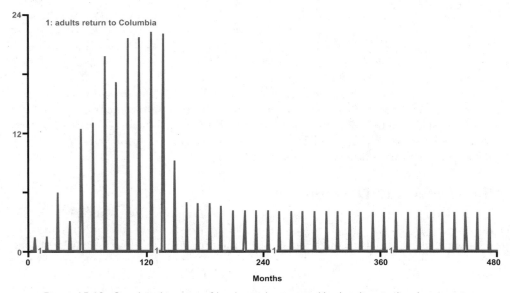

Figure 15.12. Simulated impact of land use changes and hydroelectric development.

Improving the Salmon Model

This chapter describes a system dynamics model of the salmon population in the Tucannon, a small river in southeastern Washington. The model simulates the salmon's complex life cycle and the population's resilient behavior under a wide variety of conditions. The model has been used to test the impact on the salmon population of land use development in the Tucannon Valley, hydroelectric development on the Snake and Columbia rivers, and harvesting of the returning adults. This chapter demonstrates the process of building and testing that would allow system dynamics to contribute to fishery sciences and to fishery management.

Good modeling is an ongoing process of testing, improvement, further testing, and further improvements. The Tucannon model may certainly be extended and improved, as you will be challenged to do in the exercises.

Exercises with the BWeb Model

This chapter includes two sets of exercises. The first nine exercises make use of the salmon simulator on the BWeb. You'll work on the interface layer in the role of a harvest manager. The second set of exercises calls for changes in the underlying structure of the model.

Exercise 15.1. Download and verify

Download either the Vensim or Stella version of the salmon simulator. (The Stella version is shown in figure 15.13.) Read the information buttons to learn about your role as harvest manager. Run the model for 10 years with no harvesting; then set the harvest fraction to 95% to verify the results in figure 15.13.

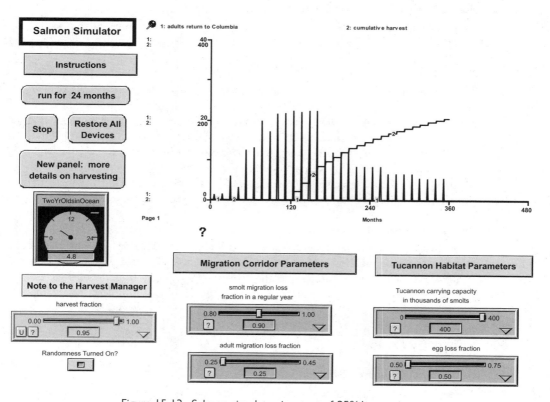

Figure 15.13. Salmon simulator in a test of 95% harvesting.

Exercise 15.2. Recover from overharvesting?

The results in exercise 15.1 match the 95% harvesting shown in figure 15.8. Harvesting at this level is simply not sustainable. What do you think would happen in the final 10 years of this simulation if the harvest fraction were reduced to zero? Change the harvest fraction and simulate the final 10 years. Do the results match your expectation?

Exercise 15.3. Verify results with four bad years

Use the simulator to reproduce the results in figure 15.4.

Exercise 15.4. Verify the stochastic simulation

Use the simulator to reproduce the results in figure 15.6. (You should see the wide variations in the number of adults, but you may not see a match in each and every year because of differences in random number generation.)

Exercise 15.5. Verify the MSY

Table 15.3 suggests that the MSY is around 75% under pristine conditions. Experiment with 70%, 75%, and 80%. Which harvest fraction gives the greatest cumulative harvest by the end of the simulation?

Exercise 15.6. Verify performance with developed conditions

Run the simulator for 120 months, then make the changes in table 15.4 to verify you get the results in figure 15.12.

Exercise 15.7. Sustainable harvest with developed conditions

Repeat the simulation in exercise 15.6 but introduce 25% harvesting starting in the 240th month. What is the annual harvest? Is it sustainable?

Exercise 15.8. Test your instincts as the harvest manager

The previous exercises provide the preparation for the main use of the simulator—to test your ability to deliver a sustainable harvest under challenging conditions. This is a group exercise, so you will work with one or two classmates. Their job is to create the challenging conditions by making changes in the parameters for the Tucannon habitat and the migration corridor. Your job is to control the harvest fraction to obtain a large and sustainable harvest. This is an open-ended exercise that invites experimentation and discussion. A good way to conclude the exploration is to describe the most challenging situation you faced and how you were able to obtain a sustainable harvest.

Exercise 15.9. Discussion exercise

Read about the Fish Banks model-based game (BWeb). Then describe how your results with the salmon simulator compare with the experience of people playing the Fish Banks game. How does you situation as the harvest manager differ from the open-access fisheries that experience overfishing and what Garrett Hardin (1968) describes as "the tragedy of the commons"?

Modeling Exercises

Exercise 15.10. Equivalent losses test

The pristine conditions call for 90% smolt migration loss, 35% first-year ocean loss, and 10% second-year ocean loss (see table 15.2). The net survival from these sources of mortality would be 5.85%. Some say these assumptions place too much loss in the migration stage and too little in the ocean stages. New losses with the same net survival could be 53.2% in migration, 75% in the first ocean year, and 50% in the second ocean year. Add new sliders to the interface to make it easy to change the ocean loss fractions. Run the model for 240 months with the original losses. Then change to the new loss fractions for the final 240 months. Document your results with a time graph of the adults returning to the Columbia.

Exercise 15.11. Third year in ocean

Draw the changes required in the flow diagram in figure 15.1 if you are told that the salmon spend 3 years in the ocean rather than 2. The total life cycle is now 5 years rather than 4. Suppose all other population parameters were the same as in figure 15.1, and suppose the loss fraction during the third year in the ocean is negligible. Figure 15.3 showed around 22,000 returning adults each year in the previous model. What do you expect to see with the new model?

Exercise 15.12. Expand the model for a third year in ocean

Build the model described in exercise 5.11 and simulate it to see the size of the equilibrium population. Does the simulation match your expectation?

Exercise 15.13. Decade-long variations in first-year ocean losses

Ocean conditions can change dramatically, especially if warm-water currents reach the mouth of the Columbia. (Warm water disrupts the cool-water upwelling that supports the growth of food organisms for juvenile salmon.) Some experts say that the ocean conditions can be unfavorable for many years in a row (perhaps a decade or so). Simulate the first 10 years with this loss at 35%, the next decade at 70%, and the remainder of the simulation at 35%. Document your results with a time graph of the number of adults returning to the Columbia.

Exercise 15.14. Limit on redds

The Tucannon has a limited number of sites suitable for redds. Bjornn (1987) thought the river could support around 3,300 redds. Add this limit as shown in figure 15.14.

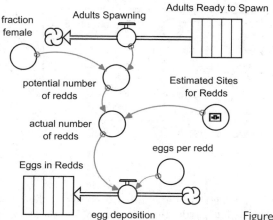

Figure 15.14. Adding a limit on sites for redds.

The actual number of redds uses the MIN function to take the smaller of the potential number and the estimated number of sites. Add the estimated number of sites to the interface layer. Set the potential number of sites to a high value (say, 10,000). Then gradually lower the input until you see a change in the results. When does this limit start to take effect in the model?

Exercise 15.15. Is rapid recovery still possible?

Run the model with the potential number of sties sites set low enough to limit the size of the population. Then repeat exercise 15.2 to see if the population will recover from 95% harvesting.

Exercise 15.16. Re-create student model of river restoration

Figure 15.15 shows a student addition to the salmon model to represent spending on river restoration. The student is an expert on river restoration, and his goal was to make the carrying capacity an endogenous variable that would respond to the amount of restoration spending. The 50 miles of river is represented by three stocks, with the initial values set for 25 miles of degraded river and 25 miles of mature, restored river. The restoration spending is controlled by a slider on the interface layer, and the restoration cost is $275,000 per mile. The annual maturation rate is 0.10/yr, and the carrying-capacity-per-mile parameters were estimated at 1,000 for degraded miles; 6,000 for restored miles; and 8,300 for mature miles. Build this sector and add it to the salmon simulator. The Tucannon carrying capacity will be 233,000 smolts at the start of the simulation. Control the restoration spending to restore 25 miles of degraded conditions. Add a stock to accumulate the restoration costs, and check that the total turns out to be nearly $7 million. Run the model for 480 months, and check that all 50 miles are in the mature, restored status by the end of the simulation. The Tucannon carrying capacity should be 415,000 at the end of this test.

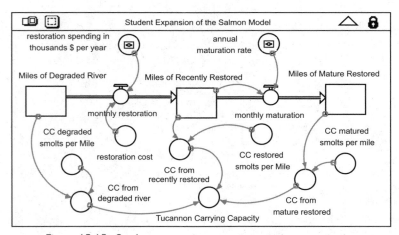

Figure 15.15. Student expansion to represent river restoration.

Exercise 15.17. Simulate the benefit of river restoration

Use the new model to simulate harvesting from the restored river. Spend the money needed to restore the river in the first half of a 480-month simulation. You should see an increasing carrying capacity and an increasing number of adults returning to the Columbia. When their number reaches 20,000, begin harvesting at 80%. The simulator will show the value of cumulative harvest if each fish is valued at $40 for the Northwest economy. How long will it take for the value of the harvested fish to exceed the $7 million spent on restoration?

Exercise 15.18. Taking advantage of the smolts simulator

The first edition of *Modeling the Environment* (BWeb) describes a model of the salmon smolts migration to the ocean. The model runs for 40 days to simulate the progress of 10 million hatchery smolts released into a Snake River above Lower Granite Dam (the first of eight dams between the smolts and the ocean). Around half are expected to reach the reservoir behind Lower Granite Dam. Figure 15.16 shows illustrative results for the remainder of the journey. Download the smolts simulator, and experiment with simulations to learn the effect of low-water and high-water years. Then experiment to learn the pros and cons of placing the smolts in a barge to speed their journey to the ocean. Now think of these findings as suggestive of migration loss fractions that would be used in the salmon simulator. Add information buttons to the salmon simulator interface (figure 15.13) to provide advice on values of smolt migration loss fraction under different conditions. Then use the new interface to see if a sustainable harvest is possible under difficult conditions for the smolts migration.

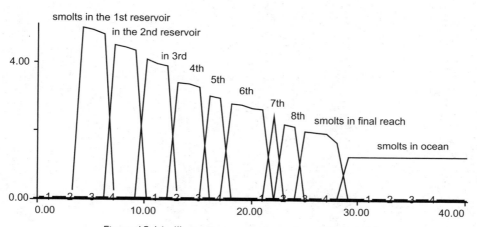

Figure 15.16. Illustrative results from the smolts model.

Exercise 15.19. Discussion: Should we merge the two models?

Exercise 15.18 uses the smolts simulator as a stand-alone model to give insights on how to set one of the inputs in the longer-term model. But why use two models if we could create equivalent results with one model? Perhaps it makes better sense to merge the two models. Describe what you would have to do to merge the smolts model with the model of the salmon life cycle. Then describe what you would do to test that the merged model is working correctly. Conclude the discussion with your views on whether merging the two models would be more productive than the two-model approach in exercise 15.18.

Further Reading

- Depletion of major fish stocks is a problem across the world (Clark 1985; Botkin and Keller 1998; Iudicello, Weber, and Wieland 1999).
- System dynamics has been used to help understand fishery problems (D. L. Meadows, Fiddaman, and Shannon 1993; Moxnes 2000; Dudley 2008), especially the problems of overfishing in open-access fisheries. Some of these applications are described on the BWeb.

Chapter 16

Managing a Feebate Program for Cleaner Vehicles

Imagine you have been asked to design an incentive program to promote the sale of cleaner vehicles in your city. And if you take on this job, you will be asked to manage the feebate program. This is quite an offer, and you are wondering whether to accept it.

Your city is huge, with over a million new cars sold each year. Most of them burn gasoline and release harmful pollutants into the airshed. Your job is to shift the sales toward cleaner vehicles. The marketplace has been changing rapidly, and consumers now have dramatically different choices, including electric vehicles (EVs), hybrid-electric vehicles (HEVs), and vehicles fueled with compressed natural gas (CNG). You are convinced that purchase price incentives are the key to shifting consumer choices. You believe a large rebate could encourage the purchase of electric cars and other alternative-fueled vehicles. But your program is suppose to be self-financing, so how do you pay for the rebates?

One answer is feebates. A fee would be imposed on the purchase of new vehicles with high pollution, and the revenues from the fees would cover the cost of rebates for purchasing alternative-fueled vehicles. Imagine that you have proposed the feebate idea, and the top officials are willing to go with your judgment. But they remind you that millions of new cars are sold each year, so the cash flow for rebates could be enormous. They want to know if you can guarantee that the cash coming in from fees will balance the cash flow out for rebates. And they ask what you propose to do if the cash flows are not in balance.

These are difficult questions, and you aren't sure how to answer them. Wisely, you turn to system dynamics to build your understanding. System dynamics has its roots in control theory, so you know it is well suited for problems of controllability. You identify yourself as the initial client for a modeling study. If you can design a good policy for managing feebates, you will want to use modeling to explain your policy to the top officials. For their benefit, as well as your own, it is useful include a control panel to promote interactive simulation and discussion.

This chapter leads you through the development of such a model. Supporting information and the Vensim and Stella models are available on the book's website, the BWeb. The chapter concludes with a set of exercises to let you experiment with fees and rebates. A second set of exercises challenges you to expand and improve the model.

Background on Urban Air Pollution

Air pollution in America's urban areas became a highly publicized issue in the 1960s. The first emission standards were established in 1965. They were followed by the Clean Air Act of

1970. It established stricter standards to take effect in 1975 in order to protect human health. Unfortunately, the air in many of our urban areas remained unhealthy more than two decades after the original 1975 target date.

The most visible sign of the continuing failure is the smog in our dirtiest cities. Smog is tropospheric ozone. It is formed by a photochemical reaction between hydrocarbons (HCs) and the oxides of nitrogen (NO_x) in the presence of heat and sunlight. Gordon (1991, 62) reported that "more than 100 of the cities in the U.S. are choking on smog, and roughly half of all Americans live in areas that exceed the ozone standard at least once a year." The air pollution problems in America's urban areas have persisted for many decades, even though the pollutants themselves are quickly dissipated in the atmosphere. The persistence of urban air pollution stems from our continued use of polluting technologies on the ground.

One of the biggest polluters is the automobile. Gordon (1991) reported that there were 187 million cars and light trucks or vans on the road in 1989, roughly 1.5 vehicles for every working American. The transportation sector was said to be responsible for two-thirds of the nation's carbon monoxide (CO) emissions and around 40% of the nation's HC and NO_x emissions. Transportation was also responsible for 30% of the nation's carbon dioxide (CO_2) emissions. The tailpipe emissions from a conventional gasoline vehicle include HC, NO_x, and CO. The HC and NO_x emissions are the precursors to ozone. CO is a heavier pollutant that becomes highly concentrated in transportation corridors and causes headaches and stress on the heart.

Each of these pollutants is subject to emissions standards set by the state agencies. If the population of vehicles continues to grow, the responsible agencies will have to adopt more and more expensive measures to lower emissions sufficiently to bring the airshed into compliance. Some air districts have issued extremely detailed plans with hundreds of control measures, along with their costs per pound of emission reduction. The marginal control measures in such plans are an indication of the cost associated with emissions over the life of a conventional vehicle (CV). The BWeb explains the cost of a CV in a polluted airshed in southern California in the late 1990s:

115 pounds of HC @ $8 plus

133 pounds of NO_x @ $12 plus

1,346 pounds of CO @ $5

for a total cost of over $9,000.

In other words, a conventional vehicle's lifetime emissions would force the air district to call on industries, municipalities, and residents to spend over $9,000 to achieve compensating reductions elsewhere in the airshed.

Box 16.1. Joe questions the $9,000 estimate

Joe raises his hand to ask about the hidden emissions for an EV. He suspects the EV batteries would be recharged each night so the vehicles would be ready for the next day's commute. That electricity could come from burning fossil fuels, so there would be emissions to the atmosphere. Shouldn't these emissions be counted against the EV?

This is a good question, one that is frequently asked in the classroom. And Joe is right about the electricity generation—it would typically occur during nighttime hours. For the electric system in the western United States, the marginal generators would be fueled by natural gas. The BWeb explains that the gas-fired generation would add some NO_x to the atmosphere, thus contributing to the formation of smog. When the NO_x emissions are evaluated at $12/lb, they could add $300 to the cost of the EV. The relative value of an EV would be reduced from $9,000 to $8,700.

> Many students have the same concern as Joe because they have become cautious about new technologies. But for some reason, students seldom ask the corresponding question about the existing vehicles: *What about the hidden emissions for a CV?* These are the emissions associated with the production of gasoline at the refinery and the distribution at the retail station. These emissions are often forgotten, but they are not hidden. Indeed, they are often in plain sight, especially if you are filling your gas tank on a hot day. The extra emissions add around $1,300 to the cost of the CV (BWeb).
>
> The cost of a CV (relative to an EV) is now $9,000 minus $300 plus $1,300 for a total of $10,000. $10,000 is a round number that is easy to remember. It will serve as a target for the feebate modeling in this chapter.

The Value of Electric Vehicles

The increased attention on EVs makes them a good point of focus for this chapter. We'll pay close attention to the comparison of an EV with a CV. The attributes for both vehicles are taken from the mid-1990s, as explained on the BWeb. With these conditions, it appears that a feebate policy should reflect the $10,000 estimate of the value of an EV relative to a CV.

The Case for a $10,000 Feebate

Feebates are appealing because they would work to supplement the market forces that create the supply and demand for vehicles. From the answer to Joe's question, a CV would impose an extra $10,000 in cost on the airshed. The $10,000 in extra cost is largely ignored by consumers. With feebates, however, this cost could be brought to the consumer's immediate attention by setting the fee and rebate to total $10,000. For example, the CV fee might be set at $2,000 and the EV rebate at $8,000. Consumers would be free to chose whichever vehicle best meets their needs, and manufacturers would be free to produce the most profitable mix of vehicles. The market would now operate with feedback to the participants on the environmental impacts of their decisions.

Feebates are also appealing because they would promote new technologies without requiring the program administrator to play favorites. Feebate programs could be implemented in a fuel-neutral and a technology-neutral manner. Each vehicle could be evaluated solely in terms of its emissions in the airshed, and the feebates could be adjusted accordingly. Any clean vehicle would qualify for a rebate; any dirty vehicle would be subject to a fee.

The Challenge of Feebates

Feebates are appealing, but could they be managed in a pragmatic manner? Suppose that the CVs and EVs are the only vehicle choices, and we elect to set the CV fee at $2,000 and the EV rebate at $8,000. We hope that the fees collected from the sale of CVs will provide the financing for the EV rebates. We announce the feebate for a given model year and wait to watch the market response. Perhaps the market will respond with 80% CVs and 20% EVs. If 80 out of 100 buyers purchase the CV despite the $2,000 fee, we would collect $160,000 in fees. The other 20 buyers receive an $8,000 rebate, so total rebates would be $160,000. This example shows the ideal situation—the cash flows are in balance, and the program may be described as self-financing. Self-financing is another reason for the popularity of feebates—they could be operated in a revenue-neutral manner so as to avoid the "tax label" (which is unpopular in the United States).

Now, consider a slight variation on the ideal situation. Suppose the market responds with a 70%/30% mix of CVs/EVs. For every 100 new-car purchases, we would collect $140,000 in fees, but we would be paying $240,000 in rebates. The program would be $100,000 out of balance for every 100 new cars sold. Think of the consequences with 1 million new cars sold every year—the feebate program would be out of balance by 1 billion dollars per year!

At this point, you are probably thinking that you could avoid this problem by setting the feebate to account for the 70%/30% market shares. You might keep the fee at $2,000 but lower the rebate to $4,667. Or you might set the fee to $1,000 and the rebate at $2,333. Both these examples abandon the goal of keeping the sum of the fee and rebate at $10,000. But suppose our focus is to get the cash flow into balance. These two examples aim to do so by setting the rebate at 2.33 times the fee. This makes sense because the CV market share is expected to be 2.33 times larger than the EV market share. However, these examples presuppose that you can accurately forecast the 70%/30% market shares.

Unfortunately, there is little reason to believe that you could develop a reliable forecast of market shares. Your efforts would be hampered by major uncertainties in describing consumer behavior and equally important uncertainties in describing the attributes of the vehicles to be offered for sale. If we are to seriously consider a feebate program, we must face the fact that the administrator will not have a crystal ball to forecast future market shares.

Since the cash flow will be frequently out of balance, it makes sense to operate with a balancing account. We'll start with $100 million in the account, and we expect to see the balance grow in years when fees exceed rebates. And there will also be years when the rebates exceed the fees, so the balance will decline. We wish to know if the fees and rebates could be adjusted to control the fund balance reasonably close to zero. This is the main purpose of the model. Along the way, it will also shed light on the mix of vehicles in operation and their emissions.

Read and Verify

The model is explained below as if you are working along on your own computer. A good way to learn is to read the step-by-step instructions and immediately verify that you can reproduce the results on your computer. The modeling is in Stella; Vensim versions of the models are on the BWeb. We are dealing with five types of vehicles, so it makes sense to take advantage of Stella's arrays feature. You'll use the array editor to create a dimension called *V*, with five elements for each of the vehicle types.

The Vehicles Sector

Figure 16.1 shows the portion of the model devoted to the accumulation of vehicles in the airshed. These variables have been combined using Stella's sector tool. The conveyor stock is assigned to cars in operation. The stock is increased by sales and reduced by retirements. The retirements are based on a 10-year lifetime for all five types of vehicles. Cars are measured in millions; sales and retirements are in millions of cars per year. Let's assume that there are 10 million cars at the start of the simulation, and that all of them are CVs. The initial value of this stock should be set to spread the 10 million evenly across the conveyor (see exercise 14.1 for how this was done with insects).

The sales is an array flow that is the product of the total sales and the market share for each of the five types of vehicles. Figure 16.1 shows the market shares as a ghost variable because it is part of another sector. You can test the vehicle sector by specifying the market shares, as shown in table 16.1. Set the total sales to begin at 1.6 million/yr and to grow exponentially at 0.047/yr. Set the model to simulate from 2000 to 2016 with DT at 1/4 year. You should have total sales of 3.4 million vehicles/yr by the end of the simulation. The total vehicles in operation will grow from 10 million to nearly 27 million during the simulation. You should see 1.55 million EVs in operation at the end of the simulation.

Now, set the annual travel for a single car at 10,000 miles/yr. The annual travel is the product of the number of cars and the annual travel for one car. The simulation begins with 10 mil-

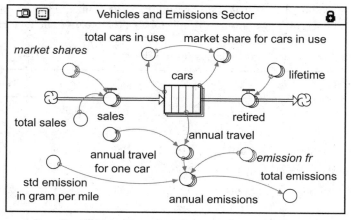

Figure 16.1. The vehicles and emissions sector.

Table 16.1. Vehicle information.

Vehicle type	Element name	Emissions fraction	Market share
Conventional	CV	1.00	52%
Electric	EV	0.00	6%
Hybrid-electric	HEV	0.37	8%
Compressed natural gas	CNG	0.42	22%
Alcohol	AL	0.81	12%

lion CVs, so annual travel is 100 billion miles/yr. The standard emissions refers to the 0.25 gram of HC emissions expected from a CV per mile of travel. Annual emissions from 100 billion miles/yr would be 25 billion grams/yr. This is the same as 25,000 metric tons/yr, so we will measure annual emissions in thousands of metric tons of HC per year. The emissions fractions for the five vehicle types are listed in table 16.1. The electric vehicle is a ZEV, so its fraction is zero. The HEV is the next cleanest, with HC emissions at 37% of the standard value assigned to a CV. AL stands for a vehicle fueled by alcohol. In the 1990s conditions simulated here, such a vehicle would have been fueled by methanol with an emissions at 81% of the standard value for a CV.

You should be able to build the vehicles sector from the information given so far. Run this sector on its own, and check that total emissions grows to just over 50,000 metric tons/yr by the end of the simulation.

Market Shares from a Discrete-Choice Model

Consumer choice between different types of vehicles could be influenced by a wide range of vehicle attributes, such as purchase price, color, add-on options, and the dealer's reputation. For this chapter, we'll draw on a vehicle choice modeling by researchers from the University of California (Bunch et al. 1992). The researchers modeled consumer choice based on six attributes that would cover the main differences between conventional and alternative vehicles. Table 16.2 shows the six attributes, along with the values assumed for this chapter. The first two attributes deal with the cost to purchase and operate the vehicle. The purchase prices range from $15,000 for a CV to $27,000 for an HEV. (Remember, these are numbers from the mid-1990s.) Fuel cost ranges from 2.7¢/mile for a CNG to 6.4 ¢/mile for a HEV.

Table 16.2. Attributes of five types of vehicles.

Attributes	CV	EV	HEV	CNG	AL
Price	$15,000	$25,000	$27,000	$20,000	$18,000
Fuel cost (¢/mile)	4.7	5.3	6.4	2.6	7.2
Range (miles)	450	100	200	200	250
Emissions fraction	1.00	0.00	0.37	0.42	0.81
Fuel availability	1.00	0.50	1.00	0.25	0.20
Horsepower	121	65	85	115	134

Now, suppose we are dealing with consumers who focus solely on the costs. The first two attributes would tell the whole story. We might combine these two attributes into a total effective cost of each vehicle. To illustrate, suppose we know that consumers weight future fuel costs as if they expect to use their vehicle for 60,000 miles of total discounted travel. In this case, the total effective cost of a CV would be $15,000 plus $2,820 (i.e., 60,000 miles @ 4.7¢/mile). The total effective cost would be $17,820. We might then compute the corresponding cost of each of the other vehicles and compare the costs. But what would we do with the comparison? Would we award 100% of the market to the vehicle with the lowest cost? This might make sense if the winning vehicle were thousands of dollars less expensive. But what if it were only a few dollars less expensive?

It's not immediately clear how we should calculate market share, even when we concentrate solely on the cost attributes. The market share calculation is made even more complicated by the presence of nonmonetary attributes like horsepower. Most readers are familiar with horsepower, and most would prefer a car with more power (everything else held equal). Data on consumer purchases of cars with different horsepower might tell us how much more a consumer is willing to pay to buy a more powerful vehicle. We have such data for conventional vehicles, but we lack data for the alternative vehicles. And we also lack sales data on how consumer choice is shaped by attributes like range and emissions. How are we to proceed if there are no sales data to support a theory of consumer choice among vehicles?

The researchers from the University of California dealt with this problem with a mail-back survey of 700 people in Southern California (Bunch et al. 1992). The people were presented with choices among many different types of vehicles whose attributes spanned a wide range of values. Their choice of vehicles was then explained with a statistical model that has proven useful in discrete choice. The model is known as a *multinominal logit model* (BWeb). The market share for a particular vehicle (v) among five competing types of vehicles uses the exponential (EXP) function as follows:

$$MS_v = \exp U_v \bigg/ \sum_{i=1}^{5} \exp U_i$$

The U is sometimes described as the utility of a vehicle. This is an abstract term that is not easily interpreted. But, for now, think of greater utility as meaning greater overall attractiveness and a higher market share. Table 16.3 shows the Stella equations used to implement the multinomial logit equation. The variables are grouped in a market shares sector, as shown in figure 16.2. The utility for each vehicle is the sum of utilities from the six attributes. The co-

Table 16.3. Stella equations for the market share calculation.

market_shares[V] = numerator[V]/denominator
numerator[V] = exp(U[V])
denominator = ARRAYSUM(numerator[*])
U[V] = U1[V]+U2[V]+U3[V]+U4[V]+U5[V]+U6[V]
U1[V] = coef_1*purchase_price[V]/1000
U2[V] = coef_2*fuel_cost[V]
U3[V] = coef_3A*(range[V]/100) + coef_3B*((range[V]/100)^2)
U4[V] = coef_4A*emission_fr[V]+ coef_4B*emission_fr[V]^2
U5[V] = coef_5A*fuel_availability[V]+coef_5B*fuel_availability[V]^2
U6[V] = coef_6*horse_power[V]
coef_1 = -.143
coef_2 = -.175
coef_3A = 2.06
coef_3B = -.303
coef_4A = -3.08
coef_4B = 1.53
coef_5A = 2.24
coef_5B = -.956
coef_6 = .00796

efficients were estimated statistically from the stated preference survey. A single coefficient means that the stated preferences were well explained with a linear relationship (e.g., utilities associated with price, fuel cost, and horsepower). Two coefficients were used when a nonlinear expression provided a better explanation (e.g., for utilities associated with range, emissions fraction, and fuel availability).

The multinomial logit model is well established in estimating discrete choice with multiple attributes (BWeb). Indeed, it has been called the "workhorse" of choice models (Louviere, Swait, and Hensher 2000, 13). The equations are suitable for statistical estimation, but they do not lend themselves to easy interpretation. Inserting these equations into a spreadsheet will help us to experiment with the numbers. Tables 16.4 and 16.5 show a spreadsheet (BWeb) to focus on

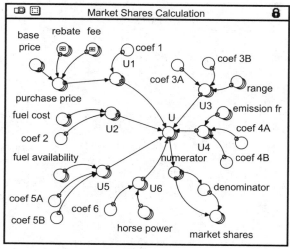

Figure 16.2. The market shares sector.

competition between a CV and an EV. (The other three vehicles are ignored for clarity.) The utility of a CV is calculated as 0.86391. The utility of the EV turns out to be a negative number: −1.3471. This may look bad for EVs, but they would still capture some of the market:

$$MS_{EV} = 0.2599/(2.3724 + 0.2599)$$

which turns out to be 10%. A 10% market share turns out to be quite plausible if the competition is limited to CVs and EVs. Indeed, a 10% EV market share figured prominently in the ZEV debate during the 1990s (BWeb).

Table 16.4. Spreadsheet check for the CV.

Attributes	Coefficients			Utilities
Price (thousands of $)	−0.143	15.0	U1(CV)=	−2.145
Fuel cost (¢/mile)	−0.175	4.7	U2(CV)=	−0.8225
Range (hundreds of miles)	2.060	4.5		
Range (hundreds of miles) squared	−0.303	20.3	U3(CV)=	3.13425
Emission fraction	−3.080	1.0		
Emission fraction squared	1.530	1.0	U4(CV)=	−1.55
Fuel availability	2.240	1.0		
Fuel availability squared	−0.956	1.0	U5(CV)=	1.284
Horsepower	0.00796	121.0	U6(CV)=	0.96316
U (CV) =				0.86391
e^U				2.372419

Table 16.5. Spreadsheet check for the EV.

Attributes	Coefficients			Utilities
Price (thousands of $)	−0.143	25.0	U1(EV)=	−3.575
Fuel cost (¢/mile)	−0.175	5.3	U2(EV)=	−0.9275
Range (hundreds of miles)	2.060	1.0		
Range (hundreds of miles) squared	−0.303	1.0	U3(EV)=	1.757
Emission fraction	−2.080	0.0		
Emission fraction squared	1.530	0.0	U4(EV)=	0
Fuel availability	2.240	0.5		
Fuel availability squared	−0.956	0.3	U5(EV)=	0.881
Horsepower	0.00796	65.0	U6(EV)=	0.5174
U (EV)=				−1.3471
e^U				0.259993

The spreadsheet makes it easy to conduct multiple experiments with different assumptions on the vehicles. If we increase the range of an EV to 200 miles, for example, the EV market share increases from 10% to 26%. Returning the range to 100 miles and lowering the price of the EV to $17,000 will deliver the same result—the market share increases to 26%. These experiments tell us that an extra 100 miles of range is worth $8,000. This is one of the trade-offs implicit in the responses of the 700 people who participated in the stated-preference survey. You can experiment with the spreadsheet to find other trade-offs that were in the minds of the participants in the survey. The reasonableness of these trade-offs is what makes the market share calculations suitable for the feebate modeling exercise. You should be able to build the market share sector from the information provided so far. Run the model, and you should see the market shares listed in table 16.1.

Cash Flow and the Control Panel

Figure 16.3 shows the third sector of the model. The balance in the feebate fund is initialized at $100 million. The fund is fed by fees collected, and it is drained by rebates paid. The third flow is the interest earned or charged. The interest rate is 0.10/yr, which means that the fund will accumulate interest earnings at 10%/yr. However, if the fund goes into the red, there will be interest charges at the rate of 10%/yr. The fee and rebate are marked with

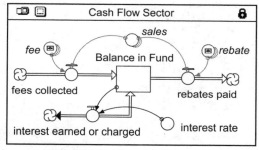

Figure 16.3. The cash flow sector.

the slider icon, so you know these will be set on the control panel. The sales is a ghosted variable from the vehicles sector. The equations for the this final sector are listed in table 16.6.

Table 16.6. Stella equations for the cash flow sector.

Balance_in_Fund(t) = Balance_in_Fund(t - dt) +
        (fees_collected + interest_earned_or_charged - rebates_paid) * dt
INIT Balance_in_Fund = 100
fees_collected = fee[CV]*sales[CV]+fee[AL]*sales[AL]
interest_earned_or_charged = Balance_in_Fund*interest_rate
rebates_paid = rebate[EV]*sales[EV]+rebate[HEV]*sales[HEV]+rebate[CNG]*sales[CNG]
interest_rate = .10

The feebates model is formed by combining the three sectors in figures 16.1, 16.2, and 16.3. The downloadable model also contains some "display" variables that make time graphs on the control variable easier to understand. Figure 16.4 shows the control panel of the model. which may be downloaded from the BWeb. An example of a display variable is the "Target is zero by the year 2016."

The simulation in figure 16.4 begins with no fee or rebate. The market shares will match the values in table 16.1, and the balance in the fund will grow from $100 million to $148 million owing to interest earnings. The simulation advances one year at a time on the assumption that the feebate manager can elect to change feebates on an annual basis. This simulation imposes a $2,000 fee on CVs and allows an $8,000 rebate on EVs starting in 2004. (There are no fees or rebates on the AL, CNG, or HEVs in order to keep the example simple.) These incentives cause the CV share of the new sales to fall from 52% to 40%. The EVs market share increases from 6% to 18%. This is a good result so far. You have tripled the market share for EVs, and the sum of the fee and rebate is $10,000, so consumers feel the environmental value of an EV.

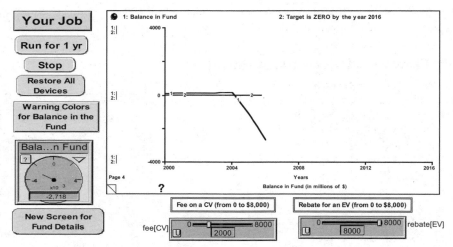

Figure 16.4. Control panel results with a $2,000 fee and an $8,000 rebate.

There are no changes in vehicle attributes or in fuel costs during the simulation, so these new market shares will remain in place as long as you keep the fee and rebate fixed. These incentives are held in place for two years in this simulation, and the fund is deeply in the red. The "Balance in Fund" warning light has turned yellow to draw your attention to the fact that the current balance has fallen below $2 billion in the red. Indeed, the current balance is a negative $2,718 million at this point in the simulation. This policy would put the state $2.7 billion in the red in just two years!

The control panel is designed with the idea that the feebate manager should be given considerable latitude in controlling the fund. The specific limits are plus or minus $4 billion. For example, if you let the fund go more than $4 billion in the red, Stella will ask for your resignation. And if you let the fund accumulate more than $4 billion, you will be fired for diverting too much money out of the economy. These are arbitrary limits. Perhaps you will find that you can control the cash flows quite successfully. In this case, you could reassure the top officials that the fund balance will be maintained within plus or minus $1 billion. However, it may turn out that large swings in the fund balance are inevitable. In this case, you should warn the top officials that even the best fund manager will need wider latitude in managing the fund.

These are some of the possible conclusions to be drawn from the exercises. This chapter is written as a work-along chapter in which you learn by building and experimenting. You have come a long way if you have constructed the feebate model on your own computer. But much of the learning is waiting in the exercises. They will guide you through a typical process of experimentation and model improvement. By the end of the exercises, you should know whether to accept the job as feebate manager.

Exercises with the BWeb Model

This chapter includes two sets of exercises. The first five exercises make use of the BWeb model. You'll work on the interface layer to simulate the existing model under a wide variety of conditions. The second set of exercises calls for changes in the underlying structure of the model. You'll make the changes on the model layer and then simulate the new model to learn if you can achieve better control of feebates.

Exercise 16.1. Verify the Stella model

Download the Stella model and verify you get the market shares in table 16.1 without fees or rebates. Verify that total emissions grow to just under 50,000 metric tons by 2016. Then verify the results in figure 16.4.

Exercise 16.2. Verify the Vensim model

The BWeb provides a Vensim version using the PLE (free) version of the software. The PLE does not support arrays, so separate variables are needed for each type of vehicle. Run the model for the entire simulation to verify the market shares in table 16.1. Then verify that total emissions grow to just under 50,000 metric tons/yr.

Exercise 16.3. Feebate policy continued

Re-create the results in figure 16.4 with the Stella model. Then continue the simulation for another two years with the same feebate values to learn how Stella can fire a feebate manager. Then re-create figure 16.4 and adopt a new feebate starting in 2006. Aim for a fee and rebate that bring the fund close to zero by 2016. And see if you can accomplish this with the sum of the fee and rebate fixed at $10,000.

Exercise 16.4. Explore feebates policies

Experiment with any combination of fees and rebates over the 16-year time period, and aim to bring the balance close to zero by the end of the simulation. Your main goal is to avoid getting fired, so be sure to keep the fund balance within the prescribed limits. Your other goals are to lower total emissions as much as possible by 2016. It would also be desirable (but not necessary) to have the sum of the CV fee and the EV rebate fixed at $10,000. And if you impose a fee on AL vehicles, set the fee to reflect the AL emissions factor. Similarly, take the emission factors for CNG and HEVs into account when allowing rebates for these vehicles. This is an open-ended exercise. The only restrictions are that the fees and rebates cannot exceed $8,000. Also, you will simulate one year at a time, so these incentives must remain in effect for at least one year. Otherwise, you are free to devise any policy that will do the best job in lowering emissions. Remember that you are practicing for the job of feebate manager, so you need a feebate policy that can be explained to the top officials.

Exercise 16.5. Feebates strategy paper

Write a strategy paper about your best feebate policy. The paper should be addressed to the top officials who have asked you to design the feebate program. The paper should inform them on the likely swings in the feebate fund over time. It should also make them aware of the reduction in emissions that might be achieved by 2016. Explain the conditions under which you would accept the job as feebate manager.

Modeling Exercises

Exercise 16.6. Endogenous feebate control

The fees and rebates are exogenous inputs controlled by sliders on the interface (the control panel). Now that you have arrived at a good policy, it is useful to convert the fees and rebates to endogenous variables. The challenge is to alter the model so that these incentives respond continuously over time to the changing conditions in the market. With endogenous incentives, we could run the model on autopilot. The fees and rebates would be adjusted automatically, freeing our time to explore a wide range of simulations with changes in assumptions about vehicles and consumers. Figure 16.5 illustrates one example of how this might be done. This is a causal loop diagram, but there are no loops unless we count the dashed lines. These are added to depict the decisions by a student who watched the balance in the fund to set the CV fee and the EV rebate. If the fund became too big, the student lowered the CV fee and raised the EV rebate. This policy would create two negative feedback loops. They are labeled "fee control" and "rebate control," but the question marks warn us that this only one possible way of thinking about feebate control. Your job is to implement your own feebate policy so it can be tested on autopilot.

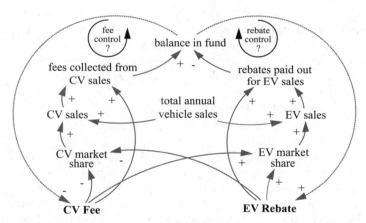

Figure 16.5. Dashed lines for possible fee and rebate control.

Exercise 16.7. Run the new model on autopilot

Run your new model with constant conditions to verify that you get similar results to your best policy in exercise 16.4. (Your results will be somewhat different since the fees and rebates are now subject to continuous adjustments.)

Exercise 16.8. Stability test with better EVs

Introduce a major change to make EVs more attractive. For example, you might increase the range of an EV from 100 to 150 miles in the year 2012. Then simulate the model on autopilot to see if your feebate policy can control the fund within the prescribed limits.

Exercise 16.9. Feebate control with different environmental values

The $10,000 estimate of the environmental value of the EV is from the mid-1990s. If the estimate were repeated today, it could be much lower owing to lower emissions from the newer CVs. On the other hand, the value of an EV would increase if we include the cost associated with CO_2 emissions from the CV. Did your endogenous feebate policy maintain the sum of the fee and rebate at $10,000? If so, use your model to test the same feebate policy with the sum of the fee and rebate maintained at $5,000.

Exercise 16.10. Fees and rebates for all vehicles

The previous exercises focus on EVs and CVs to learn about controllability. But the producers of AL, CNG, and HEVs will ask for fair treatment, so we should search for a general policy for feebates. Suppose the AL and CV are assigned a fee because of their high emissions, and that rebates will be allowed for CNG, HEVs, and EVs. If possible, set the relative value of the fees and rebates to reflect the emission fractions in table 16.1. Run the new model with your endogenous policy on fees and rebates. Can you control the fund within the prescribed limits?

Exercise 16.11. Expand to include fuel efficiencies

The fuel cost is an exogenous input measured in ¢/mile. Expand the model to specify the fuel cost and efficiency for both the EV and the EV. Set the CV fuel cost at $1.29/gal, and the efficiency at 27.5 mi/gal. The EV fuel cost is 5.3¢/kWh, and the efficiency is 1 mi/kWh. Run the new model to verify that you get the fuel costs in table 16.2 and the market shares in table 16.1.

Exercise 16.12. Impact of lower electric rates

Use the model from exercise 16.11 to simulate the change in EV market shares if the electricity price is cut from 5.3¢/kWh to 2¢/kWh. How many more EVs are in use in the year 2016 with the lower rates?

Exercise 16.13. Impact of a gasoline tax

Use the model from exercise 16.11 to simulate the change in market shares if the state imposes a $1/gal tax on gasoline. How many fewer CVs are in use in the year 2016 with the tax? How many more EVs are in use?

Exercise 16.14. Gas tax feeds the fund

Suppose the tax revenues generated in exercise 16.13 are fed into the state fund (the fund shown in figure 16.3). This would build the balance, allowing for larger rebates for cleaner vehicles. Expand the model from exercise 16.13 to allow for gasoline tax revenues to contribute to the state fund. (To keep the calculation simple, ignore the gasoline consumed by the HEVs.) Your tax revenues will come from the CVs. The simulation begins with 100 billion miles of annual CV travel @ 27.5 mi/gal. The gasoline consumption is 3.64 billion gal/yr. A $1/gal tax would generate $3.64 billion/yr, so it could quickly push the fund above $4 billion. To avoid getting fired, remove the $4 billion limit in the Stella model.

Exercise 16.15. Are taxbates easier or more difficult to control?

Let's coin the term *taxbates* to stand for a program to promote the sale of cleaner vehicles through rebates financed by a gasoline tax as well as fees. Experiment with the model from exercise 16.14 to find an appropriate combination of a gasoline tax, CV fee, and EV rebates to lower the emissions in the airshed. How do the financial challenges of operating a taxbate program compare with the problems of a feebate program?

Exercise 16.16. Interface for a taxbates control model

Expand the interface in figure 16.4 by adding a new slider to control the size of the gasoline tax to be used in a taxbate program. Add any new screens and buttons that will help a newcomer use your model to explore the dynamics of a taxbate program. Test your best taxbate program with the stability test of exercise 16.8.

Exercise 16.17. Revised version of the strategy paper

Revise the strategy paper from exercise 16.5 to include what you have learned from the endogenous modeling of feebates and the experiments with a taxbate policy. Conclude with the conditions under which you would accept the job a fund manager.

Chapter 17

Modeling Pitfalls

A pitfall is a concealed danger for an unwary person. A modeling pitfall is a danger that is concealed from a person who has not learned the problems that can arise in modeling. This chapter describes some pitfalls for readers who are using system dynamics for the first time. We'll start with some simple pitfalls, dangers than can be easily avoided once we are made aware of them.

Simple Pitfalls

Some pitfalls are easily avoided. But they can turn into big problems if we are not watchful. An example is the non-negative option for stocks in the Stella software. This option should be turned off, as you learned in exercise 2.9. Forgetting to turn this option off will leave your model vulnerable to complications, like those demonstrated with the first epidemic model in chapter 8. This pitfall is easily avoided by turning to the File/Default Settings command (see the book's website, the BWeb). You can tell Stella to make sure that all of your stocks will go negative if your flows force them to do so. Negative results for a physical stock means that you should reformulate one of the flows.

One reason physical stocks go negative is that we forget to include a connector from the stock to a flow that drains the stock. Think of the extreme case where the stock falls to zero; the outflow should also fall to zero. This requires a connection between the stock and the flow. The simplest example is to control the outflow with a rate (e.g., the death rate in figure 3.1). Control of the outflow can be more complicated, as in the connection between the stock and the evaporation flow in the Mono Lake model (figure 5.9). If you don't see such a connection between the stock and the outflow, the model could well generate erroneous behavior. This is a surprisingly frequent problem among new readers, so the book shows concrete examples (e.g., the first model of Mono Lake [figure 5.4] and the first model of the epidemic [figure 8.2]).

However, there may be occasions when a connection to the outflow is missing. An example is the other outflow (figure 5.9) that drains water out of Mono Lake. This is a small exogenous flow, so we should check if its value is reasonable. The equilibrium diagram is useful here. The flows in figure 6.1 indicate that the other outflow can remain an exogenous input without causing problems. Another occasion appears with a conveyor stock, as when there is no visible connection from the students in school to the graduation of students in figure 14.17. This is an apparent problem, not a real problem; the connection to the outflow is hidden in the formulation of the conveyor. And finally, the connection is missing in some displays in the book, as in the figure 6.2 diagram for the hydrologic cycle. These diagrams are for display, not for simulation. So you do not need to worry about the missing connection.

Another pitfall involves misuse of the biflow, a flow that can operate in two directions. You first saw the biflow in exercise 3.5, and you know that the Stella biflow will have two arrowheads. Beginning students sometimes forget to ask for the biflow when there could be a flow in both directions. And they sometimes draw the flow with the arrows pointed in the wrong direction. These mistakes are easy to avoid once you remind yourself to be watchful of biflows.

The Troublesome IF THEN ELSE Function

A pitfall for some modelers is overreliance on the IF THEN ELSE function, the logical function described in chapter 14. The function was used to simulate the operation of a furnace or an air conditioner. There was a clear rule to be evaluated, and there were two distinct outcomes based on the evaluation. The function made sense because we were describing a single device, not the average operation of all the air conditioners or furnaces in a city. But this situation is seldom the case in system dynamics models, because we are typically simulating the average performance of a population, not the actions of each individual in the population. Nevertheless, many students seem drawn to this function, especially if they are still becoming acquainted with stocks, flows, and feedbacks.

The Mono Lake model will illustrate the problems with this function. The export policy in figure 5.14 calls for a gradual change in export as the lake's elevation varies between 6,380 and 6,390 feet. This policy is easily simulated using Stella's graphical function (~). If you did this exercise correctly, the export will gradually rise to around 37 KAF/yr and remain at that value for the rest of the simulation. However, some students are drawn to the IF THEN ELSE function. An example of a student equation could be:

export = IF (elevation>6385) THEN 100 ELSE 0

This calls for an abrupt change in export with the apparent target at 6,385 feet. Figure 17.1 shows the results. There is no export for the first 10 years, and the elevation climbs gradually toward 6,385 feet. The export jumps to 100 KAF/yr in the year 2000, but it stays there for less than a year. The elevation quickly falls below the target, and export is returned to zero. This causes the elevation to grow above the target, and we see export jumping back to 100 KAF/yr. These oscillations continue in this manner for the remainder of the simulation. This pattern is

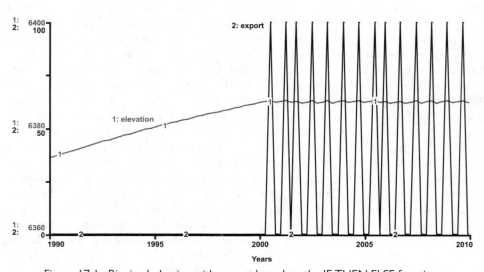

Figure 17.1. Ringing behavior with export based on the IF THEN ELSE function.

sometimes called *ringing*, a term for the unwanted oscillations in a poorly designed electrical system. Ringing is a bad sign in electrical circuits, and it's a bad sign in the Mono Lake model.

At this point, you might be tempted to improve on the student equation. Maybe you can think of an IF THEN ELSE formula to match the export in figure 5.14. You could use the IF to check for conditions below 6,380 feet; between 6,380 and 6,390 feet; and above 6,390 feet. With multiple conditions, you would need to insert a second IF THEN ELSE inside the first function. These are called *nested functions*. Nested functions are real pitfalls that should be avoided. Not only are your equations prone to errors, but the complex collection of parentheses will leave your equations difficult for others to understand. This pitfall is easy to avoid in the Mono Lake case; we simply use a nonlinear graph (~) to represent the water exported from the basin.

Let's now consider an IF THEN ELSE example that you haven't seen before. Imagine a city with 1 million cars traveling 25 miles/day with a fuel efficiency of 25 mi/gal. The normal gasoline use would be 1 million gal/day. Figure 17.2 shows a model to simulate actual use of gasoline over an 8-day period with an interruption in refueling in the second day. The behavior is shown in figure 17.3. The refueling is 1 million gal/day until time exceeds 2 days. Then it's cut by 75%. (This can be done with the IF THEN ELSE function without problems.)

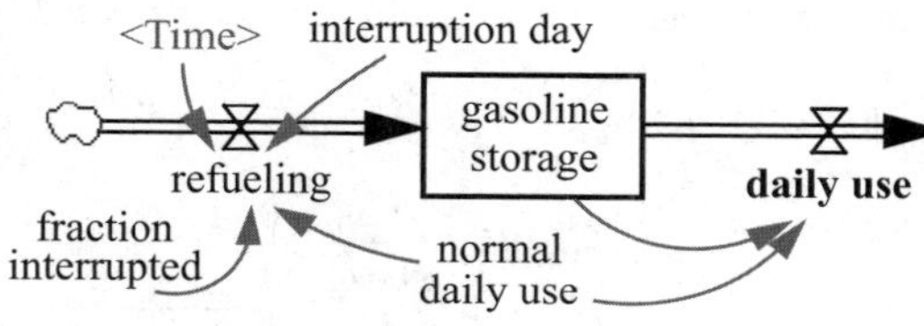

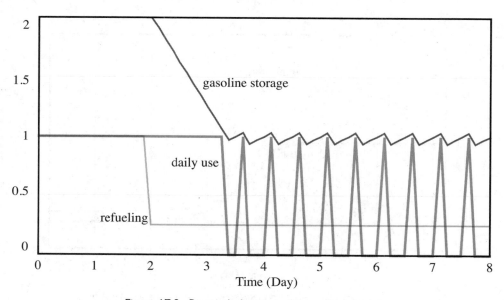

Figure 17.2. Gasoline model with an interruption in refueling.

Now what equation would you write for the daily use? It should match the normal daily use when there is plenty of storage. But it could fall to zero if there is low storage in the tanks. A typical student equation would be:

Daily use = IF THEN ELSE (gasoline storage>normal daily use, normal daily use, 0)

Figure 17.3. Ringing behavior in the gasoline model.

If we simulate the model with 2 million gallons of initial storage, we see the results in figure 17.3. Gasoline storage remains at 2 million gallons until the interruption occurs near the end of the second day. The storage is then drawn down because refueling is only 25% of normal daily use. Daily use falls to zero midway through the third day and then jumps back to normal use immediately thereafter. The model then falls into a ringing pattern that continues for the remainder of the simulation. The ringing behavior is characteristic of a poorly formulated model. We need to introduce extra variables to represent our theory on how drivers would reduce daily use of gasoline. One theory is that drivers eliminate unnecessary travel as the days of storage get too low. Figure 17.4 shows a Vensim lookup to represent a reduction in the fraction of trips as the days of storage falls too low. Figure 17.5 shows the new results.

Gasoline storage remains at 2 million gallons for the first two days, the same as before. Storage falls during the third day, but consumers are still traveling the same as before. They reduce the number of trips in the fourth day, and daily travel falls to 75% below normal by the sixth day. The system has reached a new equilibrium with refueling and daily use in balance at 0.25 million gal/day. This theory indicates that consumers would maintain around 0.33 million gallons of storage. This is minimal storage carried forward as insurance to help with essential travel in the future.

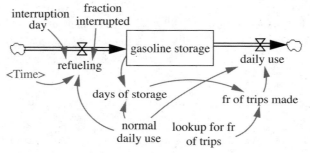

Figure 17.4. Rethinking the daily use when refueling is interrupted.

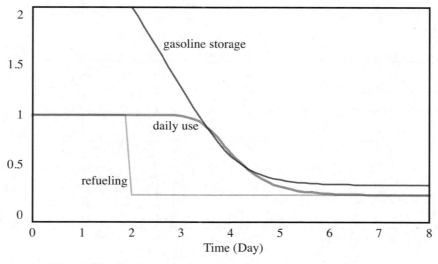

Figure 17.5. Gradual reduction in gasoline use following a disruption.

The Mono Lake and gasoline examples illustrate the general approach for avoiding the IF THEN ELSE. We think of the real-world process that controls the flow, and we add extra variables to represent the adjustment in the flow. We will most likely invoke a graphical function (~) in Stella or a lookup in Vensim to create a gradual adjustment. The gradual adjustment makes more sense than the discrete adjustments that occur with the IF THEN ELSE function.

Ringing Is a Sign of Poorly Formulated Equations

The previous examples show that ringing can appear from the inappropriate use of the IF THEN ELSE function. Ringing can also occur when equations are poorly formulated for a variety of reasons. But poorly formulated equation may escape attention if we do not test the model immediately. Therefore it is useful to check the model by scanning all the variables for ringing. Vensim and Stella make this easy to do. Simply visit each of the variables and graph their behavior. If you see ringing, take a closer look. Remember to distinguish between ringing and schooling. A model with schooling behavior will show dramatic variations that might be mistaken for ringing. An example is the salmon adults in migration in figure 15.9. This pattern arises from the salmon migrating as a school of fish—they appear in the river each spring, and their counts jump abruptly. This is not ringing, so you don't need to reformulate the model.

The Trouble with High-Turnover Stocks

High-turnover stocks can be a serious pitfall in models. The material stays in these stocks for only a brief moment before flowing out again. A very small value of DT is needed to simulate the outflows accurately. A low DT can mean slow simulations, and slow simulations can inhibit interactive simulation. If we include such stocks, we lose the opportunity for interactive simulation, one of the main advantages of system dynamics. We'll illustrate this problem with a model of the global hydrologic cycle in figure 17.6. The stocks are measured in thousands of cubic kilometers (TCkm); the flows are in TCkm/yr. The equations are shown in Table 17.1. Initial values of the stocks were shown previously in figure 6.2, with many of the values from Schlesinger (1991, 346) and rounded off for clarity. The stocks of water in the atmosphere are the high-turnover stocks. Schlesinger drew attention to this storage because it amounts to only 13 TCkm, an amount that he describes as tiny.

But enormous quantities of water move through the atmosphere each year. The difficult flows are shown by the dark arrows in figure 17.6. The precipitation over oceans is typically 380 TCkm/yr, around 40 times the stock. Precipitation over land is typically 110 TCkm/yr,

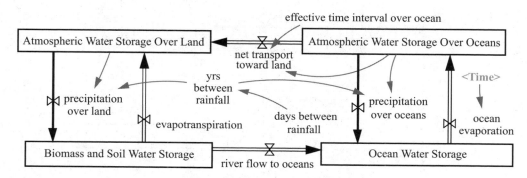

Figure 17.6. Model of the global hydrologic cycle.

Table 17.1. Equations to test the hydrologic cycle model.

precipitation over land =	Atmospheric Water Storage Over Land/yrs between rainfall
precipitation over oceans =	Atmospheric Water Storage Over Oceans/yrs between rainfall
evapotranspiration =	70
ocean evaporation =	IF THEN ELSE (Time>3 , 520 , 420)
river flow to oceans =	40
days between rainfall =	9
effective time interval over ocean =	0.25
net transport toward land =	Atmospheric Water Storage Over Oceans/effective time interval over ocean
yrs between rainfall =	days between rainfall/365

also around 40 times the stock. These large flows are calculated by dividing the stock by the average time between rainfall. We assume 9 days or 0.0246 years between rainfall. The river flow to the ocean and the evapotranspiration are constant at the equilibrium values. The ocean evaporation will increase from 420 to 520 TCkm/yr after the third year. We simulate the model with various values of DT to see if we get accurate results. The simulation with DT at 0.015625 is shown in figure 17.7. (This value is 1/64 year because the software developers recommend DT in powers of ½.) The precipitation is in equilibrium at 380 TCkm for the first 3 years; then it adjusts quickly to the new equilibrium when there is an exogenous increase in the evaporation from the oceans. The thick curve is the accurate result, but it requires 64 steps to simulate a single year. A model for 100 years would require over 6,000 steps. But if we raise the value of DT, we see ringing behavior shown in figure 17.7.

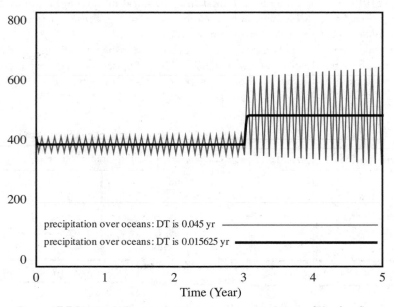

Figure 17.7 Ringing behavior due to inaccurate simulation of the fast flows.

Should We Get Rid of the High-Turnover Stocks?

It seems we are in a unfortunate situation. The last thing we need is ringing, so you might conclude that we should use a small value of DT and live with the slow simulations. Perhaps there are some models that simply cannot be simulated rapidly?

We do not face this difficult trade-off with the hydrologic model. We can have fast simulations and numerical accuracy. The key is to avoid the high-turnover stocks, the stocks of water in the atmosphere. We know the water in the atmosphere is a central part of the global cycle, so it must appear in the model. But it does not have to appear as a stock. A better approach is to use an auxiliary to represent the atmospheric storage based on their likely equilibrium values. We know that the flows will adjust atmospheric water quickly, as shown in figure 17.7. So we estimate the values of the stock by asking how the flows would balance themselves. This requires some algebra, and algebra is made easier with shorter variable names. Let:

YrRain = years between rainfall
AtSL = Atmospheric Storage over Land
PL = AtSL/YrRain = precipitation over land
NTL = net transport toward land, and
ET = evapotranspiration

If the flows are in equilibrium, we know that

PL = NTL + ET

We now have two equations for PL, the precipitation over land. These two equations must give the same result. So we know that:

NTL + ET = AtSL/YrRain

Our goal is to find the value of atmospheric storage over land, so we rearrange to get:

AtSL = YrRain*(NTL + ET)

The atmospheric storage can be found by an algebraic combination of three variables, so there is no longer a need to integrate a stock. A similar approach may be used for the atmospheric water storage over ocean. We find the value of this stock that would guarantee that ocean evaporation is balanced by the sum of net transport toward land and the precipitation over oceans. The new equations are shown in table 17.2.

Table 17.2. Auxiliary equations for the atmospheric water storage.

Atmospheric Water Storage Over Land =
    yrs between rainfall*(evapotranspiration+net transport toward land)

Atmospheric Water Storage Over Oceans =
    ocean evaporation/[(1/yrs between rainfall)+(1/effective time interval over ocean)]

Vensim uses *auxiliary variables* to allow algebraic combinations of other variables. Stella calls these *converters*, and the converters are normally circles. Vensim auxiliaries are shown in this book without an icon. If we stick with these choices, the diagram lacks the visual match with textbook images of the hydrologic cycle. To remind ourselves that water is stored in the atmosphere, we could assign a rectangular shape to the auxiliaries, as shown in figure 17.8. The rectangles are in gray to alert us that they are not really integrated like stocks. The note explains that we use algebra rather than integration to find the water storage in the atmosphere. The dark connections in bold are cosmetic as well; they remind us of the flows in figure 17.6. The reformulated model in figure 17.8 gives accurate results with DT = 1/4 year. A 100-year simulation would now require only 400 steps, so we will have rapid simulations. And if, for some reason, the days between rainfall were smaller, there is no need to reduce the value of DT.

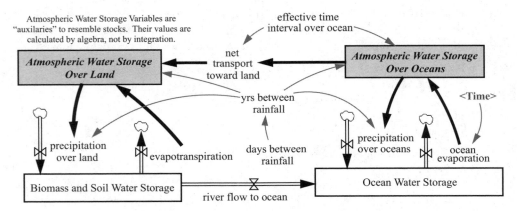

Figure 17.8. The high-turnover stocks are replaced with auxiliaries.

A Surprising Pitfall: Models as Calculators

Some students with a strong mathematical background become excited about the complex calculations that can be performed with Stella and Vensim. Those with training in differential equations recognize that system dynamics models are a collection of first-order differential equations. The software programs find numerical solutions and are sometimes viewed as powerful calculators. Students with background in differential equations are sometimes eager to design models to make visual connection with the examples in their mathematics classroom (see figure 17.9). This is a highly condensed version of the Mono Lake model from figure 5.9. A student brought this model to me one day, asking for help in writing the equation for the flow. His goal was to replicate the Mono Lake model in a form that reminded him of a differential equation. He named the flow dW/dt and assigned short names to all the inputs to the model. His challenge was to write the equation for dW/dt. He then planned to verify that the condensed model would give the same results as figure 5.10. It turned out that he was unable to write a "clean" equation for the flow, and he felt disappointed about his math skills.

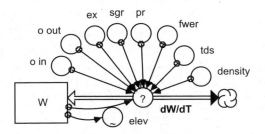

Figure 17.9. A condensed version of the Mono Lake model.

The point of this story is not the student's failure to write the equation; it's his thought that it would be useful to try. His reasoning boiled down to the view that system dynamics models should be designed to resemble models from classes in differential equations. He also viewed models as calculators, and he believed we should strive for compact calculators that do not take up much space on the screen.

This student story illustrates a view held by some with strong training in mathematics. They sometimes design models to strengthen the correspondence between the image on the screen and the image in a math textbook. This is an unproductive and distracting activity, and it takes valuable time away from the more important connections that should be made. For models to be useful with the client, we should aim for connections between the client's ideas and the variable on the screen. It is also unproductive to view a system dynamics models as a calculator. Models certainly perform complex calculations, but they are devices for communication, not just for calculation. Effective communication requires a lot of effort and extensive interaction with the group of people working on the project. Both Vensim and Stella provide many features to aid in communicating the structure and results of a model, and the BWeb provides suggestions for improved communication in the modeling process.

Easy Checks to Spot Modeling Problems

The pitfalls described in this chapter are small, technical problems of model formulation. They are easily fixed once you are aware of them. We can make ourselves more aware by conducting several checks. For example, we can check that the non-negative option for stocks has been turned off. Another simple check is to scan all the variables to see if there is ringing. Ringing is usually a bad sign, so the ringing variable will lead you to the flows that need to be reformulated. Another simple test is to cut DT in half and repeat the simulation. We should see the same results. If we don't, we should cut DT in half and simulate again. (There will be occasional exceptions to this rule, as explained on the BWeb in the model of the Tsembaga clan and their pig festivals). If we do find that DT has to be smaller, we should look for the variables that change in the DT test. These variables may involve a high-turnover stock, and we might consider replacing it with auxiliaries.

Some models have formulation problems that do not appear in the normal tests. For example, the results may appear reasonable over an interval from 2010 to 2030, and the team has performed many tests over this 20-year interval. A simple test is to run the model somewhat longer, say to the year 2035. This test can reveal formulation problems that do not surface during the normal time horizon.

Some experts include units consistency on the list of easy checks for a model. Both Vensim and Stella provide tools for automatic checking of unit consistency. These checks are easy to do once we invest the time to write the units into the appropriate window with each new equation. Experienced modelers all agree that units must be internally consistent, and many recommend automatic checks with the software. I believe we will learn more about the model if we check the units ourselves, as explained in appendix A.

The final suggestion for checking a model is to create a control panel to make sensitivity testing easy. Create two or three graphs with the principal variables in view. Then add sliders to the screen. You might select 5 to 10 inputs based on your own intuition as to which inputs are most likely to be important. Then take the time to experiment with the inputs and check for plausibility of the results. Be sure to experiment with a wide range of values and to try some extreme tests with all of the inputs pushed to their upper or lower limits. Informal sensitivity testing from a control panel is a fun way to learn about a model and its problems. However, it may not cover all the inputs, and it may miss the effect of simultaneous changes. Comprehensive sensitivity testing can provide a more thorough check, as explained in appendix D.

Exercises

Exercise 17.1. Verify gasoline model

Build the model in figure 17.4 and verify that it gives results similar to figure 17.5.

Exercise 17.2. IF THEN ELSE discussion

The gasoline model in figure 17.4 uses the IF THEN ELSE function to create the 75% reduction in the refueling flow. But we recommend against this function to simulate the daily use. Why do you think this function should be permitted for the inflow and not the outflow?

Exercise 17.3. Verify the hydrologic cycle model

Build the model in figure 17.6 and verify the results in figure 17.7.

Exercise 17.4. Ringing with the DDT model

Chapter 22 describes models of DDT accumulation. Build the DDT model (BWeb) that uses a stock for DDT in the air. This is a high-turnover stock, so we expect ringing in the precipitation flows. Verify the original results with DT = 0.02 year. Then test the model by doubling DT and simulating again. For what value of DT do you first see ringing?

Exercise 17.5. Population model

Build the model in figure 17.10 with a 50-year adult interval, a 15-year youth interval, and 0.25 year for the infant interval. The births are fixed at 10 million persons/yr and the simulation begins with 0.5 million infants, 150 million young, and 500 million adults. Set the infant mortality rate to 0.20/yr, and simulate for 20 years with DT = 1/16 year. What are the equilibrium values for infant deaths and survivals?

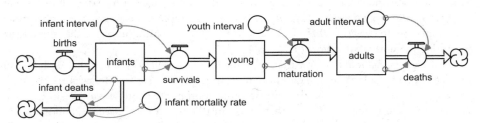

Figure 17.10. Infants are the high-turnover stock in a population model.

Exercise 17.6. Ringing in the population model

Double the value of DT in exercise 17.5, and repeat the simulation. Do you see ringing? If not, double DT again and repeat the simulation. For what value of DT do you first see ringing?

Exercise 17.7. Population model without the infants stock

Build the population model in figure 17.11 with the same parameters as in exercise 17.5. The stock of infants has been removed, but you will write the appropriate equation for the expected number of infants. You need to write an equation for infant survivals as well. Then simulate the new model to verify that you get the same results as in exercise 17.5.

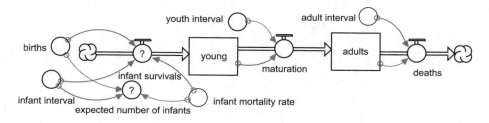

Figure 17.11. Population model with the stock of infants removed.

Chapter 18

Introduction to Cyclical Behavior

Many systems exhibit intriguing cyclical behavior. You will probably recall the first time you felt the rhythmic beating of your heart. Or perhaps you remember the first time you saw ocean waves crashing on the shore. These and other oscillatory systems are fascinating to observe, but they are difficult to understand.

This chapter demonstrates how system dynamics may be used to improve our understanding of cyclical behavior. Improved understanding is needed in a wide variety of professions— from economics to population biology. The economist benefits from a better understanding of cycles in a major industry like real estate; the biologist benefits from a better insight into the cycles in wildlife populations. These examples are shown in coming chapters. But first we will illustrate cyclical behavior with the familiar flowers and sales models from chapter 7. The first illustration shows oscillations imposed by outside forces. The remaining examples are more interesting. They show oscillations that arise from inside the system.

Cycles Imposed by External Forces

Recall the model that generated S-shaped growth in the area of flowers. Figure 18.1 shows a new version of this model to allow for changes in rainfall. Let's assume that rainfall varies around a mean of 20 inches/yr with a sinusoidal pattern. The amplitude of the sine wave is 15 inches/yr, and the period is 5 years. This pattern can be represented with Stella's built-in SINE WAVE function. The intrinsic growth rate will be 100%/yr with optimal rainfall of 20 inches/ year. The nonlinear graph (~) represents the decline in the growth rate when rainfall deviates

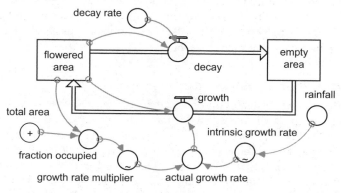

Figure 18.1. Flowers model with variable rainfall.

from the optimum. Let's use the graph to show a decline to 0.6/yr if the rainfall falls to 10 inches/yr or rises to 30 inches/yr. The intrinsic growth rate will fall to zero if the rainfall falls as low as zero or rises as high as 40 inches/yr.

Figure 18.2 shows the simulated area of flowers, along with their growth and decay. The variations in rainfall cause large swings in the growth but only minor swings in the area. The general pattern could be described as S-shaped growth with a superimposed cycle. But the variations in rainfall do more than simply add a cyclical variation to the previous results. The area of flowers takes longer to reach quasi-equilibrium conditions, and the equilibrium values are somewhat lower than in chapter 7. These changes make sense because the flowers enjoyed optimum growth conditions in chapter 7, but they experience variations around the optimum in the new model.

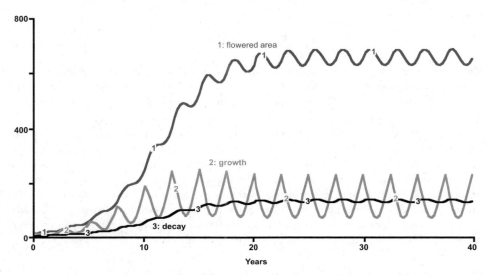

Figure 18.2. Oscillations in the area of flowers from external changes in rainfall.

This first example reveals some general conclusions that may apply to all systems, not just to flowers. First, we see that cycles imposed from outside can be transformed and moderated. In this example, a 5-year cycle in rainfall is changed to a 2.5-year cycle in the growth of flowers. The simulation demonstrates that a large cycle in flower growth would persist, but there would be only small cycles in the area covered by flowers. It appears that the system acts to buffer the impact of variable rainfall. The flower example is typical of most systems—the buffering takes place at the stock in the system.

Oscillations from Inside the System

Figure 18.3 shows a new version of the flowers model. There are 1,000 acres of suitable area, and the fraction occupied will lead to a reduction in the growth rate. But this reduction is no longer immediate. The new twist is a lag before a change in the fraction occupied is translated into a change in growth rate. You know from chapter 14 that the SMOOTH function can be used to represent the delayed effect of information. Most lags can be represented by either first-order or third-order smoothing of information (see figures 14.12 and 14.13). Let's select first-order smoothing for the first simulation. This choice makes sense if some of the impact is felt immediately, but the total impact is spread over a longer time interval. The lag time is 2 years; the intrinsic growth rate is 75%/yr, the decay rate is 20%/yr, and the graphical function (~) for the growth rate multiplier is the linear shape (figure 7.4) that created logistic growth in chapter 7.

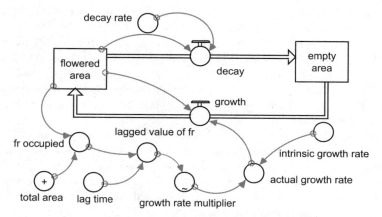

Figure 18.3. Flowers model with a lagged effect on the growth rate.

We now simulate the model with all of these assumptions fixed in time. There are no variations in rainfall as in the previous example, and there are no other disturbances. You might expect the model to show S-shaped growth under these conditions, but figure 18.4 shows a much different pattern. The flowered area grows during the first 12 years, eventually covering more than 900 acres. But the flowered area begins to decline after the 12th year, and the decline frees up more space for growth in the future. By the 18th year, the growth exceeds decay, and the flowered area increases once again. The variations in growth eventually fade away, and the system reaches dynamic equilibrium. The dampening is relatively strong, and the oscillations have almost disappeared by the end of the simulation. If undisturbed, the system will find its way to an equilibrium with 734 acres of flowers.

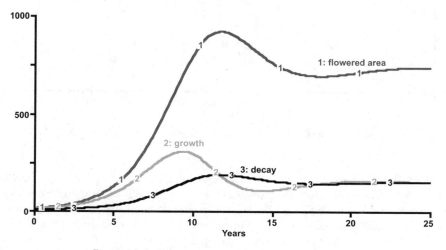

Figure 18.4. Damped oscillations in the flowers model.

Damped Oscillations

This pattern is called *damped oscillations.* Damped oscillations are quite common, with the most familiar example probably being the pendulum. It will show oscillations over time, but they will be damped owing to the frictional forces that take the energy away from the swinging pendulum. The oscillations in figure 18.4 are not caused by any external factors, and they did not appear in chapter 7. This means they must be caused by the lag in the way the growth rate reacts to how the flowers have filled up the suitable area. The new model assumes that the flowers do not immediately sense the effect of their congestion, so their growth pushes the area past the equilibrium value. Once the full effect of the congestion is felt, the growth falls below the decay, and the area shrinks over time. The lag applies on the way down as well as the way up, so the flowers do not immediately sense the extra space that has been made available. Their area falls below the equilibrium value before growth is sufficiently strong to push the area toward the equilibrium value.

Since the lagged effect causes the oscillations, we would expect that shorter lags would lead to smaller oscillations. Figure 18.5 shows four simulations to confirm this idea. All four simulation use an intrinsic growth rate of 75%/yr and first-order smoothing of the fraction of area occupied. The lag times range from 2 years to 0.75 year, and the comparison graph shows smaller overshoots with the shorter lag times. The fourth simulation with the 0.75-year lag time is the most extreme example since it shows no oscillations whatsoever. This pattern is called *overdamped oscillations.* This may strike you as a curious term since there are no oscillations in this simulation. We use the term *overdamped* when a system has the potential to oscillate, but the oscillations are not present under current assumptions.

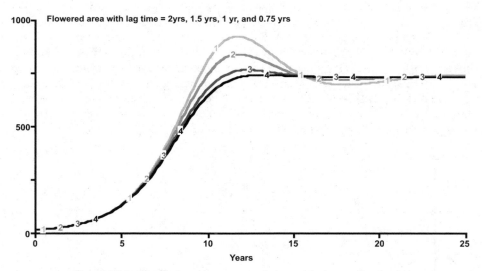

Figure 18.5. Oscillations dampen out faster with shorter lag times.

The overshoot and oscillations in the flowers model are caused by a lag in the feedback signal that would normally cause the growth to come into smooth accommodation with the limited space for flowers. The longer the lag, the greater the tendency for the simulation to overshoot the equilibrium and to oscillate. The extent of the overshoot is also influenced by the rate of growth as the flowered area shoots past the equilibrium value. The faster the growth, the greater the overshoot. Figure 18.6 shows five simulations to confirm this idea. All simulations use first-order smoothing with a 1-year lag time, but the intrinsic growth rates range from 40%/yr to 80%/yr. The first simulation (with 40%/yr) requires about 30 years to find an equilibrium area of around 500 acres. There are no oscillations in this particular case, so this is another example of overdamped behavior. The fifth simulation (with 80%/yr) shows the greatest overshoot. The flowered area grows to nearly 1,000 acres before feeling the effect of the congestion. This simulation finds its way to equilibrium of around 750 acres in around 20 years.

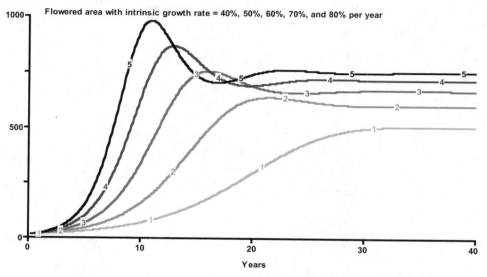

Figure 18.6. Oscillations dampen out faster with lower growth rates.

These experiments show that a system with S-shaped growth can turn into an oscillatory system owing to a simple lag in the feedback effect on growth rate. We've learned that longer lag times lead to larger overshoot and greater oscillations, and we've learned that more rapid growth can also lead to larger overshoot and greater oscillations. The oscillations shown so far are either damped or overdamped. To see more volatile patterns, we turn to the sales company model.

Oscillations in the Sales Model

Chapter 7 used a sales model and a flowers model to demonstrate that systems with similar structure generate similar behavior. So you might expect that oscillations could arise in both systems if we were to introduce a lag in the information feedback that limits the size of each system. The sales company simulation begins with 50 people and shows S-shaped growth to around 750 people. The growth slows in a gradual manner as the sales personnel experience reduced effectiveness when they crowd into the same area. Figure 18.7 shows a new version of the sales model to allow a lag in the effect of the number of salespersons on their effectiveness. The new relationship is emphasized by the arrow marked with a \\ to remind us of the delay. The new model assumes that it takes time for the congestion in salespersons to translate into a reduction in their individual effectiveness. A third-order SMOOTH is used to represent the lag based on the thought that the effect is not felt immediately after an increase in the size of the sales force. The lag time is set at 1 year; the fraction to sales is increased to 0.55; and all other parameters are the same as in chapter 7.

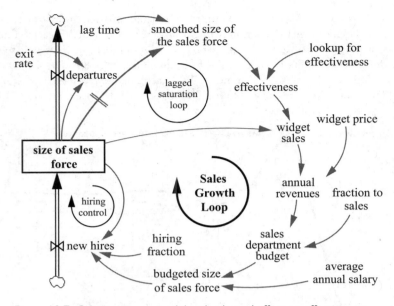

Figure 18.7. Sales company model with a lagged effect on effectiveness.

The new model generates the oscillations shown in Figure 18.8. The company is simulated to grow from 50 to over 950 and to oscillate around an average value of 825. The growth is quite rapid during the first 5 years of the simulation. (As long as each person continues to sell two widgets a day, the company will continue to rapidly expand the sales force.) New hires peak around the 7th year and then fall quickly to zero. With 20% of the sales force departing each year, the total sales force declines to below the previous peak. This allows for an improvement in the effectiveness, and the company finds itself in a more favorable financial position. New hires increase around the 12th year and push the sales force back over 825. The company will experience another decline in effectiveness, a decline in the sales department budget, and

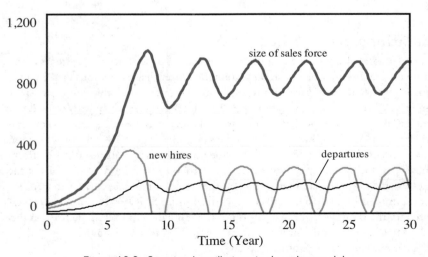

Figure 18.8. Sustained oscillations in the sales model.

a subsequent drop in new hires. Figure 18.8 shows that this oscillatory pattern would continue for the remainder of the simulation.

The oscillations in the sales model arise for the same reasons as in the flower model. The size of the sales force shoots past the equilibrium value since the signal on loss of effectiveness is not felt immediately. By the time the sales force experiences the loss of effectiveness from too many salespersons, the company has budgetary problems and new hires have been reduced to zero. With departures continuing at 20%/yr, the company will shrink in size, and effectiveness will eventually improve. But the lagged effect is felt on the way down as well as on the way up. So the company declines to below the equilibrium value before it begins to rebuild the sales force. Figure 18.8 shows the initial overshoot followed by five booms in hiring. If you study the peaks in hiring, you will see no dampening over time. Simulations that persist year after year are called *sustained oscillations*. (An example of sustained oscillations is a pendulum with no frictional forces or wind resistance to cause the oscillations to dampen out over time.)

Summarizing Oscillatory Patterns in a Parameter Space Diagram

The flowers and sales model show that the introduction of a simple time lag can create oscillations in a system that previously exhibited only S-shaped growth to equilibrium. Experiments with these models will show that the oscillations are greater when the lag times are longer or if the system is growing rapidly when it approaches the equilibrium value. The oscillations seen so far are *sustained, damped*, and *overdamped*. These arise in models with only a small number of input parameters. In the case of the flowers model, there are two important parameters: the length of the lag time and the intrinsic growth rate. With only two parameters, we could record the various patterns of oscillations in a two-dimensional *parameter space diagram* (as shown on the BWeb). The sales model contains many more parameters, but we would expect that the combined effect of the sales company rules could be translated into the equivalent of the intrinsic growth rate. If this were done, a two-dimensional parameter space diagram could help one appreciate the oscillatory tendencies of the sales model (BWeb).

Marking the Information Delays

These examples reveal the importance of delays in information feedback. Systems that were found to be quite stable in chapter 7 now show oscillations caused by the introduction of a time lag. This is a general tendency, as noted by D. H. Meadows (2009, 54). She observed that a simple delay in a balancing feedback loop can make a system more likely to oscillate. Because of the importance of time lags, it is useful to mark their location in model diagrams.

Figure 18.7 shows how this is done with model diagram for the sales company. Notice the \\ symbol on the arrow from the sales force to the smoothed value of the sales force. And for emphasis, the saturation loop has been renamed the "lagged saturation loop."

A similar symbol appears in the causal loop diagram for the flowers model in figure 18.9. Notice the // symbol on the arrow from the fraction occupied to the growth rate. This arrow is part of the negative feedback loop that controls the growth of flowers when they occupy a large fraction of the suitable area. Notice that the lagged value of the

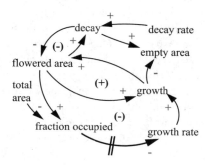

Figure 18.9. Adding a // to mark the time lag in the flowers model.

fraction occupied does not appear in this diagram. The new variable may be omitted in the interest of clarity (the // tells the story.) A comparison of figures 18.7 and 18.9 reveals that the time lags have been introduced in the same place in both systems. They both appear in the negative feedback loop that brings the system into balance. In both cases, the time lag can cause the system to shoot past the balance point and begin to oscillate.

The Importance of Delays

These examples reveal the importance of delays in understanding dynamic behavior. Both the flowers system and the sales system were quite stable in chapter 7. But the introduction of a single time lag in a control loop was sufficient to induce oscillatory behavior in this chapter. These examples teach us that relatively simple systems can exhibit qualitatively different cyclical behavior. You've seen examples of sustained, damped, and overdamped oscillations, but there are many more. Some systems are so unstable that the oscillations grow larger and larger over time. (Think of the first time you tried to bounce a basketball.) These are called *growing oscillations*. No system can grow forever, so a system with growing oscillations eventually encounters a limit. When the system comes into accommodation with the limit, the oscillations can fall in repeating pattern called a *limit cycle*. (The beating of your heart is a limit cycle.) We'll discuss examples of *growing oscillations* and *limit cycles* in later chapters. More complex cycles are described on the BWeb.

All the cycles have been shown in deterministic simulations. There were no outside disturbances of any kind, so the oscillations may be said to have originated from inside the system. Stochastic simulations assume random changes in external conditions. Some of the exercises will help you see how the random disturbances can change the pattern of oscillations.

Living in a World with Delays

The examples in this chapter illustrate useful rules of thumb for living in systems with delays. We have learned that delays in the feedback signal can be responsible for overshoot and subsequent oscillations. If we can shorten the delay, we may achieve more stable behavior. Simulations with these models will also show that the tendency for overshoot is greater when a system is approaching its limit at high speed. If the delays in the systems are fixed, slowing the growth may be the only way to avoid the overshoot and the subsequent oscillations.

These findings should make sense from your own experiences. Delays are all around us, and they pose dynamic problems. However, with experience, we have learned to manage our activities. Think of your experiences at the dinner table. Food ingestion is part of a complex physiological process involving the body's signal that replaces the feeling of hunger with the feeling of satiation. There are delays in this signal, so it is possible to overeat if we eat rapidly. However, with experience, we have learned to avoid the overshoot by eating slowly. A similar overshoot involves the intake of alcohol. There are delays in the body's absorption of alcohol. The first-time drinker will not be familiar with these delays, imbibe too much alcohol, and suffer the effects of the overshoot. These are familiar systems, and we have learned our own rules of thumb for living with the physiological delays.

But how do we manage new systems that are unfamiliar to us? We lack the everyday experiences, so there are no rules of thumb for dealing with time lags. This is where modeling can contribute. Models can help us build understanding of the effect of time lags. We can then experiment with the model to develop rules of thumb for the management of environmental systems.

Exercises

Exercise 18.1. Sales model verification

Build the sales model in figure 18.7 and verify the sustained oscillations shown in figure 18.8. Then change the length of the simulation from 30 years to 300 years. Is the company still oscillating at the end of the simulation?

Exercise 18.2. Add a trainees stock to the sales model

Expand the sales model to include a stock of trainees. The stock is fed by the flow of new hires and drained by a flow of trainees joining the sales force. Set the joining flow as the number of trainees divided by a training delay, and set this delay to 1 year. Document your results with a time graph of the size of the sales force and the flows of new hires and departures.

Exercise 18.3. Impact of training delay on the cycles

The training stock introduces a material delay in the "lagged saturation loop" shown in figure 18.7. The total delay around this loop is longer than before. So perhaps the additional delay will cause larger oscillations. Conduct three simulations with the training lag at 0.5, 1.0, and 1.5 years. Document your results with a comparative time graph of the size of the sales force. Does a longer training delay make the system more volatile?

Exercise 18.4. A stochastic simulation

Use the sampled noise structure from figure 14.8 to create a random value for the intrinsic growth rate. Set the noise interval to 1 year, the low value to 0.74, and the high value to 0.76. This will create a new random number every year, but the intrinsic growth rate will be essentially the same as the deterministic simulation. (Make sure you are using a first-order SMOOTH with a 2-year lag and verify that you get a different value for the intrinsic growth rate each year.) Your simulation should show damped oscillations in the area of flowers similar to that in figure 18.4.

Exercise 18.5. Expand the magnitude of the noise

Repeat exercise 18.4 with the low value at 0.70 and the high value at 0.80. Document your result with a comparative time graph showing the area of flowers from this simulation and from the simulation in exercise 18.4.

Chapter 19

<hr>

Cycles in Real Estate Construction

Environmental problems challenge us to understand business systems as well as environmental systems. Human systems are interconnected with the environment in complex ways that are difficult to understand. Interdisciplinary models can help us build our understanding, but the challenge is to include the human and the environmental parts of the system within the same model. We can prepare to meet this challenge by becoming familiar with models of business system. This chapter introduces examples of business models that deal with cyclical behavior. The principal example deals with the construction cycle in the real estate industry.

Real estate was selected for this case study because the problems will be familiar to many readers. You may have followed the problems from the news coverage during the past two decades. Real estate prices in many regions of the United States were soaring in the 1990s, and there was a boom in construction. But prices have fallen during the first decade of the 21st century. As of 2009, many regions are experiencing falling prices and a decline in construction. The bust phase of the construction cycle has been very difficult for families, businesses, real estate developers, and the lending industry.

This chapter describes simple models to help us understand the cyclical pattern in construction of commercial office space. The models demonstrate how construction cycles can arise from a combination of the long lead time for construction and the investment decisions of real estate developers. The model descriptions are brief; supporting information is available on the book's website, the BWeb.

The real estate models are good exercises to practice the modeling of a business system. Additional business models are described in the exercises at the end of the chapter. One exercise deals with inventory cycles in the aluminum industry, an industry with a huge appetite for electricity and a tremendous potential for energy savings from recycling. The other exercise involves the Beer Game, a multiplayer board game that illustrates the challenges of inventory control in a supply chain.

Background on Boom and Bust in Real Estate

The recent news coverage of the housing industry has made real estate problems a focus of attention. But boom and bust is nothing new for this industry. Indeed, the long history of real estate is dominated by a series of exuberant building booms and subsequent busts. The graph in figure 19.1 is based on Homer Hoyt's detailed account of Chicago real estate values (Hoyt 1933). The chart shows land values, new construction, and business activity, all scaled in percentage of variation from a normal value. Population surges occurred in Chicago several times during the century. These surges were an important factor, but Hoyt believed that the key to the boom-and-bust pattern was the way investors reacted to the population surges. Hoyt (1933, 387) observed that developers did not normally react in time to prevent land values from increasing far beyond the increase in population. The high prices then led to an exuberant response:

> *Developers scramble to build at many locations around the city, and a great many men work secretly and independently on a great variety of structures in many sections of the city. There is no central clearing house to correlate the impending supply of buildings with the probable demand, so that when all these plans came to fruition, an astonishing number of new structures had been erected.*

This overreaction sets the stage for the bust:

> *Gross rents fall, and net rents fall even faster.*
> *Land values plummet, and foreclosures are everywhere.*

Hoyt concluded his book by speculating that the "real estate cycle itself may be a phenomenon that is confined chiefly to young or rapidly growing cities." This may have been a hopeful view in 1933, but boom and bust has continued in cities like Dallas and Boston in the 1970s, as explained in texts by DiPasquale and Wheaton (1996) and Sterman (2000). And the most recent evidence of boom and bust appears across different regions of the United States in the 1990s and into the first decade of the new century.

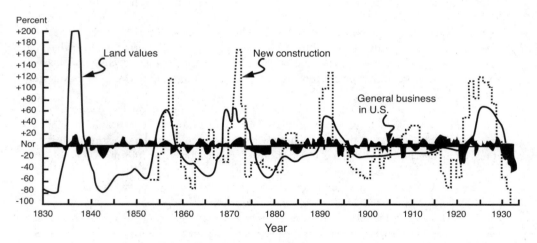

Figure 19.1. Real estate values in Chicago.

Reference Mode

Figure 19.1 gives a general idea for the target pattern for the modeling. There is a tendency for bursts of construction to appear every 15 or 20 years. The construction tends to appear after a major increase in real estate values, and it continues after prices have peaked and are on a downward trend. The construction boom leaves the city overbuilt, and prices are depressed for many years. Figure 19.1 shows boom and bust, but the cycles do not appear in clockwork fashion. However, they do return again and again, and they cause land values to swing dramatically. Land values can be 50% above normal in some years and then 50% below normal only five years later. The graph also shows that the cyclical pattern is asymmetric, with the bust period lasting around twice as long as the boom period.

Chicago was a rapidly growing city in the 19th century, but the introductory models in this chapter will work with a simpler situation. We will start by simulating a market with a fixed demand for office space. We know from chapter 18 that delays in key feedback loops can lead to cyclical behavior, so we will pay particular attention to the delays in office space development. Developers have a lot to do to translate an idea into a new building ready for occupancy. They could face a year for assessment, another year for design and procurement, and perhaps 2 years for construction. Understanding the effect of the 4-year delay could be the key to understanding the cyclical pattern of construction.

Figure 19.2 assigns separate stocks for assessment, design, and construction stages. The final stock represents the space that is available for occupancy. We'll assume an average life of 50 years before the space is subject to demolition. We'll assume a large city with the need for 100 million square feet of office space. (The space could accommodate 1 million workers if each required 100 square feet.) The flows will be measured in millions of square feet per year. Demolition is 100 million square feet divided by 50 years, or 2 million square feet/yr. It's often useful to start a model in equilibrium, so each of the flows will be 2 million square feet/yr.

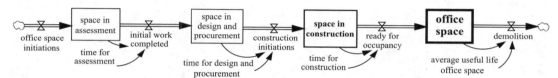

Figure 19.2. Stocks and flows for office space development and construction.

First Model: Cycles in Rent from External Factors

Figure 19.3 shows a simple model to get started. The first step is to simplify the development and construction process by focusing only on the construction phase. This simplification makes sense if there are many developers who have completed the advance work on sites and projects. Let's assume that there are many preapproved projects ready for construction if developers see profitable conditions. If this is true, the effective lead time can be reduced to a 2-year construction process. The completions flow is defined as the office space under construction

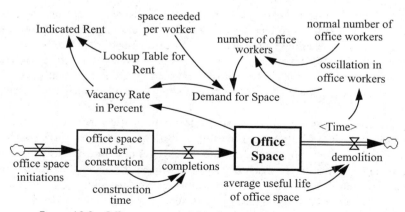

Figure 19.3. Office space model with a variable demand for space.

divided by the 2-year construction time. This model is initialized with 4 million square feet under construction. And office space initiations is constant at 2 million square feet/yr.

The demand side of the model will be the source of dynamic behavior. The model assumes a city with 0.9 million workers, each of whom requires 100 square feet of rental space. The normal demand for space is 90 million square feet. The 100 million square feet can accommodate this demand with 10 million square feet to spare. We have a vacancy rate of 10%, which is a common value in the city. The rents are measured in $/month per square foot, and we'll assume $30/month when the vacancy rate is at the normal value of 10%. If vacancies fall below 10%, the rent can rise dramatically.

Let's test this simple model with a sinusoidal variation in the number of office workers. Figure 19.4 shows a test with a 13-year cycle in the number of office workers (the top curve in the graph). The office space is fixed at 100 million square feet, so the office worker variations lead directly to variations in the vacancy rate and the indicated rent. The rent is the middle curve in the graph; it peaks in the 3rd, 16th, and 28th year. These peaks coincide exactly with the peaks in the number of office workers. This simple example is used to illustrate a cycle that originates from outside the system.

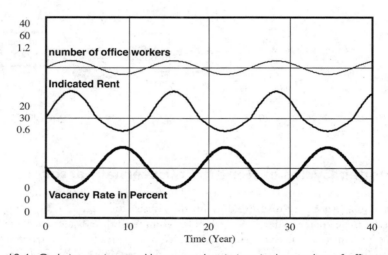

Figure 19.4. Cycle in rent imposed by external variations in the number of office workers.

External factors can certainly influence the real estate market, as explained by Hoyt. But Hoyt's description teaches us to look inside the real estate industry for a full understanding of the cycle. It was the decisions by the developers that were responsible for the repeating pattern of boom and bust. This first model focuses our attention elsewhere. Office space is totally constant, and the variations in rent are attributed to an external oscillation in the demand for space. At this stage, you might be wondering why we are off to such a poor start in the modeling.

This first model is presented here to illustrate a point of view that often dominates discussions among experts who are not accustomed to the feedback perspective on dynamic behavior. The natural tendency is to attribute the changes in market conditions to external conditions. This inclination is particularly strong when experts are making short-term forecasts of market conditions (BWeb). Our goal is different. We are looking for an explanation of the long-term pattern of boom and bust in construction. The next model turns our attention to the construction decisions of the developers.

Construction Responds to the Rent

Figure 19.5 shows a model incorporating the developers' response to rent. The oscillation in the number of office workers has been removed, and we have 0.9 million workers until some extra workers are added to disturb the initial equilibrium. The indicated rent is a nonlinear function of the vacancy rate, the same as in the previous model. Let's assume that the actual rent takes some time to fully respond to changes in vacancies, and we use a 2-year first-order adjustment process for the rent to follow the indicated value. Developers are assumed to pay close attention to the rent, but it is not clear whether they base project initiations on the current rent, the past rent, or a forecast of future rents. The model in figure 19.5 assumes that developers follow the past rents over a 2-year interval to ascertain the trend. The trend is then used to look ahead 2 years to prepare a forecast of the rent about the time that new projects would become available for occupancy. The forecasted rent is compared to the total levelized cost, which is set to $28 per square foot per month. A higher profitability forecast will lead to a higher fraction of construction initiated. The industry is assumed to have a maximum capacity of 20 million square feet of annual initiations.

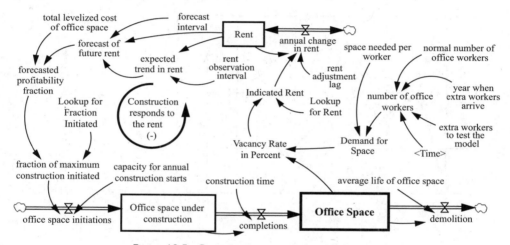

Figure 19.5. Construction responds to the rent.

Figure 19.6 shows a test simulation with the market in equilibrium for the first 5 years. The rent is constant at $30 per month, which is sufficient for developers to initiate construc-

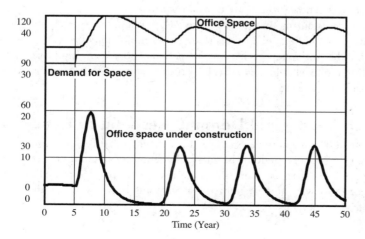

Figure 19.6. Construction cycle in the second model.

tion on 2 million square feet/yr. This is exactly the amount needed to counter the demolition of older buildings. The equilibrium is disturbed by the arrival of 0.05 million workers in the 5th year. They increase the demand for space by 5 million square feet, cutting the vacant space in half. The step increase in office workers sets the model into motion. The response takes the form of a nonlinear oscillation in construction, with peaks in construction in the 7th year and again in the 23rd, 34th and 45th years. We have a construction cycle with a period of around 11 years. The variations are highly volatile, with construction during the booms peaking around five times the average construction.

Figure 19.7 shows the changes in the vacancy rate and the rent in the second model. The simulation begins with the vacancy rate at 10% for the first 5 years. This rate is cut in half when the extra workers arrive in the 5th year. The rent increases quickly thereafter, and the developers' forecast of the rent increases even more. Vacancies increase substantially during the next few years owing to the first building boom. The vacancy rate is over 20% in the 11th year; the rent is well below the starting value; and the forecasted rent is even lower. Rents begin to increase again around the 15th as the rate of demolition exceeds the new office space completions.

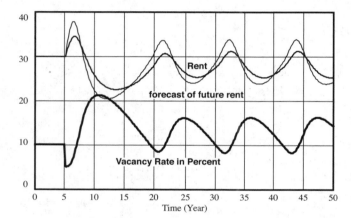

Figure 19.7. Vacancy and rent in the second model.

This second model shows the general pattern of a construction cycle. The developers' costs are fixed, and the demand for office space is fixed after the 5th year. The simulation teaches us that a long cycle in construction could arise totally from the decision-making process of the developers. The key to the cycle is the developers' tendency to continue building during the period of high prices. They are simulated to start many projects, far more than are needed to accommodate the extra workers. These projects are under construction for 2 years, and they do not exert an influence on rents during that time. Meanwhile, developers are watching the rent and preparing a trend forecast of the rent 2 years ahead. This combination of assumptions leads to a pattern of boom and bust that repeats itself with a period of approximately 11 years. The target pattern called for a somewhat longer cycle. Let's turn to a third model to see if we come closer to matching the reference mode.

Third Model: Both Supply and Demand Respond to Rent

The next model assumes that office workers are employed by companies that pay attention to rents in addition to the many other costs of doing business. Higher rents can induce the companies to make some changes in the space per worker. But the ability to respond is limited, and the companies do not necessarily want to change their way of doing business unless higher rents appear to be a permanent problem.

This combination of assumptions is implemented in the demand response loop in figure 19.8. The new model assumes that companies observe the rent over a 5-year interval. (This observation time is represented by a first-order SMOOTH, with the delay in the information marked by the \\ symbol in figure 19.8.) The observed rent is compared to a standard rent of $30/month to obtain the fractional increase. Economic models often summarize demand response with a price elasticity of demand. We'll follow this practice with a rent elasticity of –0.2. This means that a 100% increase in rent will lead to a 20% reduction in the office space needed per worker. A reduction in the demand for office space will lead to an increase in the vacancy rate and a subsequent decline in the rent. This tells us that the demand response loop generates negative feedback in the system. However, when compared to the supply response, the demand response tends to be slow (owing to the 5-year observation interval) and weak (due to the –0.2 elasticity).

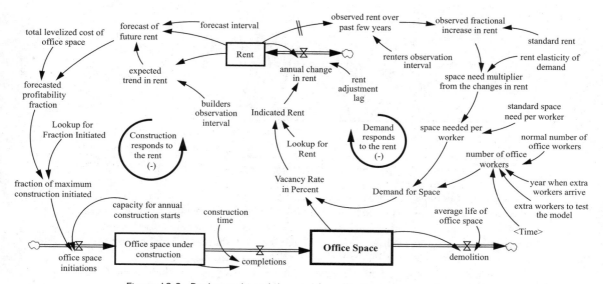

Figure 19.8. Both supply and demand for office space respond to rent.

Figure 19.9 shows the response of the new model to the step increase in the number of office workers. The demand for space increases with the arrival of the new workers in the 5th year. The graph shows minor variations in the subsequent demand in the remainder of the simulation. The effect of the demand response on the larger system is evident from a comparison of the construction in this graph with the construction in figure 19.6. The comparison reveals the construction cycle is slightly less volatile in the new model. The inclusion of a demand response has moderated the construction cycle slightly, but the overall pattern is much the same as before.

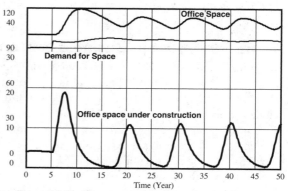

Figure 19.9. Construction cycle in the third model.

A Sensitivity Test

The third model contains a wide variety of parameters that are highly uncertain. It's useful to experiment with changes in parameter estimates to see their impact on the construction cycle. Figure 19.10 focuses on the length of the forecasting interval. The graph compares the office space under construction in two simulations. The thin curve is the previous result from figure 19.9. It assumes that developers look at the trend in rent and use that trend to forecast rent 2 years into the future. The thick curve is from a simulation with the forecasting interval set to zero. In other words, the developers base their profitability assessment on the current rent rather than on a forecasted rent.

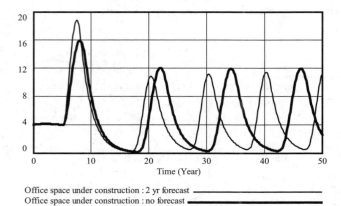

Office space under construction : 2 yr forecast ————————
Office space under construction : no forecast ████████████

Figure 19.10.
Testing the importance of the forecasting interval.

The two simulations show the same construction during the first 5 years. Then the extra workers arrive, and the model responds with an initial boom in construction. The initial boom is larger in the simulation with a 2-year forecast. This is to be expected, because the developers are extrapolating the uptrend into the future. Their rent expectations are higher, and they initiate more construction than if they were looking only at the current rent.

The first boom is triggered by the arrival of the extra workers, but the next boom arises from the subsequent decision making by the developers. Forecasting leads developers to get started earlier with the second boom. There is overbuilding, but it is less extensive than the new simulation. The new simulation (with developers focusing on the current rent) shows a longer delay before they initiate construction in the second boom. Once they get started, there is a larger boom in construction. This sets the stage for a longer interval before the start of the next building boom. The period of the new construction cycle is around 13 years. This is more in line with the reference mode established at the start of the chapter.

Boom and Bust in Business Systems

From the recent news coverage of real estate, you might conclude that the problems of boom and bust are unique to the real estate industry. But real estate is only one of many industries that become trapped in cycles of boom and bust. Other industries with cyclical problems include aircraft construction, shipbuilding, and semiconductors. The commodity industries have their own version of investment cycles. Examples include agricultural commodities (e.g., pork, beef, coffee); forest products (e.g., pulp, paper, lumber); and metals (e.g., aluminum, copper, zinc). Indeed, the examples of cyclical behavior in business systems are so numerous that a listing could span the alphabet, from aircraft to zinc, as explained by Sterman (2000, 792).

Ideas from previous chapters proved useful in the development of the real estate models in this chapter. Insights from the real estate models are transferable as well. Capital-intensive industries with long lead times are especially vulnerable to boom and bust. The electric power industry is an example. The regulatory rules for this industry were changed substantially in the 1990s, and problems of blackouts and price spikes followed shortly thereafter in the western United States. System dynamics has been used to help us understand the dynamics of the restructured power industry (BWeb). The models revealed that construction of power plants would fall into a cycle of boom and bust for fundamentally the same reasons as seen in commercial real estate (Ford 2002). This analysis helps us understand the reasons for the California electricity crisis of 2000–2001 and whether the crisis conditions could return in the future.

Exercises

Exercise 19.1. Real estate construction cycle

You can experiment with the real estate models by taking advantage of the BWeb materials. A good way to learn is to build the models on your own and then compare what you have done with the downloadable models. You can then use the models to conduct additional sensitivity tests. The BWeb provides additional models to focus on the developers' tendency to continue building when there is an extraordinary amount of office space already under construction. This tendency is key to the overbuilding, which sets the stage for the bust phase of the boom-and-bust cycle.

Exercise 19.2. Supply chain challenges in a classroom game

The BWeb provides a set of exercises connected with the product distribution game, known as the Beer Game. The game provides a fun way to learn about system modeling and systems thinking. (The Beer Game is probably the most popular pedagogic device in the field of system dynamics.) The first version of the game was created in the 1960s to demonstrate principles of supply chain management. Players aim to meet customer demand by ordering product in the multistage supply chain in figure 19.11. Their goal is to minimize inventory costs, but to still maintain sufficient inventory to fill incoming orders.

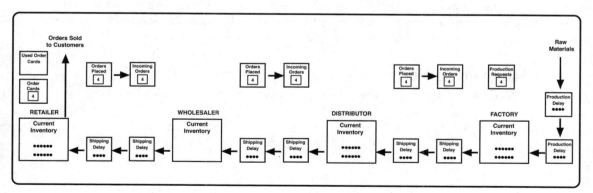

Figure 19.11. Initial conditions for the Beer Game.

The game has been played by teams of undergraduate students, graduate students, teachers, and business executives. The intriguing result is that all teams end up with problematic behavior. This comes as quite a surprise to the players, since there are no disruptions of supply (e.g., owing to strikes or factory problems), and the factory can deliver any amount of beer, no matter how large the orders. But there are delays for the beer to travel through the supply chain. Dealing with these delays is quite a challenge, even for seasoned business leaders participating in executive education programs. The BWeb provides instructions for ordering the game, and it explains a series of models to simulate the problematic behavior that arises in the game.

Exercise 19.3. Inventory cycles in the aluminum industry

Aluminum smelting is a commodity industry with highly volatile prices. Figure 19.12 shows the key stocks and flows in a model from the first edition of *Modeling the Environment* (BWeb). The model generates a limit cycle in aluminum prices and smelter utilization owing to the delays in the supply and demand loops (figure 19.13).

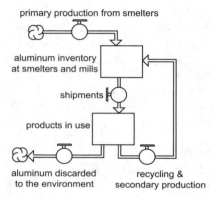

Figure 19.12. Main stocks and flows in the aluminum model.

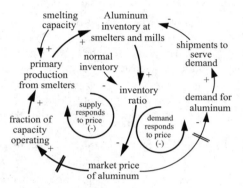

Figure 19.13. Supply and demand respond to price in the aluminum model.

The model operates with a fixed amount of smelting capacity (because new smelters are often linked to electricity development rather than aluminum prices). The cycle in aluminum prices is said to be an inventory cycle, since prices rise when inventories are low, and they fall when inventories are high. The BWeb materials provide an extensive collection of exercises to practice simulating the cycles in this industry.

Acknowledgment

• The graph of real estate values in Chicago in figure 19.1 is adapted from Hoyt (1933).

Chapter 20

Cycles in Predator and Prey Populations

The relationship between predators and their prey has always occupied a special place in the minds of ecologists. Indeed, there has probably been more written on the subject of predator-prey interactions than on any other single topic in ecology (Matson and Berryman 1992). Ricklefs (1990, 403) describes "predation as a clean demographic event that readily lends itself to modeling." Some predator-prey systems can be brought into the laboratory and subjected to experimentation. The experiments reveal dynamic patterns ranging from stable oscillations to violently unstable behavior.

The great interest in predator-prey systems makes them an ideal case for system dynamics. This chapter demonstrates how system dynamics may be used to examine the interaction between predators and the deer herd on the Kaibab Plateau in northern Arizona. This particular deer herd is famous for a major irruption early in the 20th century. This chapter looks at the possible dynamics of the deer and predator populations if the predators had not been exterminated in the early years of the century.

Background on Predator-Prey Cycles

One of the most famous predators is the lynx, a medium-sized cat that can prey on the snowshoe hare. The lynx-hare interactions are well known because of the value of their pelts to the Hudson Bay Company (Elton 1933). Company records document the long history of the ups and downs in the sale of pelts, and they show a peak in the number of lynx pelts every 9 to 10 years. If we assume that the number of pelts is a good measure of the relative size of the population, one would conclude that these two populations have oscillated in a cyclical manner for over 100 years.

Odum (1971, 191) describes the hare-lynx cycle as an example of cycles "which involve regular oscillations or cycles of abundance with peaks and depressions every few years, often occurring with such regularity that population size may be predicted in advance." Odum explains that the best-studied examples among the mammals are cycles with a 9- to 10-year or a 3- to 4-year periodicity. The hare-lynx is a "classic example" of a 9- to 10-year oscillation. The shorter 3- to 4-year cycle is said to be characteristic of many northern murids (lemmings, mice, voles) and their predators (such as the snowy owl).

But predator-prey populations to do not always interact in a stable manner. Odum describes experiments by Pimentel (1968) in which houseflies and parasitic wasps were first placed together in a limited culture system. The intriguing results arose from newly associated populations that were brought from the wilds and inserted into the controlled laboratory experiment. These experiments showed that the populations would oscillate "violently." The population of flies increased sharply, followed by a sharp increase in the population of wasps. Their interactions were highly unstable, leading to a crash to zero. However, the experiments also showed that it was possible for the flies and wasps to interact in a stable manner. Test populations derived from colonies in which the two species had been associated for two years could coexist in the experimental environment. Odum (1971, 222) believes that these experiments are instructive for natural systems where there are

> hundreds of similar examples which show (1) that where parasites and predators have long been associated with their respective hosts and prey, the effect is moderate, neutral, or even beneficial from the long-term view and (2) that newly acquired parasites or predators are the most damaging.

What Is the Reference Mode?

Little is known about Kaibab populations prior to the start of the 20th century. Nevertheless, it's important to specify a reference mode. Let's draw the target pattern as cyclical behavior in the populations with a period of around 10 years. We would expect to see a peak in the deer population followed a few years later by a peak in the predator population. But how would the cycles change over time? For example, should the reference mode show sustained or damped oscillations?

Indian artifacts indicate that the deer and predators have coexisted on the Kaibab Plateau for many decades. So it makes sense to assume the populations have had ample opportunity to coexist with one another. Let's set the target pattern as oscillations that persist over time. The model should exhibit either damped or sustained oscillations in the absence of random disturbances. And the cycles should persist over many years in stochastic simulations. We should expect oscillations with a period of around 10 years.

Background on the Deer and Their Predators

The Kaibab deer is a Rocky Mountain mule deer. They attain reproductive maturity at about one and a half years of age. Twins are very common, and occasionally triplets are seen. Russo (1970) reports that there is little information available on the total size of the mule deer population prior to 1906. He cites Rasmussen's (1941) estimate of around 4,000 deer at the turn of the century. You'll learn in chapter 21 that sheep and cattle were introduced and that some of the livestock were taken by predators. But we'll focus our attention on the cycles in deer population prior to the introduction of livestock. The deer population could be represented by stocks in the model, but how many stocks make sense? Perhaps there should be separate stocks for the fawns, the adults, and the senescent. And perhaps we should assign separate stocks for males and females.

A model with so many stocks might be needed for some purposes, but the best way to start is with a much simpler model. Some veterans remind themselves to "keep it simple, stupid" (KISS). Let's follow this advice by assigning a single stock to the entire deer population. The flows will be births, deaths due to old age, and deaths due to predation. (We can ignore migration, since the Kaibab is an isolated plateau.)

The Kaibab was home to populations of coyotes, bobcats, mountain lions, and wolves at the start of the 20th century. Data on the size of the predator numbers are lacking, so estimates of the different populations will be highly uncertain. The principal predator is the mountain lion (also known as a cougar). Following the KISS advice, we will adopt a single stock to represent the combination of predator populations. But how many predators makes sense when we start the model? Ricklefs (1990, 436) describes surveys of a variety of predator-prey systems. The prey populations are much larger than the predator populations, but the prey/predator ratio can vary greatly from one study to another. A typical ratio for the Kaibab situation is around 80 prey for each predator. So we will proceed on the assumption that there were around 50 predators coexisting with the 4,000 deer at that start of the 20th century.

Thinking about Equilibrium

Although the populations are seldom in equilibrium, it's helpful to become familiar with the numbers by imagining an equilibrium situation. Figure 20.1 shows an example with 50 predators and 4,000 deer. The deer population is subject to a flow of births, deaths, and predation. If each predator were able to kill 40 deer per year, predation would be 2,000 deer per year. If the deer live around 14 years (absent predation), the death rate would be 0.07/yr, and the deaths would be 280/yr. Let's suppose that 50% of the deer are female; that around 70% of the females are fertile; and that the average litter size is 1.6. These factors combine to give a birth rate of 0.57/yr, and there would be 2,280 births per year. The deer population is in equilibrium.

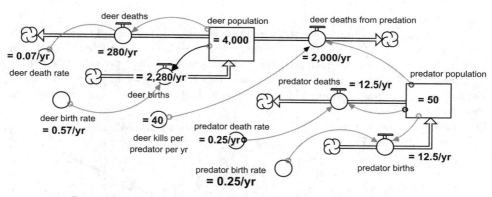

Figure 20.1. Imagine an equilibrium with 50 predators and 4,000 deer.

The predators' birth rate and death rate are set at 0.25/year, and this population is also in equilibrium. The death rate makes sense if the average predator lives 4 years. (A cougar could live 10 years or longer with ample prey, so the 4-year life span represents a difficult situation.) Cougar females can become fertile after about 2 or 3 years. They have a litter every other year, and the average litter is 3 kittens (Armstrong 1987). With a life span of 4 years, only around a third of the females would be fertile. This combination of assumptions leads to the birth rate of 0.25/yr.

Figure 20.1 begs the question of whether the predators would be able to achieve 40 kills per year per predator. The success of predators is thought to depend on the density of prey. With higher density, the cougars are better able to find the prey and attack in a successful manner. The Kaibab Plateau is one of the largest and best-defined "block plateaus and one of the very few that is bounded on all sides by escarpments and slopes which descent to lower lands" (Rasmussen 1941). The density of prey at the start of the 20th century is thought to have been 4,000 deer spread over 800,000 acres. So we will start with a density of 5 deer/1,000 acres. With higher density, deer kills per predator should increase. And with higher deer kills, we should see an increase in the predator birth rate and a reduction in their death rate. This combination of assumptions is implemented in the initial model

An Initial Model

Figure 20.2 shows an initial model with the deer kills depending on deer density. The diagram also shows that we have simplified the deer flows by combining the births and deaths into a bi-flow called "deer net births." The deer net birth rate would be fixed at 0.50/year.

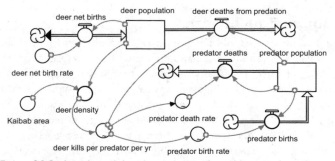

Figure 20.2. Initial model with predation depending on deer density.

The diagram shows three converters with the ~ symbol. Each of these is represented by a nonlinear relationship. Figure 20.3 shows that higher deer kills are possible at higher deer density. At the right extreme, each predator is assumed to kill 60 deer per year if the density is 10 deer/1,000 acres. This is somewhat greater than 1 kill per week, which is what will be assumed as the satiation limit. The midpoint is highlighted to remind us of the equilibrium discussion. This model assumes 40 kills/yr per predator if the density is 5 deer/1,000 acres. The opposite end of the graph shows zero kills when there are no deer to be found.

The end points describe the extreme situations. At one extreme, there is satiation; at the other, there are no kills if there are no deer. But what about the shape of the graph between the extremes? Do you think the shape is reasonable? At this point, we should remind ourselves that the predator is a very elusive animal, so there is little direct information to help us specify the particular shape. You should view figure 20.3 as simply one possible description of the preda-

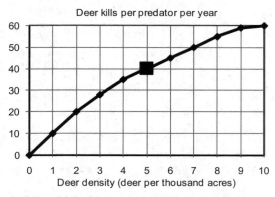

Figure 20.3. Predation depends on deer density.

tors' success at different densities. This particular shape corresponds to a combination of *Type I and II functional forms* used in studies of predation (Watt 1968; Taylor 1984; Pratt 1995; Hastings 1997).

Figure 20.4 shows nonlinear relationships for the predators' birth rate and death rate. The diagram highlights the equilibrium values discussed previously. We will assume that the two rates are 0.25/yr when the kills are 40 deer per year per predator. The right side of the graph represents the situation with high kills. The death rate is 0.10/year (which corresponds to a life span of 10 years). With such a long life, around 75% of the females would be fertile, and the birth rate would be 0.55/yr. The predators' growth rate would be 0.45/yr. This rate of growth is nearly as high as the deer growth rate. We are dealing with two populations with the potential to respond strongly under favorable conditions. Absent predation, the deer population can grow at 50%/yr; with plenty of prey, the predator population can grow at 45%/yr.

The left side of figure 20.4 shows the difficult situation with low kills. If the kills per predator fall to 10 per year, for example, the death rate would increase to 50%/yr (corresponding to a life span of only 2 years). With such a short life, cougars would not live long enough to have kittens, and the birth rate would be zero. With zero kills, the predators would be forced to hunt for inferior prey and would suffer the high death rate of 70%/yr.

You should be able to build and test the initial model from the information given so far. The simulation will start with 4,000 deer that experience net births of 2,000 deer/yr. The predation will be 40 kills per predator times 50 predators, so the deer population

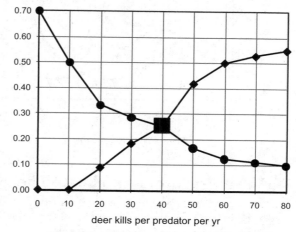

Figure 20.4. Predators' birth and death rates.

will be in equilibrium. The density will remain at 5 deer/1,000 acres; deer kills will remain at 40 per predator per year; and the predators will experience a birth rate and death rate of 0.25/yr. The predators will also remain in equilibrium. The simulation will show that both populations are in equilibrium; but do you think the equilibrium is stable? One way to find out is to start with slightly different values for the initial population.

Simulating the Initial Model

Figure 20.5 shows the results if we start with 40 predators. Time starts in the year 1900; this is strictly for convenience, and the simulation does not represent events on the Kaibab Plateau during this time period. Instead, it shows the first attempt to describe what might have happened had the wild populations not experienced the disrupting effects of hunting, trapping, and livestock grazing. Figure 20.5 shows that the deer population grows initially because the predators are not numerous enough to keep them in check. As the deer density increases, the predators become more successful in predation, and their population grows during the first 6 years. The larger population of cougars then causes the deer population to decline. The predator population peaks around the year 1906 and begins to decline as well.

Figure 20.5 shows both populations declining during the interval from 1908 to 1912. By then, the predator population is low enough to allow the deer population to grow again. It reaches 14,000 by the year 1921. These high numbers lead to higher success by the predators, so their numbers begin to grow rapidly after 1916. By the year 1924, the predators soar off the chart. This large population is simulated to drive the prey population to zero. This simulation shows that the deer population would be eliminated by around the year 1924; the predator population would be eliminated a few years later.

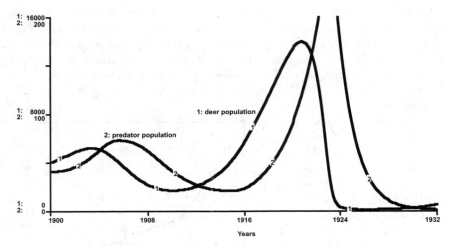

Figure 20.5. Simulation results from the initial model.

Do Wild Predators Annihilate Their Prey?

You might be wondering if these results are a coincidence of the initial values assigned to the predator population or the deer population. Or you might wonder if the unstable behavior could be attributed to the high birth rate of the deer or to the high birth rate of the predators. These sensitivity tests are left as exercises. The tests will reveal a consistent pattern of unstable behavior. In each simulation, the deer population will eventually be driven to extremely small numbers. You might then see a mathematical rebound, but wild populations cannot survive when their numbers fall below a minimum viable number. The sensitivity tests will all show the same general result—the two populations interact to create explosive oscillations that would lead to their extermination.

These results raise an important question, one considered by Watt (1968, 134):

> *Why don't predators annihilate their prey? The big cats and other large predators are impressively efficient machines of destruction, so one would assume that they would denude the landscape of food, and then die from starvation.*

Watt explains that predators do not normally hunt the prey population to zero. Rather, they have been observed to select those individuals from the prey population that have the least chance of escape and are easiest to catch. (i.e., the young, the old, and the weak). Pratt (1995, 76) describes some predators as killing nearly all of a prey in excess of some threshold number. If the prey density falls below the threshold, the predators would no longer find it profitable, and they switch to a different prey. Pratt alerts us to the possibility that the threshold level may be determined by the availability of prey hiding places and the prey's social behavior.

The importance of hiding places is confirmed in various experiments in which prey are provided with additional refuge or in which extra time requirements are imposed on the predators. Hastings (1997, 66) describes Huffaker's (1958) experiments where predatory mites search for prey mites on 40 oranges in a grid pattern. By arranging the oranges in different patterns, Huffaker was able to obtain both stable and unstable cycles in the mite populations. These results suggest that the landscape of the Kaibab Plateau is the key to stability. If the deer are better able to find refuge at low density, they could avoid the extermination dynamic. And since we think the deer and their predators have coexisted for many years, it makes sense to revise the model.

Revising the Model

From the previous discussion, you might think that we should expand the model to include the effect of prey hiding places, prey social behavior, threshold levels, and distinctions between strong and weak prey. Perhaps we should add a variable to account for the number of hiding places, another variable for the threshold level, and still another variable to account for the social behavior that allow the deer to avoid predation. The preceding discussion might lead one to expand the deer population as well. Rather than one stock, we could include three stocks to simulate the deer age structure. The expanded model could show young deer maturing to mature deer and mature deer aging to older deer. It could then be used to simulate predators that concentrate on the young and old deer.

Adding these many factors would greatly increase the complexity of the model. At this stage, it would be more useful to learn if the combined effect of these factors could be represented in an implicit fashion. Let's experiment with a change in the shape of the predation curve to account for a landscape that allows refuge for the deer at low density. Figure 20.6 shows an example with the same end points as before. The satiation limit is 60 kills/yr per predator, and there are zero kills if the density is zero. The two graphs also share the same result with 5 deer/1,000 acres. The new shape in figure 20.6 shows the largest differences at low density. With a density of 2 deer per 1,000 acres, the new assumption is for 8 kills per predator per year. This is well below the 20 kills with the previous shape. This difference may be taken as a sign that deer are better able to find refuge and that the predators may be switching to alternative prey. The new shape shows a steeper response as the density increases from 2 to 5 deer per 1,000 acres. This new shape corresponds to the *Type III function* often used to describe predator behavior (Watt 1968; Taylor 1984; Pratt 1995; Hastings 1997).

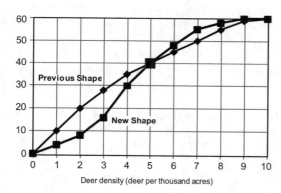

Figure 20.6. Alternative shapes for the predation function.

Figure 20.7 shows the simulated behavior of the new model. We begin with 4,000 deer and 100 predators, so the populations are far from equilibrium at the start of the simulation. The large predator population causes a decline in the deer population during the first few years, and the predator population declines quickly thereafter. By the year 1905, there are only 36 predators on the plateau. Their low number allows the deer population to grow, and it reaches a peak of around 5,000 by the year 1908. The higher deer density improves the predator situation, and their numbers grow to a peak of around 70 shortly after the year 1911. The predator population peaks again in 1922, and for a third time in 1932. These peaks occur around 2 years after peaks in the deer population We have a damped cycle with a period of 10 years. As the cycles fade in magnitude, the populations find their way to the equilibrium values discussed previously. There would be 4,000 deer and 50 predators.

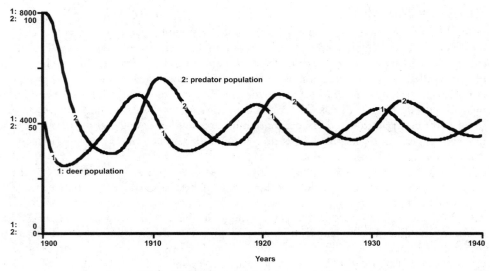

Figure 20.7. Simulation with the new predation function.

Further Sensitivity Testing

These two simulations point to the importance of the predation function, especially the kills assumed for low values of the deer density. Table 20.1 lists the two shapes simulated so far. Let's continue the testing with the third shape in the table. It shares the same extreme points; there are no kills at zero density and 60 kills at high density. The third shape also shares the same result at the midpoint. The sigmoid pattern is more pronounced with the third shape, so the kills are lower at low density. The prey have greater refuge, and we would expect to see the third shape lead to a more damped pattern of oscillations.

Figure 20.8 confirms this expectation. The simulation starts with 100 predators and 4,000 deer, the same initial values as the previous simulation. But this new simulation shows the two

Table 20.1. Deer kills with different predation functions.

Deer density	1st shape	2nd shape	3rd shape
0	0	0	0
1	10	4	0
2	20	8	2
3	28	16	10
4	35	30	25
5	40	40	40
6	45	48	50
7	50	55	58
8	55	58	60
9	59	60	60
10	60	60	60

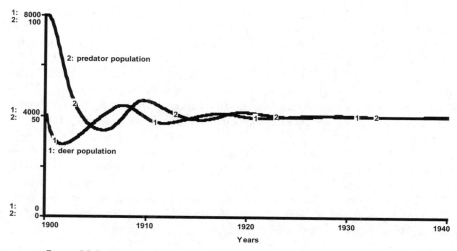

Figure 20.8. Heavily damped oscillations with the third shape for predation.

populations drawn quickly toward the equilibrium values. After only 20 years, we see around 4,000 deer and 50 predators. This is a very heavily damped system.

Stochastic Simulation

Random factors can sometimes alter the pattern of behavior compared with the deterministic simulations. You've seen examples in figure 14.6 (flowered area) and figure 15.6 (salmon). The Kaibab deer and the predator population were also exposed to many changes that are not represented in the simulations shown so far. If we take their effects to be random, we can repeat the simulation to learn if randomness will alter the pattern of oscillations. This is done by using the sampled noise structure in figure 14.8 to represent the deer net birth rate. The values are uniformly uncertain from a low of 0.3 to a high of 0.7 with a 1-year sampling interval. You may interpret this randomness to mean that the deer are exposed to "good years" and "bad years" that are not explained in the model. Figure 20.9 shows results when predation is

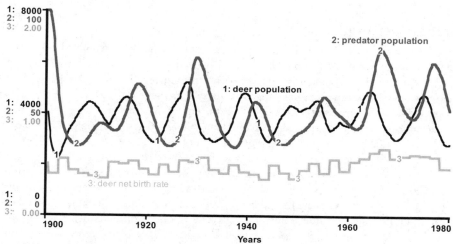

Figure 20.9. Stochastic simulation with predation governed by the second shape.

controlled by the second shape. The deer net birth rate varies each year around the normal value of 0.5/yr. The effect of the randomness is apparent by comparing the new results with figure 20.7. The previous simulation shows the populations close to equilibrium in about 40 years. The new simulation runs for 80 years, and it leaves the impression that the populations could be varying forever. This simulation teaches us that randomness can change a damped system into one with oscillations that continue indefinitely. The oscillations are much more irregular, but some fundamental patterns remain. For example, the peaks in predator populations appear approximately every 10 years, and they appear around 2 years after the peaks in the number of deer.

Figure 20.10 shows the corresponding results when predation is controlled by the third shape. These new results may be compared with the deterministic simulation in figure 20.8. It showed a heavily damped system that found its way to the equilibrium within 20 years. The stochastic simulation shows that the variations in populations are confined to a relatively small range, but there is no sign that the populations will reach the equilibrium values. This simulation also leaves the impression that the variations will continue forever.

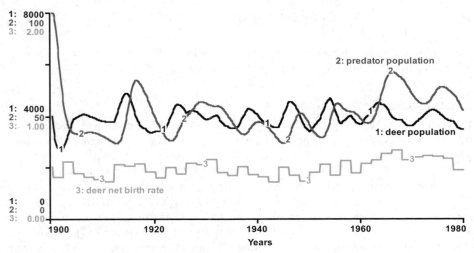

Figure 20.10. Stochastic simulation with predation governed by the third shape.

Further Testing

The stochastic simulations show the general pattern of behavior that we defined at the outset of the modeling exercise. At this point, we have a plausible explanation of the dynamics that might have occurred if the predators had not been exterminated. The model matches the reference mode, so it makes sense to conduct sensitivity tests to see if the pattern of oscillations are robust. These tests are among the exercises at the end of the chapter.

Policy testing is normally the final step in the modeling process. This chapter does not include policy simulations, but you can explore an example of policy testing on the book's website. The BWeb provides exercises to focus on the stability of the predator-prey system if wildlife managers allow hunting to remove an excess number of predators. The question is how much hunting could be permitted before the intrusion eliminates the inherent stability of the system. The BWeb also provides exercises to expand the model to include a second prey population or a second predator population. These new populations will then be in competition with the populations in the existing model. The challenge is to learn if the expanded model confirms the competitive exclusion principle. And the BWeb provides exercises for disaggregating the stock of deer to represent the fawns, the adults, and the elderly. The challenge is to represent the predators' preference to prey on the young and elderly population.

Conditions on the Kaibab Plateau changed dramatically near the turn of the 20th century. Thousands of sheep and cattle were introduced in the 1880s, and predators were largely exterminated by 1920. These changes led to an irruption of the deer population and a subsequent die-off due to starvation. The story of the Kaibab irruption is the subject of chapter 21.

Exercises

Exercise 20.1. Verify annihilation results

Build the model in figure 20.2 and simulate it with DT = 0.125 year. Remember to start with 4,000 deer and 40 predators. Document your results with a time graph to match figure 20.5.

Exercise 20.2. Sensitivity to initial number of predators

Set the initial number of predators to 45 and repeat the simulation in exercise 20.1. Do you still get annihilation? Set the initial number of predators to 48 and simulate again. Do you still get annihilation?

Exercise 20.3. Sensitivity to initial number of deer

Set the initial number of predators to 50 and the initial number of deer to 4,100. Do you get annihilation? Set the initial number of deer to 3,900 and simulate again. Do you get annihilation?

Exercise 20.4. Verify damped oscillation

The model in exercise 20.1 can also generate damped oscillations. Change the predation function to match the second shape in table 20.1 and simulate with DT = 0.125 year. Remember to start with 4,000 deer and 100 predators. Document your results with a time graph to match figure 20.7.

Exercise 20.5. Size of DT

Repeat the previous simulation with DT = 0.25 year. Then repeat the simulation with DT = 0.0625 year. Do you get the same results? What is your recommended value of DT for numerical accuracy?

Exercise 20.6. Sensitivity to the deer net birth rate

Ricklefs (1990, 507) suggests that "higher prey production" could tend to destabilize predator-prey models. Test whether this applies to the model from exercise 20.4. Repeat the simulation with the deer net birth rate at 0.5/yr. Then conduct simulations with the net birth rate at 0.6/yr, 0.7/yr, and 0.8/yr. Document your results with a comparative time graph of the deer population. Do you see more volatile oscillations in the new simulations?

Exercise 20.7. Name the loops

Figure 20.11 shows six feedback loops in the model diagramed in figure 20.2. Verify that the loops are correctly labeled as positive or negative feedback. Then assign names to each of the loops that summarize their role in the system. (Think of short names, like the names in figures 15.5, 9.18, 9.19, and 23.7.)

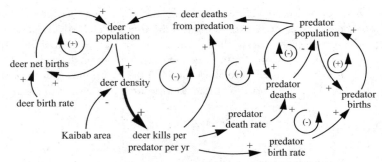

Figure 20.11. Causal loop diagram for the predator-prey model.

Exercise 20.8. Causal loops and predation

The bold arrow in figure 20.11 draws our attention to the link from density to deer kills. The + means that higher density leads to more deer kills per predator per year. It also means that lower density leads to fewer kills per year. This + label makes sense when we use the first shape for the predation function. Does it make sense for the second function? Does it make sense for the third function?

Exercise 20.9. Discussion of feedback

A causal loop diagram often provides insight on the dynamic behavior of the system (as when a positive loop alerts us to look for exponential growth). Does the causal loop diagram in figure 20.11 alert you to whether the predators and prey populations will end up annihilating one another?

Exercise 20.10. Verify the point attractor

Repeat exercise 20.4 to get damped oscillations. Then add a scatter graph to verify the results in figure 20.12. The number of deer is on the x axis, and the number of predators on the y axis. The simulation begins at (4000,100) and ends up at (4000,50). These diagrams are sometimes called *state-space diagrams*. This diagram shows the model's progress toward the equilibrium point. It's as if the system is attracted toward the center point in the diagram. Scientists sometimes refer to this point as the *point attractor* of the system (Gleick 1988, 134).

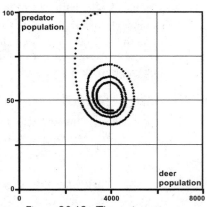

Figure 20.12. The point attractor.

Further Reading

- For a historical perspective, look to Lotka (1925) and Volterra (1926) for examples of modeling before the use of computer simulation. They were the first to use a mathematical model to describe cycles in predator prey populations. Their models were restricted to a form amenable to an analytical solution. This limited analysis led to populations that oscillated in a closed cycle.
- Swart (1990) examines and extends the classic Lotka-Volterra model.
- Ricklefs (1990, 416) discusses the oversimplification of closed cycles.
- The deer and predators of the Kaibab Plateau are described by Russo (1970).

Chapter 21

The Overshoot of
the Kaibab Deer Population

The overshoot is the fifth dynamic in the panel of six shapes in figure 21.1. It resembles S-shaped growth in the early stage, but the system does not achieve a smooth accommodation with its limited resources. Systems that shoot past their limits are vulnerable to a collapse if the resources are damaged by the excessive growth. You've read about overshoot in urban populations in chapter 1 (Forrester 1968) and in fisheries in chapter 15. The overshoot pattern in coupled human-natural systems often arises from excessive development. Examples include too many boats in a fishery; too many livestock on grazing land; and too much irrigation infrastructure in a watershed (Cavana and Ford 2004). An understanding of the propensity for overshoot is crucial if we are to achieve sustainable management of human and natural systems.

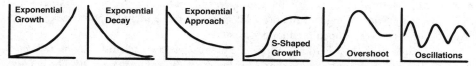

Figure 21.1. Six shapes to represent the dynamic patterns simulated in the book.

This chapter describes the overshoot in the deer population of the Kaibab Plateau. This overshoot has been attributed to human disturbance in the forest ecosystem. The Kaibab story has been told and retold, interpreted and reinterpreted, as scientists search for the best lessons from this event at the turn of the 20th century. The Kaibab deer herd is described in chapter 20, and we'll build from the previous modeling. The purpose is to illustrate the iterative nature of the modeling process. This is a long chapter that does not reach its destination immediately. Along the way, you will come to appreciate the process of iterative modeling.

Background on the Kaibab Overshoot

The ecological term for a sudden increase in an animal population is *irruption*. The term can also refer to a sudden increase followed by a decline. For example, Caughley (1970) defines a population irruption as an "increase in numbers over at least two generations, followed by a marked decline." This chapter uses the words *irruption* and *overshoot* interchangeably.

The overshoot of the Kaibab deer herd is widely known among ecologists, primarily because of the initial report by Aldo Leopold (1943). He believed the Kaibab irruption to be the first of a series of irruptions "which have since threatened the future productivity of deer ranges from Oregon to North Carolina, California to Pennsylvania, Texas to Michigan." He reported having found no record of a deer irruption in North America "antiquating the removal of deer predators. Those parts of the continent which still retain the native predators have reported no irruptions." Leopold interpreted this as circumstantial evidence to support "the surmise that removal of predators predisposes a deer herd to irruptive behavior" (Leopold 1943). He viewed the irruption as the result of human intervention that upset the balance of natural forces in the ecosystem. His views were explained in the essay "Thinking Like a Mountain" (Leopold 1949). His views had a strong intuitive appeal at the time (Zeoli 2004), and their appeal continued for the remainder of the century (Botkin 1990). Indeed, Leopold's interpretation of the Kaibab irruption was accepted and taught in textbooks on general and animal ecology (Alee et al. 1949; Andrewartha 1961; Odum 1971). Since that time, the Kaibab story has become more widely known as scientists interpret the cause of the irruption (Lack 1954; Kormondy 1969; Caughley 1970; Odum 1971; Botkin 1990).

This chapter uses system dynamics to help us understand the dynamics of this important historical event. The goal is to explain the irruption of a deer population following the extermination of predators. The modeling draws principally on the ideas first put forward by Aldo Leopold.

The Kaibab Plateau

The Kaibab Plateau is located on the north rim of the Grand Canyon. It extends around 60 miles north and south and is around 45 miles wide at its widest point. Rasmussen (1941) describes the area as one of the largest and best-defined "block plateaus and one of the very few that is bounded on all sides by escarpments and slopes which descend to lower lands." The plateau appears to the distant eye as an isolated tableland. John Wesley Powell called it the Kaibab, a Paiute Indian name meaning "mountain lying down." The plateau supports a wide variety of vegetation, with dramatic changes at different elevations. The vegetation types include shrubs, sagebrush, grasslands, piñon-juniper, and spruce-fir. The piñon-juniper woodlands provide the winter range for the deer. The summer range includes ponderosa pine and spruce-fir forests as well as open mountain grasslands.

The Kaibab deer are a Rocky Mountain mule deer, known scientifically as *Odocoileus hemionus*. Their reproductive and life characteristics are described in chapter 20. The herd was

Sketch 21.1.
Deer in Kaibab country.

thought to be around 4,000 deer at the start of the 20th century. Although information on the deer population is sparse, there are extensive records on the livestock that were first introduced to the plateau in the 1880s. Russo (1970) estimated large numbers initially, around 20,000 cattle and 200,000 sheep. There was extensive overgrazing and subsequent changes in the composition of vegetation (Zeoli 2004, 2). However, by 1907, there were only around 8,000 cattle and 10,000 sheep on the plateau, and the livestock numbers declined gradually thereafter.

The Kaibab Plateau was also home to populations of coyotes, bobcats, mountain lions, and wolves. Data on the size of these predator populations are lacking, but we do have information on predator kills (Russo 1970, 126). During the interval from 1907 to 1923, for example, estimates of predator kills included 3,000 coyotes, 674 mountain lions, 120 bobcats, and 11 wolves. It is believed that the wolf was exterminated in these years, and that the mountain lion population was greatly reduced. The predators were indigenous to the Kaibab Plateau. Nevertheless, their removal by hunting and trapping was based on a general consensus that predator control was beneficial. Russo (1970, 127) observes that "it was in reality a blanket policy that may have grown from an idea of protecting desirable wildlife by eliminating its enemies."

Sketch 21.2.
Exterminated predators.

The deer population grew rapidly around the time that the predators were removed. By 1918, there was recognition that the large number of deer was beginning to influence the condition of the forage. Russo (1970) explains that continued bad reports in 1920 and 1922 led to the formation of a special committee, and a reconnaissance party spent 10 days on the plateau in August 1924. They reported that it was common to see over 100 deer in a day's drive. One member reported seeing over 1,000 deer along a 26-mile highway leading to the rim of the Grand Canyon. The committee was not sure of the size of the population, but "all local witnesses examined placed the number of deer in the Kaibab Forest at not less than 50,000." Rasmussen (1941) estimated the 1924 deer population at around 100,000.

The reconnaissance party observed that the forage "can only be characterized as deplorable, in fact they were the worst that any member of the committee had ever seen." Forage conditions were "far from desirable in every respect. No new growth of aspen was located, and all the trees were highlined." The committee observed that white fir, which is commonly eaten

Photo 21.1 Skirted trees were a common sight on the Kaibab Plain.

by deer only under stress of food shortage, showed effects of recent and heavy use. Skirted trees became a common sight, as indicated in photo 21.1. "Any spruce or fir tree within reach was nipped and fed upon." The reconnaissance party found the deer to be in deplorable condition: "in nearly every case the outline of the ribs could be easily seen through the skin."

Russo (1970) describes a major die-off during 1924–1928. He cited one report that 75% of the previous year's fawns died during the winter. Total deer losses during these years were estimated at around 4,000, owing to a combination of starvation, hunting, and predators. Kormondy (1969, 96) describes the deer population falling by around 60% during two successive winters. He adds: "By then, the girdling of so much of the vegetation through browsing precluded recovery of the food reserve."

These difficult conditions prompted local groups to initiate some extraordinary but futile efforts to rescue the deer. Russo described one "as an event that sounds more like fiction than fact." Over 100 people loaded down with cowbells, tin cans, and other noisemakers formed a line on foot and horseback, determined to drive the deer to the south rim of the Grand Canyon (where they would be collected and transferred to better range). The noisemakers knew their effort was futile when they reached Saddle Canyon. By this point, "there were no deer in front of the men but thousands of deer behind them." Another extraordinary, but futile effort was the removal of fawns. Organizers signed contracts to deliver over a thousand fawns to private parties, but most fawns died during the first few days in the so-called fawn farms.

By 1928, government hunters were deployed to reduce the size of the deer population. The deer slayers took to the field in December of 1928 and killed over 1,100 deer. The government program was highly controversial and discontinued in the following year. During all this time, the policy of hunting and trapping predators continued. During 1927, predator "control measures" eliminated 403 coyotes, 111 wildcats, and 11 mountain lions. Russo (1970, 46) describes all of this with a sense of incredulity:

> *Paradoxical situations? Here the deer are dropping dead from starvation by the hundreds and outwardly every effort is being made to reduce the population because the range is in poor condition . . . but, the predator is still controlled.*

The year 1930 was a year of extra summer rainfall, and Russo reports that the deer enjoyed a good growth of "weeds, grass, and mushrooms" and "deer were reported in good condition throughout the year." By 1932, the deer population was estimated at around 14,000. The range was recorded to be "in better condition than it has been in a great many years." And one of the Forest Service game reports declared that the number of deer "appears to be about right for the range" (Russo 1970, 50).

This background is sufficient for the first step in the modeling process ("A" is for "get Acquainted with the problem"). The next step is to define the dynamic problem that is the focus of the modeling.

Step 2: Be Specific about the Dynamic Problem

The Kaibab irruption is an unusual topic because we are looking back in time. We can't change what has happened, and our understanding will be constrained by the limited information from the turn of the 19th century. However, the Kaibab story can contribute to general understanding of ecological systems and their reaction to human disturbances. The purpose of the model is to add to the understanding of the dynamics of an important event widely discussed in ecology.

You know that the best way to be specific about the dynamic problem is to draw a reference mode. It serves as a target pattern of behavior, and it usually corresponds to one of the six shapes in figure 21.1. A sketch of the Kaibab reference model is shown in figure 21.2. The time interval is from 1900 to 1940. The initial deer population is around 4,000. The population is assumed to be relatively constant in first decade of the century. The population then grows rapidly from 1910 to 1924, perhaps reaching a peak of around 100,000. The subsequent die-off is thought to have occurred in the late 1920s, perhaps as rapidly as 60% in just two years. By the 1930s, conditions on the plateau had improved, and the population was thought to be around 14,000.

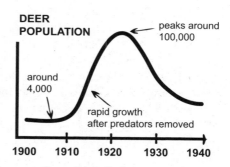

Figure 21.2. Reference mode.

Notice that the time graph is sketched by hand, and that the vertical axis is not labeled. The sketch is not a compilation of precise estimates. Quite the contrary. It is simply a rough drawing to depict a possible pattern of behavior based on the accounts by Russo, Rasmussen, and others. Our goal is to explain a population pattern that remains stable during the initial years and grows rapidly when predators are removed from the system. The population is expected to peak at anywhere from 50,000 to 100,000 and to die off rapidly owing to starvation. The model should help us understand the interactions that produced this overshoot.

It's also useful at this stage to specify a policy variable. Predator hunting was certainly the consensus policy at the turn of the 19th century, so the number of predators should be viewed as the primary policy variable. However, hunting of deer is the current management policy in the region, so deer hunting will be introduced as an additional policy variable at the end of the chapter.

First Model: Keep It Simple

Figure 21.3 shows the stocks and flows in a model with a single stock variable. We are starting as simply as possible, remembering the KISS ("keep it simple, stupid") advice from chapter 20. The model is implemented in Vensim; the book's website, the BWeb, provides equivalent models in Stella. We'll start by taking advantage of parameter values from chapter 20. There will be 4,000 deer at the start of the simulation; their numbers will be in equilibrium, with 2,000 net births/yr and 2,000 deaths/yr from predation. The predation will be achieved by 50 predators with kills of 40 deer/yr. The deer kills are found from a nonlinear lookup to match the second shape in chapter 20. The predator population was endogenous in chapter 20, but it makes more sense to treat the predators as exogenous in the new model. Their numbers will be forced to zero with Vensim's graphical lookup. In this case, time is the input to the lookup, and we set the number of predators to decline from 50 to zero during the period from 1910 to 1920.

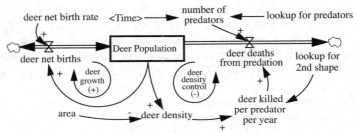

Figure 21.3. First model of the deer population.

Figure 21.3 shows additional information to help identify the feedback loops in the model. The positive feedback is labeled with the (+) and named to remind us that it will lead to exponential growth in the deer population. The negative feedback loop involves density-dependent control of the deer population. You can follow the causal arrows in the diagram around each loop, with the help of the + or − signs at the tip of each arrowhead. Such diagrams are sometimes called *hybrid-diagrams*, since they combine the regular model diagram with symbols normally found in a causal loop diagram. This diagram is sufficient to show the feedbacks, so there is no need for a causal loop diagram.

Figure 21.4 shows the simulation results of the first model. The deer population is held in check at 4,000 for the first 10 years. The predators are lowered to zero during the next 10 years, and the deer population grows rapidly. It doubles to 8,000; doubles again

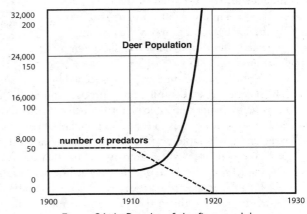

Figure 21.4. Results of the first model.

to 16,000; and doubles yet again to 32,000 before the end of the decade. This explosive growth is caused by the unfettered growth rate of 50%/yr. The population soars off the chart before 1920, and it is not coming back. This is definitely not the reference mode.

Second Model: Add the Forage Requirement

The second model (figure 21.5) will keep track of the forage requirements and the forage available on the plateau. According to Vallentine (1990, 279), the mule deer requires roughly 23% of an animal unit equivalent (AUE) of biomass. The AUE is the dry matter that would be consumed by a 1,000-pound nonlactating cow (about 12 kilograms/day). This translates to an annual consumption of 1,007 kilograms/yr. Let's round this off to 1,000 kilograms/yr, which is the same as one metric ton (MT) per year. The forage required for 4,000 deer would be 4,000 MT/yr.

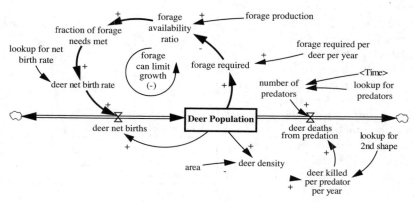

Figure 21.5. Second model of the deer population.

It is the new growth that is nutritious and consumed by the deer. Let's use the term forage production to represent the new growth produced each year. We'll assume 40,000 MT/yr, 10 times what the deer require. The estimate is based on an instinctive assumption that the plateau is able to produce far more forage than the small deer herd would require. The forage availability ratio is the ratio of forage production to the forage required. As long as the ratio exceeds 1, the fraction of forage needs met will be 1. If the forage availability ratio falls below 1, the fraction of the forage needs met will fall as well. If the availability falls to 50%, for example, the deer are only able to satisfy 50% of their forage requirement.

Table 21.1 shows the relationship between the net birth rate and the fraction of forage needs met. With ample forage, the net birth rate is 0.50/yr, the value in chapter 20. The net birth rate declines if the deer are not able to satisfy their forage needs. The steepness of the decline is difficult to estimate since there is very little information avail-

Table 21.1. Deer net birth rate.

Fraction of forage needs met	Deer net birth rate
0.2	−0.40
0.3	−0.40
0.4	−0.20
0.5	0.00
0.6	0.20
0.7	0.40
0.8	0.45
0.9	0.48
1.0	0.50

able. (Some information is provided by deer observed in New York State [Ricklefs 1990, 337], but the range of conditions was quite small.) The net birth rate will fall to zero if the fraction of forage needs met falls to 50%. And if it falls further, the population can decline as rapidly as 40%/yr.

Figure 21.6 shows the simulation results, with the deer population scaled from 0 to 100,000. The deer population remains at 4,000 during the first decade. The predator numbers are reduced starting in 1910, and the deer population grows rapidly to around 80,000 by the year 1920. At this point, their forage requirement is 80,000 MT/yr. This is twice what the Kaibab can produce, and the faction of forage needs met is 0.5. With only half their forage needs met, the deer net birth rate is zero. Births and deaths are in balance, and the population remains at 80,000 for the remainder of the simulation.

The results in figure 21.6 are certainly much closer to the target pattern. The population grows rapidly during the decade that the predators were removed and reaches a high value similar to the numbers reported by some observers from the 1920s. But the simulation does not show the die-off. You might wonder if the problem lies in some of the parameter estimates. Perhaps we should try a different estimate of the forage required per deer? Or maybe we should change the estimate of the forage production? You can build the model and experiment with different values of these

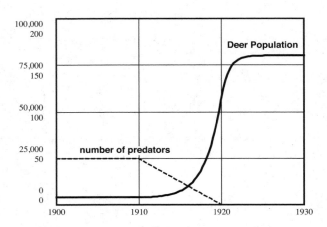

Figure 21.6. Results of the second model.

parameters. All your experiments will show the same general pattern: the population will grow rapidly after the predators are removed and come into equilibrium when the forage requirement is twice what the plateau can provide.

The problem with the second model is the structure, not the parameter estimates. We need to go beyond a constant production of forage. Let's expand the model to simulate the growth, decay, and consumption of biomass.

Third Model: Endogenous Forage Production

Forage production is our term for the growth of new biomass. We've simulated growth in biomass in the flowers model in chapter 7. Ideas from that chapter can be adapted to the Kaibab situation, as shown in figure 21.7. A single stock is used to represent the combination of the grasses, piñon, juniper, pine, spruce, cliff rose, and other vegetation found on the plateau. The stock will be measured in metric tons (MT) and initialized at a small value. We will then simulate the model to allow it to follow an S-shaped path to dynamic equilibrium. We can then take the equilibrium conditions as indicative of the Kaibab at the turn of the 19th century.

Let's assume an average life of 10 years, so the biodecay rate will be 10%/yr. And suppose the maximum biomass is 400,000 MT. If the model were to actually reach this value, the decay would be 40,000 MT/yr. If the system were in equilibrium, the new growth would be 40,000 MT/yr, a value similar to the previous model. New biomass is the biomass productivity multiplied by the standing biomass. Since the new biomass is what the deer consume when browsing, we need to subtract the forage consumption from the new biomass to get the addition to standing biomass. The forage consumption will be fixed at 4,000 MT/yr to test the model.

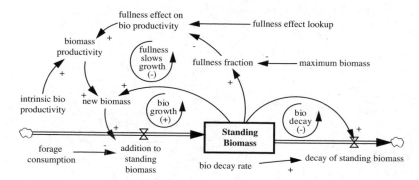

Figure 21.7. Model for S-shaped growth in biomass.

Table 21.2 shows the effect of fullness on bioproductivity. The effect is defined at 1.0 when the fullness is close to zero, so the bioproductivity is identical to the intrinsic value (which is 0.40/yr). If the standing biomass grows to fill 80% of the space, the effect will fall to 0.2. This would lower the bioproductivity to 0.08/yr. In this case, the rate of production would be insufficient to counter biodecay. So we would expect this model to fall short of 80% fullness.

Figure 21.8 shows a 20-year simulation to test the biomass model. The standing biomass grows to just over 300,000 MT. The equilibrium is achieved when the additions to standing biomass balances the decay of standing biomass. These two curves come together around the 15th year of the simulation. This test is conducted with the deer removing 4,000 MT/yr of new growth. The relative importance of the forage consumption is revealed by the small difference between the new biomass and the additions to standing biomass.

The next step is to combine the newly created biomass model with the previous modeling of the deer population, as shown in figure 21.9. The model will begin in the year 1900 with the equilibrium values found at the end of the test simulation. We know that the deer population is not going to remain constant

Table 21.2. Effect of fullness on productivity.

Fullness	Effect on bioproductivity
0.0	1.0
0.2	1.0
0.4	0.9
0.6	0.6
0.8	0.2
1.0	0.0

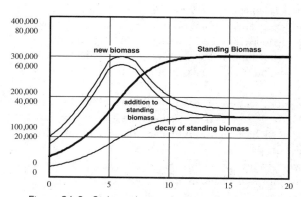

Figure 21.8. S-shaped growth in standing biomass.

at 4,000, so we need to represent forage consumption when their numbers climb to 80,000 or even higher. Figure 21.9 assumes that a fraction of new biomass is within reach of the deer. The new biomass available is compared to the forage requirement to define a forage availability ratio. As long as availability ratio exceeds 1, the deer will obtain 100% of their forage needs from the new growth. However, if the ratio falls below 1, the fraction of forage needs met will decline, and there will be a decline in the net birth rate.

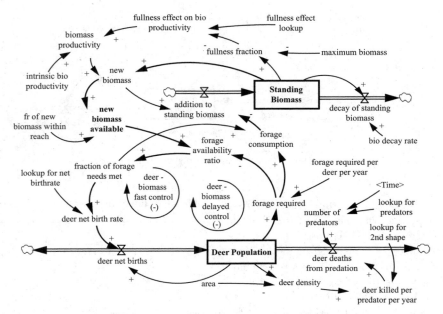

Figure 21.9. Third model of the deer population.

Figure 21.9 is a hybrid diagram to draw our attention to some of the feedbacks in the third model. Two negative loops are highlighted in the diagram. They act to control the size of the deer population. The inner loop is labeled as providing fast control. To see the negative feedback, consider how the inner loop would respond to an increase in the deer population. This would increase the forage required, reduce the availability ratio, reduce the fraction of forage needs met, lower the net birth rate, and slow the growth in the population.

The outer loop involves the stock of standing biomass so there are delays not felt by the inner loop. This outer loop is labeled as delayed control in figure 21.9. Think of how it would respond to an increase in the number of deer. There would be an increase in forage required, an increase in forage consumption, and a reduction in the additions to standing biomass. With less standing biomass, we would see less new biomass available to the deer. The availability ratio would decline, causing a decline in net births. The delayed effect is to slow the growth in the deer population.

Figure 21.10 shows that the third model comes closer to matching the reference mode. The population grows to around 90,000 by 1924 and then declines slowly for the remainder of the simulation. The decline tells us that we have an unsustainable situation, but the decline is nowhere close to the dramatic collapse drawn in the reference mode.

Figure 21.11 shows additional results in the form of a stacked graph of the forage consumption and the additions to standing biomass. The sum of these variables is the new bio-

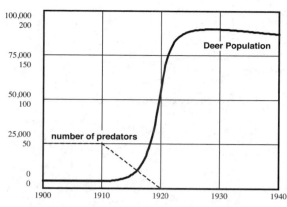

Figure 21.10. Main results of the third model.

mass produced each year. The model begins with around 35,000 MT/yr of new biomass production. The deer consume only 4,000 MT/yr, so there are 31,000 MT/yr added to the standing biomass. But conditions change dramatically around 1915. The production of new biomass increases owing to the effect of the fullness loop in figure 21.7. With more space for growth, production climbs to nearly 60,000 MT/yr. The forage consumption increases dramatically as well. It exceeds 40,000 MT/yr shortly after 1920. The deer are consuming over two-thirds of the annual production, and the remaining third is added to the standing biomass. These additions are slightly below the biodecay at the end of the simulation.

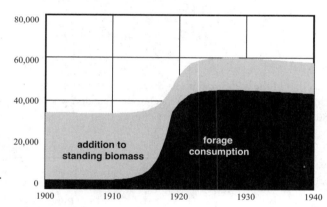

Figure 21.11. Stacked graph of forage consumption and additions to standing biomass.

The slow decline in standing biomass does not match the reference mode. We know that the deer resorted to eating older, less nutritious biomass (see photo 21.1). The next step is to add the consumption of older biomass to learn if it will cause the collapse in the deer population.

Fourth Model: Add Consumption of Standing Biomass

The consumption of old biomass is highlighted as an extra flow in figure 21.12. This flow will not take effect until the deer are unable to satisfy their forage needs from new growth. When insufficient new biomass is available, the deer will experience a need for additional forage. A fraction can be met by the consumption of older biomass.

The initial value of the standing biomass is around 300,000 MT. The current value is compared with the original value to obtain the fraction of original biomass remaining. We assume that the deer will eventually find it difficult to reach suitable older biomass (as noted in the caption to the photo 21.1). If half of the original biomass is remaining, the model assumes

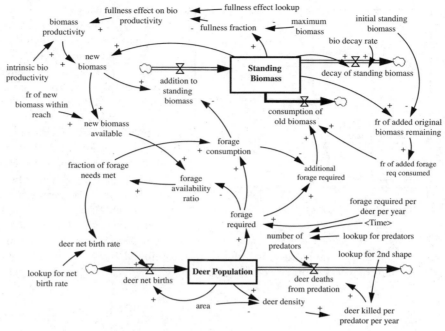

Figure 21.12. The fourth model: deer resort to eating old biomass.

that half of the additional needs are met. Figure 21.13 shows that the new model comes much closer to reproducing the reference mode. The deer population grows rapidly, reaching around 80,000 in the year 1924. The population then falls quite rapidly, reaching 25,000 by the end of the simulation.

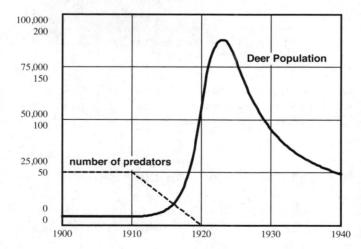

Figure 21.13. Results of the fourth model.

Figure 21.14 shows additional results in the form of a stacked graph of biomass consumption. The lower portion is in black; it represents the forage consumption. This is the new growth that the deer prefer. The gray portion is stacked on top. This is the older biomass that is consumed out of desperation. The deer first resort to the older biomass around 1919. And within a few years, over a third of their consumption is from the older biomass. Both forms of consumption decline in the second half of the simulation owing to the collapse in the deer population. This model comes close to the target pattern, but the peak population is somewhat less than the estimate from Rasmussen. The missing ingredient may be the nutritional value of the older biomass.

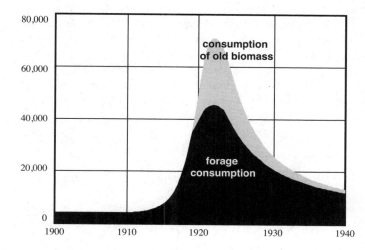

Figure 21.14. Stacked graph of the biomass consumed by deer.

Fifth Model: Nutritional Benefit of Old Biomass

Figure 21.15 shows the addition of an old-biomass nutrition factor. This was ignored in the previous model; its value is set to 25% in the new model. This means that 4 kilograms of older biomass deliver the equivalent benefit of 1 kilogram of new growth. The fraction of

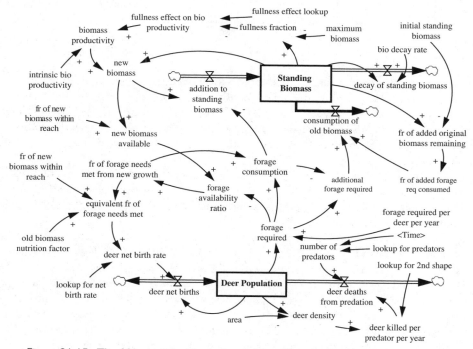

Figure 21.15. The fifth model adds some nutritional benefit from the older biomass.

added forage requirement consumed is counted in the calculation of the equivalent fraction of forage needs met. This fraction is calculated on the right side of the diagram and used on the left side. An arrow connection would be confusing, so we make use of Vensim's "shadow variable" to move a copy of the fraction to the right side of the diagram.

We know from figure 21.14 that there is around 20,000 MT/yr of older consumption when the population reaches a peak. The new model counts a portion of this as equivalent to the regular consumption, so we would expect the deer population to grow to a higher peak. And with a steeper climb to a higher peak, we should expect a steeper decline during the collapse stage of the overshoot.

Figure 21.16 confirms this expectation. The deer population climbs to just below 120,000 by the year 1923. It then declines rapidly, hitting 30,000 by 1931 and around 20,000 by the end of the simulation. This new pattern is a good match for the target pattern established at the start of the chapter.

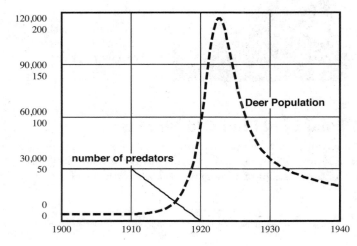

Figure 21.16. The fifth model shows the reference mode.

We now have reached a major milestone in the modeling process. Replicating the reference mode means that we have an internally consistent theory of the overshoot and collapse of the deer population. You may say that we have an endogenous theory of the irruption. Think of the exogenous inputs to this model. Except for the number of predators, all the inputs remain constant throughout the 40-year simulation. The overshoot pattern emerges from the feedback structure of the model.

Simulating the reference mode is an important accomplishment, but you might be wondering if we really needed five iterations. Perhaps the iterations have been stretched out to make each model easier for a newcomer to understand? Maybe an experienced modeler would get to the reference mode in one or two steps?

You should know that veteran modelers understand the benefit of iterative, incremental modeling. They remind themselves to keep the model simple at the outset. They limit each new addition to a single improvement. It's easy to make mistakes along the way, and the difficult challenge is finding and correcting the mistakes. An incremental process of expansion and testing, further expansion, and further testing is the best approach. It will help you find and correct your errors, one error at a time. (The last thing you need is a model with multiple errors.)

The final model uses many parameters that are highly uncertain. The next step in the modeling process is to test the sensitivity of the model to changes in these parameters.

Sensitivity Testing

The purpose of sensitivity testing is to learn if the general pattern of behavior is strongly influenced by changes in the uncertain parameters. The mechanics of a test are simple. We select an input, change the value, and repeat the simulation. If we see a similar pattern, we know that the change did not alter the fundamental tendency. Think of the panel of six shapes in figure 21.1 when we talk about the fundamental tendency. If we see an overshoot in one simulation and S-shaped growth in a sensitivity test, we know that the parameter changes the fundamental pattern. However, if the model continues to show the overshoot, we know the parameters are less important than the model structure. Models that give the same general pattern are said to be *robust*.

Figure 21.17 shows a sensitivity test to determine the importance of the nutritional factor. This is a comparative time graph, with the deer population scaled from 0 to 160,000. The population peaks at just over 80,000 when the old biomass has no nutritional value. The base case assumption is 25% nutritional value, and the population peaks at just below 120,000. If the old biomass has 50% of the benefit of new biomass, the population would peak at nearly 160,000. The three simulations show the identical result until the year 1920. This is expected, since deer do not turn to the older biomass until this time. The test with a 50% benefit shows a steeper growth followed by a steeper decline. The populations are approximately the same at the end of the simulation.

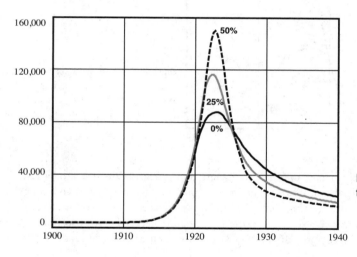

Figure 21.17. Sensitivity test for the nutritional factor.

This first test shows that the Kaibab model is insensitive to the nutritional factor. We see the overshoot pattern regardless of the value assigned to this parameter. It's certainly true that the peak population changes dramatically. We have just over 80,000 deer in one simulation and nearly 160,000 deer in another. The difference in peaks is quite dramatic in figure 21.17. But our goal is not forecasting or point prediction. We are looking for a general understanding of the tendency for overshoot. With that goal in mind, we have a robust model.

Figure 21.18 shows similar results in tests of the forage requirement per deer. The simulations assume 0.5, 1.0, and 1.5 MT of biomass per deer per year. The comparison shows dramatic changes in the peak population. It can range from just over 60,000 to nearly 240,000. But the overshoot pattern remains in all three simulations. Figure 21.19 compares simulations with three different estimates of the fraction of the new growth that is within reach of the deer. This parameter can range from 55% to 95%, and we still get the overshoot pattern. The model is robust in the face of the changes made so far.

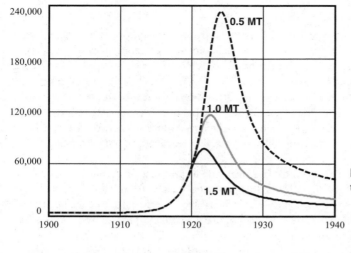

Figure 21.18. Sensitivity test of the deer forage requirement.

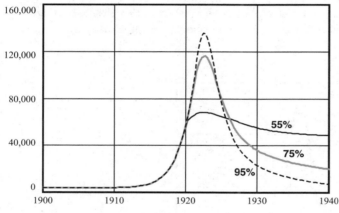

Figure 21.19. Sensitivity test of the fraction of new biomass within reach.

You can experiment with changes in other parameters, and you'll find that the model continues to give the overshoot pattern. But there are many inputs to this model, and you might be wondering if the one-at-a-time testing conducted so far is sufficient. How could we test the model to simultaneous changes in a dozen or more input parameters? This may seem like a daunting task, but you'll learn in appendix D that Vensim makes it easy to conduct comprehensive sensitivity analysis.

Let's conclude the testing with a stochastic simulation to see if random changes will alter the fundamental pattern. Figure 21.20 shows a test with random variations in the intrinsic bioproductivity. The graph compares the base case simulation with a stochastic simulation with the bioproductivity uniformly uncertain from 20% to 60%/yr. The random values are sampled and held for one year. The randomness leads to almost no change in the deer population in the first half of the simulation. This makes sense, because the forage production far exceeds the needs of the deer. But randomness leads to differences in the second half of the simulation. There are some good years in 1921 and 1922, which lead to greater production in new growth, and the deer population climbs to a higher peak. This sets the stage for a steeper decline later in the simulation. The years from 1927 to 1930 are bad years for bioproductivity, and the deer population ends up lower than before. The overall conclusion from the stochastic test is similar to the other tests—we have a robust model that continues to give the overshoot pattern under a wide range of parameter assumptions.

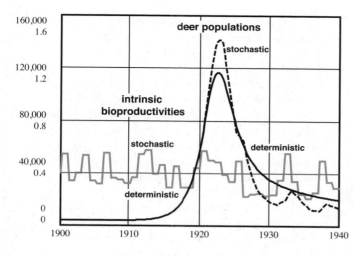

Figure 21.20. Stochastic simulation with randomness in the bioproductivity.

Policy Testing

Policy testing is the final step in the modeling process. The consensus policy at the turn of the 19th century was to remove the predators from the plateau. Russo (1970) documented the predators' kills and concluded that the killing of predators "was in reality a blanket policy that may have grown from an idea of protecting desirable wildlife by eliminating its enemies." The number of predators is an exogenous input to the model, so it is easy for you to experiment with changes in the number or timing of the predator kills. Such experiments will reveal that any reduction of the predators below the 50 needed to keep the deer in check will lead to an irruption. These results arise from the fact that predators are an exogenous input that does not react to changes in the deer population. You know from chapter 20 that the predator numbers should grow in the years following the increase in deer density. The best way to test the predator kills policy is to combine the two models. This is left for you as an exercise.

An alternative policy is deer hunting, a policy that makes sense in the absence of natural predators. Controlled hunting is common in Europe and in North America. In 1985, for example, around 7 million deer were killed in Europe and North America. The economics of hunting can be substantial. In 1976, around $2.6 billion was spent by hunters killing white-tailed and mule deer in the United States, the equivalent of around $1,000 per kill (Clutton-Brock and Albon 1992).

Figure 21.21 shows the addition of a flow to reduce the deer population due to hunting. There are two inputs—the year to initiate the program and the annual kills. The Vensim equation would use the IF THEN ELSE function. If time exceeds the year to initiate hunting, the number of deer kills is equal to the number of annual kills. Otherwise, it is zero. This model could be used to experiment with the annual kills or the year to start the program. You can experiment with this policy, and you'll discover that it is impossible to find the right value for the annual kills. Set the value too high, and the deer population will decline to zero; set it too low, and you won't prevent the irruption.

Figure 21.22 shows a different approach to hunting. The manager sets a goal for the population, and hunting is used to prevent the deer population from exceeding the goal. This policy requires measurement of the deer

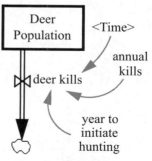

Figure 21.21. Deer kills added to the model.

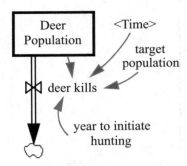

Figure 21.22. Deer hunting based on a target population.

population, so it is more complicated to administer. And it requires the agency to make a judgment about the population that is suitable for the Kaibab Plateau.

Figure 21.23 assumes that the wildlife manager specifies the hunting goal as a fraction of the deer population. We'll implement this approach in an idealized fashion. There will be no delays or errors in measuring the population, and we'll assume that hunter kills match exactly the call from the wildlife manager. If the deer population is controlled at low levels, they should have ample forage, and the net birth rate is 50%/yr. In this case, it makes sense to set the hunting fraction to 50%/yr.

Figure 21.24 shows a graph from a control panel to experiment with the model. This particular simulation shows 50% hunting initiated in 1921. The deer population is the dark curve in the graph. It is growing rapidly around 1915 to 1920. This rapid growth might trigger the implementation of the hunting program in 1921. (This year also corresponds to the time when a committee was formed to look into the difficult conditions on the plateau.) The consumption of old biomass begins around 1919 and is substantial in 1920. If a monitoring program had been in effect, the overbrowsing would have been evident by 1920. So one may think of this simulation as representing the prompt implementation of 50% harvesting after the overbrowsing had been observed.

This is a highly idealized program with no lags in measuring the population and with deer hunters delivering exactly the requested kills. But the results are disappointing. The deer population does decline, but you know from the base case that the decline was about to occur anyway owing to starva-

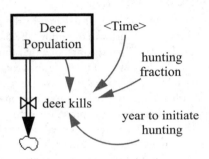

Figure 21.23. Deer kills based on a hunting fraction.

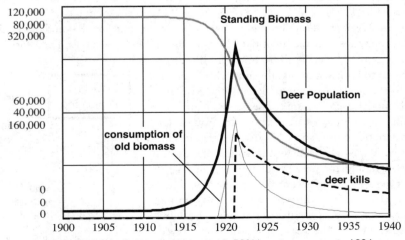

Figure 21.24. Policy simulation with 50% hunting starting in 1921.

tion. The second half of the simulation shows a continued decline in the standing biomass. It also shows a continued overbrowsing of the biomass. It appears that this hunting policy is initiated too late to deliver a sustainable situation.

You can experiment with different hunting policies. You'll find that it is possible to control the deer population with 50% harvesting if you initiate the program in advance of the signs of overbrowsing. For example, this proactive policy could lead to a deer population of around 24,000, and the Kaibab could deliver new growth well above the forage requirement of the deer.

Additional Policy Tests

At this point, we have an idealized policy that can control the deer population in the absence of predators. If we were encouraged by this result, the next set of policy tests should elaborate on the policy. For example, it's important to test whether the 50% harvest policy would control the deer population if we were to simulate the delays in measuring the size of the deer population. Also, we might wonder if the hunting policy would work given discrepancies in what the hunters harvest and the target amount. Another complication involves the weather. It is important to consider whether a policy would work with the changes in the many variables that would be influenced by random changes in the weather. We should test our best hunting policy in the presence of random variations in the intrinsic bioproductivity and the normal value of the deer net birth rate.

Another policy consideration is the lack of an explicit target for the deer population. The 50% harvesting policy leads to an equilibrium population, but the size of the population is a coincidence of when the hunting policy is started. We might make sense to expand the hunting policy so that the desired size of the deer population is an explicit policy. Readers familiar with deer hunting will know that the hunting regulations require distinctions between bucks and does and between young and old deer. The simulation of deer hunting policies would call for multiple stocks to represent the sex and age structure of the deer population.

What about the Excluded Variables?

We'll conclude the modeling process with the question that often arrives after finding an encouraging policy result. We look at the model and wonder about the many variables that have been left out. Just think of the list of variables that have been excluded from the final model. It could include each of the hunting factors mentioned previously. The excluded variables also include changes in elevation across the plateau, changes in seasons, and the impact of snowfall on biomass availability. Many variables are excluded by the high level of aggregation. For example, there are no distinctions between the cougars, coyotes, and bobcats, and there are no distinctions between different types of vegetation, such as shrubs, grasses, pine, and spruce. Furthermore, we don't distinguish between the conditions on the summer range and the winter range. Finally, and perhaps most important, the model does not deal with the cattle and sheep that occupied portions of the plateau.

Now suppose you were to draw a bull's-eye diagram for the final model. If you placed all the excluded variables around the outside of the diagram, it would be immediately obvious that there are more items left out of the model than are in the model. It would be natural to react in an anxious manner. After all, one of the excluded variables could be more important than the variables that have been simulated so far. You might worry that the computer could give the wrong answer. But if you try to add each and every variable to the model, the modeling process could go on and on. Will the model ever be big enough to deliver the right answer?

As we contemplate these questions, it's useful to remember that computer simulation is not a magic path to a right answer. Simulation modeling should be viewed as a way to gain improved understanding of the dynamics of the system. With this limited goal in mind, we should resist the inevitable temptation to add more and more variables. Indeed, we should feel proud if we arrive at a relatively simple model that can explain the reference mode. Also, it's important to remember that adding more and more variables will sometimes confuse the issues rather than illuminate them. As a practical suggestion, it's good practice to limit our initial modeling effort to a model that is similar to the fifth version of the Kaibab model. The model is sufficiently complex to generate the reference mode from inside the system. Yet it is so compact that we can see every variable and every interconnection in figure 21.15, a diagram that occupies less than one full page in the book. The Kaibab model demonstrates what you have learned from other models in this book—that it is possible to obtain important insights about dynamic behavior with a relatively small model.

Postscript on the Kaibab Story

This chapter uses the Kaibab case to illustrate the steps of modeling and the iterative nature of the modeling process. The model shows an internally consistent theory, based on the original views by Aldo Leopold (1943, 1949). But it is important that you be aware of the considerable uncertainty about what actually happened on the Kaibab Plateau.

Caughley's (1970) article from *Ecology* is useful here. He notes that irruption in ungulates (e.g., mule deer) have been initiated by a change in food or habitat. The process is terminated by overgrazing. Since the Kaibab irruption is widely known, Caughley carefully reviewed the reports by different observers and concluded that there was an accumulation of evidence that was powered by the willingness of one author after another to accept the previous author's account of the Kaibab situation. He concluded that the "cause of the irruption is more doubtful than the literature suggests." He did not dismiss the removal of predators as the primary factor initiating the irruption. Nor did he spend much time with an alternative explanation. His only reference to an alternative is that the "increase in deer numbers was certainly concomitant with reduction of pumas and coyotes, but it also coincided with a reduction of sheep and cattle." Caughley noted that two observers writing in the 1960s "considered that the increase of deer was a consequence of habitat being altered by fire and grazing, and that the reduction of predators was of minor influence." Caughley concluded that the data on the deer herd in the period from 1906 to 1939 are unreliable and inconsistent, and he believes that the many factors that could have contributed to the upsurge in the deer population are hopelessly confounded. Caughley concluded his review by asserting that the Kaibab case is "unlikely to teach us much about irruption of ungulate populations."

A more recent discussion of the Kaibab story appeared in Botkin's (1990) *Discordant Harmonies*. Botkin describes Caughley's review and seems to agree that the Kaibab case is unlikely to teach us much about the irruption. Botkin feels that there are lessons to be learned from the telling and retelling of the Kaibab story by prominent naturalists. Botkin argues that their eagerness to retell the stories reveals their paradigm (their way of viewing the world):

> It is surprising that such careful and observant naturalists as Leopold, Rasmussen, and the others who examined the Kaibab history and to whom the study of nature was important would have accepted one explanation among many when the facts were so ambiguous. Many interpretations are possible, only one of the possible stories was accepted, a story that painted a clear picture of highly ordered nature within which even predators had a highly essential role.

Botkin believes that the story of the mule deer on the Kaibab Plateau is only one of the many from the first half of the twentieth century regarding the removal of predators. He believes the information on each case is sketchy, but "to proclaim that we do not have enough information to know if an irruption of mule deer was caused by the removal of the mountain lion, or even if the irruption occurred at all, is to speak against deep-seated beliefs about the necessity for the existence of predators as well as all other creatures on the Earth."

As students learning how to build and test models, you should remember that constructing and testing a model of the Kaibab story does not make the story true. It does reveal that the story "hangs together" in an internally consistent manner. That is, the model shows that the original description of the irruption can be explained through computer simulation in which the interrelationships in the model give rise to the rapid growth and subsequent collapse of the deer population. If you embrace this account of the Kaibab case, you could then use the simulation model to test policies to manage the deer herd.

Alternatively, if you wish to challenge the original story, you might change the model to learn if the irruption could be explained in an entirely different manner. For example, you might introduce livestock onto the Kaibab Plateau and alter the assumptions to minimize the role of the predators. The new model could then be used to test whether the removal of livestock was the primary trigger of the irruption. These changes are a substantial modeling challenge, one that requires a commitment to learn from the events on the Kaibab Plateau. This challenge was undertaken by Len Zeoli (2004). His thesis suggests an alternative explanation of the irruption that emphasizes the effect of livestock overgrazing on the structure and composition of Kaibab forage. Zeoli reduced the importance of predators and emphasized the favorable foraging conditions that followed the removal of livestock. More information on his intriguing interpretation of the Kaibab irruption is provided on the BWeb.

Exercises

Exercise 21.1. Verification

You can experiment with the Kaibab models by taking advantage of the BWeb materials. A good way to learn is to build the models on your own and then simulate them to verify that you get the same results shown in this chapter. Then compare your own models with the BWeb models.

Exercise 21.2. Checking for numerical accuracy

The simulations in this chapter were generated with DT = 0.25 year. Repeat any of the simulations in the previous exercise with DT cut in half. Are the results numerically accurate? If not, what is your recommended value of DT?

Exercise 21.3 Causal loop diagram

Some of the feedback loops are shown in the hybrid diagrams in figures 21.7 and 21.9. But these diagrams do not show all the loops. Draw a causal loop diagram to reveal the vicious circle that appears when the deer resort to consumption of the older biomass. The loop should include the reduction in standing biomass, the reduction in new biomass available, and an increase in the need for still more consumption of old biomass. Your diagram should include each of the variables in figure 21.15 that appear in this loop.

Exercise 21.4. Deer births and deaths

The births and deaths of deer are combined into net births in all the models in this chapter. Expand any of the models to include separate flows for births and deaths. (You might draw on ideas from the separate births and deaths of predators in chapter 20.)

Adopt assumptions on the birth and death rates that are equivalent to the net birth rates in table 21.1. Then simulate your model to see if you get similar results to the model in this chapter.

Exercise 21.5. Model merger

Chapter 20 provides an endogenous explanation of the number of predators. Use this structure to replace the exogenously specified predators in chapter 21. Run the model without the killing of predators. Does the model show predator-prey oscillations similar to the damped oscillations in figure 20.7?

Exercise 21.6. Predator hunting policy

Expand the model in exercise 21.5 by adding predator kills. Set the kills to reduce the predator population to nearly zero during the decade from 1910 to 1920. Turn in a time graph to compare with the results in figure 21.16.

Exercise 21.7. Cumulative number of predator kills

The reported predator kills mentioned by Russo are extraordinary (e.g., 674 mountain lions killed between 1907 and 1923). Add a stock to the model from exercise 21.6 to keep track of cumulative predator kills. Document your work with a time graph of the predator population, the annual predator kills, and the cumulative kills.

Exercise 21.8. Reversal of predator hunting policy

Use the model from exercise 21.6 to test the cessation of the predator hunting policy when the predators have been cut approximately in half. This could represent a change in attitudes toward predator kills after five years or more. Document your results with a time graph to compare to figure 21.16. Does the reversal allow the predators to re-bound? Can the predators prevent the irruption in this new simulation?

Acknowledgment

•   Sketches 21.1 and 21.2 are adapted from Russo (1970).

Chapter 22

DDT in the Ocean

The birth of the modern environmental movement is often associated with the publication of *Silent Spring* in 1962. Rachael Carson used this book to describe the dangerous side effects of chemicals invented during the 1940s and 1950s. The most important was the insecticide DDT (dichloro-diphenyl-trichloro-ethane). DDT was first used in the 1940s with spectacular success, and it was viewed by many as the ideal insecticide. It has been estimated that DDT and other insecticides have prevented the premature deaths of at least 7 million people from insect-transmitted diseases such as malaria. In 1948, Paul Müller received the Nobel Prize for his work on DDT.

Carson's book appeared 14 years later. It warned of the dangers of the continued use of DDT. This chemical is especially dangerous to top carnivores such as ospreys, hawks, eagles, pelicans, and the peregrine falcon. It disrupts their hormonal regulation of calcium, leading to thin eggshells and reproductive failure. The birds take up the DDT from the fish, and the fish take up the DDT in the estuaries and the ocean. DDT was banned in the United States in 1971, 9 years after the publication of *Silent Spring*. Since that time, there has been a dramatic recovery of some bird populations, such as the brown pelican on the California coast. However, DDT is still used today despite its impact on bird populations. The United Nations World Health Organization reported worldwide usage at more than 30,000 metric tons/yr in the 1990s (Botkin and Keller 1998, 209).

Modeling of DDT Flows in the Environment

The first edition of *Modeling the Environment* described two models of the accumulation of DDT in the environment. The models demonstrated the application of system dynamics to simulate the flows of persistent pollutant through the soil, atmosphere, and the ocean. This chapter describes the main features of the models and discusses their distinctive features. The models and chapter 21 are available on the book's website, the BWeb.

This is a short, but important chapter because of the generality of the approach. System dynamics can be used to simulate a wide variety of persistent pollutants that involve slow degradation and multiple media (soil, water, atmosphere, etc.). One example is by A. Anderson and J. Anderson (1973) on the flow of mercury contamination through the environment. Other models simulate bioremediation to hasten the reduction of toxics in the environment. Examples include the work by Brennan and Shelley (1999) on lead and the work by Siegel et al. (2003) on cesium.

DDT in the Soil

Let's begin with the simple model in figure 22.1. It simulates the degradation of DDT in the soil. We'll set the degradation rate to 7%/yr and test the model with an application of 100 tons in the year 1950. You know that the degradation should remove half of the DDT in 10 years. It will then remove half of the remaining amount in the next 10 years. Figure 22.2 confirms the expected pattern. DDT in the soil jumps to 100 tons with the application in 1950. There are 50 tons remaining in the soil by 1960. This 10-year interval is called the half-life for DDT in the soil.

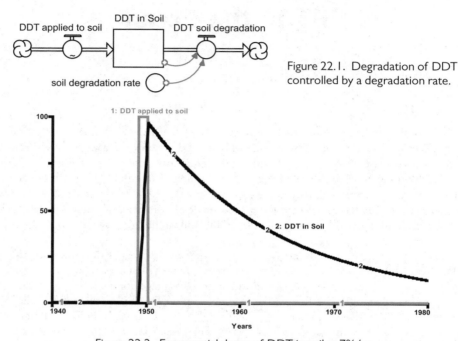

Figure 22.1. Degradation of DDT controlled by a degradation rate.

Figure 22.2. Exponential decay of DDT in soil at 7%/yr.

Randers (1973) reviewed research on DDT and noticed that half-lives are often used to characterize the decay processes in various media. He decided to make half-lives the common factor to calculate flows through the soil, water, atmosphere, and ocean. His estimates are listed in table 22.1. The half-lives for soil processes are 10 years for degradation and 2 years for evaporation. Also, an extremely small amount of soil DDT goes into solution and then into rivers. Once in the rivers, the trip to the ocean is characterized by a half-life of only 0.10 year. The ocean half-lives range from 0.3 year for excretion from fish to 15 years for natural degradation. DDT can also end up in the atmosphere, where it is quickly removed by precipitation.

Table 22.1. Half-lives used by Randers (1973).

Flows and media	Half-lives
degradation in soil	10 yrs
evaporation from soil	2 yrs
into solution from soil	500 yrs
runoff from river	0.10 yrs
precipitation from atm.	0.05 yrs
degradation from ocean	15 yrs
death of fish	3.0 yrs
excretion of fish	0.3 yrs

The DDT models (BWeb) operate with all flows governed by a half-life. Figure 22.3 shows how this may be done. The half life is set to 10 years, and the soil degradation equation is written to deliver the same decay as before.

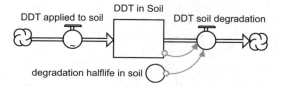

Figure 22.3. Degradation with a half-life.

The new equation is:

DDT_soil_degradation = DT_in_Soil/(1.44*degradation_half-life_in_soil)

This equation makes sense if you think of a 10-year half-life. The DDT in soil is divided by 14.4 years. The reciprocal of 14.4 years is 0.6944/yr, so the flow is essentially the same as the flow with a degradation rate set to 7%/yr.

Now imagine what would happen if the DDT applied to the soil were constant at 50,000 tons/yr. DDT in the soil would increase over time until the degradation comes into balance with the application. This equilibrium value multiplied by 7%/yr must be 50,000 tons/yr. This tells us the equilibrium value would be around 714,000 tons.

DDT Accumulation with Several Flows

But the DDT in soil is influenced by evaporation as well as degradation. The evaporated DDT enters the atmosphere, where it may reside for only a few weeks before rainfall sends some of it back to the soil. These are some of the flows simulated in the previous edition of *Modeling the Environment*. Their overall effect is summarized with the equilibrium diagram in figure 22.4. In this case, total application was set to 100,000 tons/yr, with half landing on the soil and half blowing out to sea. When these flows are continued long enough, DDT in the soil builds to an equilibrium of 215,700 tons. The largest outflow is evaporation. It sends 71,900 tons/yr to the atmosphere, but 36,600 tons/yr returns owing to precipitation over land.

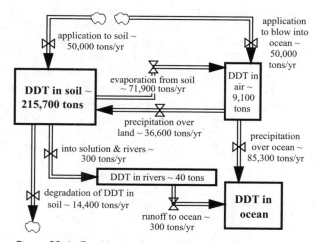

Figure 22.4. Equilibrium flows in the soils and atmosphere.

DDT Accumulation in the Ocean

Figure 22.5 draws our attention to the purpose of the model—how much DDT ends up in the ocean and the fish. These values are taken from 1980, the year when DDT in fish hit a peak of 330 tons. This may seem trivial compared to the 2 million tons in the ocean. But the concentration of DDT in fish is the key to understanding the subsequent impact on the birds. DDT builds up in the bodies of fish in a process called biomagnification. The concentration in fish is found by dividing by the weight of the fish in the ocean. The concentrations are shown in parts per hundred million (pphm) in the Stella bar chart in figure 22.6. This chart dramatizes the bioconcentration as DDT moves up the ocean food chain.

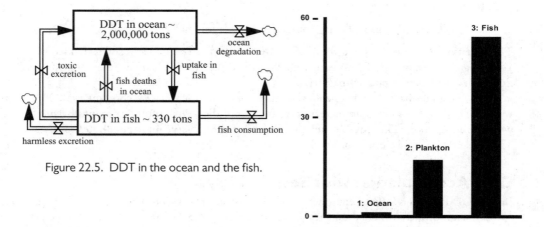

Figure 22.5. DDT in the ocean and the fish.

Figure 22.6. Peak DDT concentrations (pphm) in the ocean and in the fish.

Sluggish Response of DDT to Changes in Application

The purpose of the DDT modeling was to learn the responsiveness of the system to changes in application. Randers (1973) estimated that application peaked in the 1960s, shortly after the publication of *Silent Spring*. Rachael Carson and other scientists warned that the amount of DDT in the soil and the oceans would continue to increase for many years. Randers showed that they were right to issue these warnings. His simulations assumed that worldwide application of DDT peaked in the mid-1960s and would decline over the next three decades. The base case simulation showed that DDT in soil would reach a peak in the mid-1970s and begin to decline. DDT in the soil would be largely eliminated by the year 2000, but the DDT in the oceans was a different story. The simulations showed DDT concentrations in fish continuing to climb during the two decades following the reduction in their application. DDT in fish would begin to decline around 1980, but another two decades would be required before fish concentrations would fall below the levels observed in the early 1960s. The total lag for fish concentration to return to the values that triggered the publication of *Silent Spring* was estimated at four decades. These results are described on the BWeb along with a collection of modeling exercises.

Dealing with High-Turnover Stocks

The DDT models illustrate the challenge of simulating a system with high-turnover stocks. Figure 22.4 shows two such stocks—the DDT in rivers and in the atmosphere. The DDT in rivers holds only 40 tons, but the rapid flow to the ocean leads to an outflow of 300 tons/yr.

The atmosphere is similar. There are only 9,100 tons of DDT in the air, but the frequent precipitation causes 85,300 tons/yr to exit the atmosphere and enter the ocean.

High-turnover stocks are problematic because we need an extremely small DT to simulate the outflows accurately. This means slow simulations if are simulating over many decades. So, if we need the long time horizon for the model, we have two choices: we can either tolerate the slow simulations, or we can look for a way to eliminate the high-turnover stocks.

I believe we should avoid building models with slow simulations. Rapid simulations are essential if we are to encourage interactive simulation and discussion. Some models are slow to simulate, and their developers sometimes speak of the long computer time as a good thing (as if the long lag reveals that the model is performing a complex calculation). I believe we should think twice about building models with long execution times. In most system dynamics models, the long execution time can be traced back to high-turnover stocks. This means we need to get rid of these stocks.

You can tell from figure 22.4 that there is no problem getting rid of the DDT in rivers. We simply ignore the DDT that goes into solution since this flow to the ocean is trivial compared to the flow from precipitation. But how would we eliminate the DDT in the atmosphere? This stock occupies a pivotal position in the system. Indeed, its position corresponds to the pivotal position of the stock of atmospheric water in the hydrologic cycle in figure 6.2. The challenge is to eliminate these pivotal stocks without losing their role in shaping the key flows. Chapter 17 explains how this could be done for water in the atmosphere; the BWeb materials explain how it could be done for DDT in the atmosphere.

Systems with Sluggish Responses

The DDT models provide pragmatic examples of simulating pollutant flows through multiple media (e.g., soils, rivers, atmosphere, ocean, and fish). The models are used to address the slow response to a change in policy. For DDT, the policy option is the reduction in its application. The simulations reveal that four decades are required for the DDT in the fish to finally return to the levels that triggered the change in policy.

Perspective on this slow response is provided by comparison with the Mono Lake and Tucannon River cases from previous chapters. The Mono Lake case focuses on the responsiveness of the lake elevation to a change in the amount of water exported from the basin. Our initial expectation was for a slow and delayed response, but the model surprised us with a rapid and steep increase in elevation. The salmon population of the Tucannon River was distinctive for its incredible resilience in the face of external disturbances. With pristine conditions, the population could recover rapidly from major changes in migration conditions or in harvesting. The simulations alerted us to some delays in the population response because of the four-year life cycle. It also alerted us to the need for patience in interpreting highly variable salmon counts.

Modeling can help us anticipate the responsiveness of systems to changes in policy. The salmon case teaches us to watch for responsiveness within a four-year life cycle, perhaps longer when the response is obscured by random variations. The DDT case teaches us to brace ourselves for a four-decade response time. The problems that persist for many decades are among the most challenging of environmental problems. In my view, the most fundamentally difficult problems of the 21st century are associated with the accumulation of CO_2 in the atmosphere. This is the topic for the next chapter.

Chapter 23

━━

CO$_2$ in the Atmosphere

Carbon dioxide (CO$_2$) is a trace gas that appears naturally in the atmosphere. Atmospheric CO$_2$ acts to capture a portion of the outgoing infrared radiation and provides a warming effect that is natural and essential to life. CO$_2$, water vapor, and methane are among the greenhouse gases (GHG) whose effect is analogous to the glass in a greenhouse. They act like an atmospheric blanket to keep the earth warm.

The greenhouse effect has been occurring for millions of years. Since the start of the industrial era, however, CO$_2$ emissions from power plants, automobiles, and other human-made sources have increased substantially. These anthropogenic emissions have led to excessive accumulation of CO$_2$ in the atmosphere. The higher concentrations act to trap more of the outgoing long-wave radiation, thereby contributing to global warming and a wide variety of climate changes.

This chapter describes the potential for system dynamics to contribute to improved understanding of CO$_2$ accumulation and climate change. It begins with an overview of the main stocks and flows in the global carbon cycle. The chapter then demonstrates a simple method to explain recent projections of the atmospheric CO$_2$ in the 21st century. Such projections are highly uncertain because of our limited understanding of the feedback processes.

This chapter explains the key feedbacks and discusses the extent to which they are represented in climate models. The greatest need is for models that are more effective in linking the terrestrial, atmospheric, and oceanic systems into a coupled system. Increased coupling is especially important to represent the positive feedbacks that act to destabilize the system. These loops are responsible for the rapid warming that is evident in the geologic record, and they could lead to rapid climate change in the future. The chapter concludes with a review of the three types of models commonly used to understand climate change. The most pressing need is for improved integrative models that can help us understand the power of the destabilizing loops. System dynamics models can help meet this need.

The Global Carbon Cycle

Figures 23.1 and 23.2 show the accumulation of CO$_2$ in the atmosphere as part of the global carbon cycle. The first diagram shows the carbon flows in a visual manner. The second diagram summarizes the key stocks and flows in the system. The names in the Vensim diagram show current storage in gigatons (GT) of carbon. The flows are measured in GT/yr (with numbers rounded off for clarity).

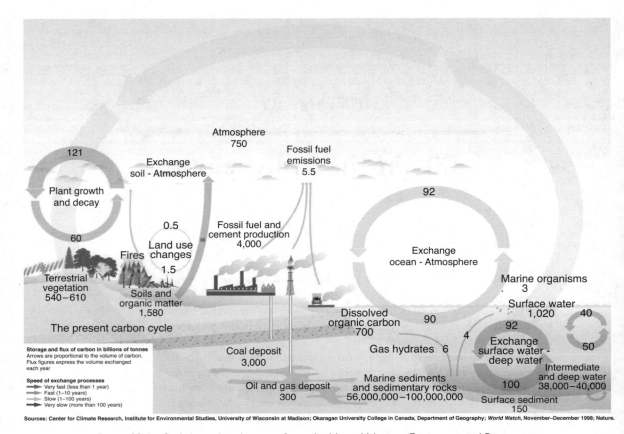

Figure 23.1. Carbon cycle schematic from the United Nations Environmental Program.

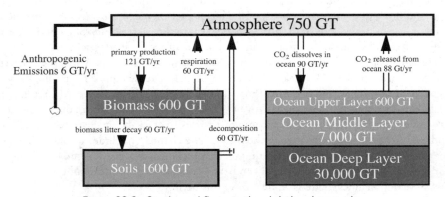

Figure 23.2. Stocks and flows in the global carbon cycle.

The many flows make it difficult to anticipate the growth in atmospheric carbon. We could simplify the situation if we focus on the anthropogenic emissions, the 6 GT/yr highlighted in figure 23.2. If this were the only flow into or out of the atmosphere, accumulation would be simple. For example, the anthropogenic load has been growing at around 1.4 %/yr. With this trend, the load would reach 7.5 GT/yr in around 15 years. During this short interval, around 100 GT would be added to the atmosphere. If anthropogenic emissions were then to remain constant at 7.5 GT/yr for 100 years, another 750 GT would be added. Atmospheric storage of CO_2 would then be over twice as large. The 750 GT corresponds to a concentration of 352 ppmv (parts per million by volume). If the atmospheric storage doubles, the concentration would double as well. Atmospheric CO_2 would be over 700 ppmv in just over 100 years.

A doubling of atmospheric CO_2 provides a benchmark for thinking about climate change. This illustrative calculation is made simple by focusing only on the anthropogenic emissions. It's as if the anthropogenic emissions that enter the atmosphere remain there forever. However, CO_2 is removed rather quickly from the atmosphere by primary production of terrestrial biomass or by absorption in the ocean's upper layer. (A typical CO_2 molecule is exchanged with the ocean in less than a year after entering the atmosphere.) But the prompt removal does not mean that the carbon is removed from the system. Rather, it circulates through the system, reentering the atmosphere at one point, exiting the atmosphere again at a later point, and returning to the atmosphere yet again further in the future. The overall effect of these circular flows is surprisingly long-lived. CO_2 released to the atmosphere today "will contribute to increased concentration of this gas and the associated climate change for over a hundred years" (Houghton 2004, 227).

Let's turn our attention to the net effect of the natural flows to and from the atmosphere. The left side of figure 23.2 shows that total of the natural flows out of the atmosphere exceeds the inflow by 1 GT/yr. This imbalance suggests that around 1 GT/yr of carbon is added to the stock of biomass or soil. This means the carbon stored in the biomass would grow over time (perhaps owing to extensive reforestation of previously cleared land). The right side of the diagram shows that the natural flow out exceeds the inflow by 2 GT/yr. The combined effects is a net outflow of 3 GT/yr. In other words, natural processes are currently acting to negate approximately half of the anthropogenic load (IPCC 2001; Socolow et al. 2004).

As the use of fossil fuels grows over time, the anthropogenic load will increase. But scientists do not think that natural processes can continue to negate 50% of an increasing load. There are limits on the terrestrial system associated with reforestation of previously cleared land (Socolow et al. 2004) and with carbon sequestration in plants and soils owing to carbon-nitrogen constraints (Gill et al. 2006). Also, the current absorption of 2 GT/yr is disrupting the chemistry of the ocean's upper layer (Socolow et al. 2004; Royal Society 2005). This contributes to problems of acidification, and it changes the balance of dissolved minerals. Higher CO_2 can reduce the concentration of carbonate, the ocean's main buffering agent, thus affecting the ocean's ability to absorb CO_2 over long time periods.

Figure 23.3 summarizes the situation. You'll recognize this diagram from the chapter 4 simulation of the accumulation of CO_2 in the atmosphere (figure 4.1). The simulation showed that the cumulative effect of these flows is a doubling of CO_2 in the atmosphere by the end of the century.

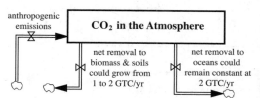

Figure 23.3. Summary of atmospheric flows.

Doubling of CO₂ in the Atmosphere

The doubling of atmospheric carbon is emphasized here because a doubling has been projected from a wide variety of climate models. Indeed, a doubling is often used as a benchmark condition, as when scientists speak of the *equilibrium climate sensitivity*, the temperature increase to be expected from a doubling of CO_2 concentration relative to the preindustrial concentration of 280 ppmv. For example, the climate sensitivity was estimated at 3°C in the recent assessment by the Intergovernmental Panel on Climate Change (IPCC 2007, 9). The IPCC warned that the climate sensitivity is highly uncertain; it could range from 2.0°C to 4.5°C.

A somewhat similar projection was obtained by Webster et al. (2003) in calculations with the climate model developed at the Massachusetts Institute of Technology (MIT). The carbon cycle results corresponded to the net flows shown in figure 23.3. The atmospheric CO_2 was projected to grow from 350 ppmv to 700 ppmv by the year 2100. Figure 23.4 summarizes the MIT projections. The mean temperature impact in the business-as-usual case is a warming of 2.4°C. For perspective on this estimate, think of the many changes in climate system in the previous century with a warming of 0.6°C. The MIT projection for the 21st century indicates four times as much warming as the world experienced in the previous century.

Figure 23.4 summarizes a statistical analysis of hundreds of simulations by showing the 5%, 50%, and 95% estimates of CO_2 concentrations and temperature impacts in the year 2100. The business-as-usual results show higher temperatures and greater variation. The mean temperature impact is 2.4°C, but it could range from 1.0°C to nearly 5.0°C. The other line connects the corresponding results if policies are adopted to reduce emissions. The general goal was to stabilize atmospheric CO_2 around 550 ppmv. CO_2 concentrations vary across a narrower range because of the stabilization policy. The most likely result is a CO_2 concentration of 512 ppmv and a global average surface temperature increase of 1.7°C. The temperature impact would range from 0.8°C to 3.2°C. The MIT analysis confirms what many models are showing—that the excessive warming can be reduced, but it cannot be eliminated. The modeling also confirms what nearly all scientists are finding—that the range of impacts will be greatly reduced by policies to lower anthropogenic emissions.

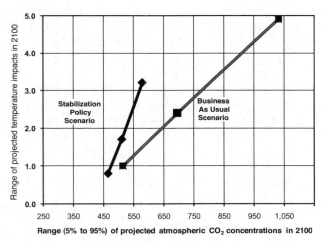

Figure 23.4. Summary of climate modeling.

The 5%-to-95% uncertainty bands portray the range of impacts from the uncertain parameters. These bands were calculated from 250 simulations with simultaneous changes in parameter estimates. The results in figure 23.4 summarize the parametric uncertainty associated with the MIT model.

Parametric Uncertainty versus Structural Uncertainty

Parametric uncertainty is only part of the uncertainty faced by scientists and policy makers. Changes in the fundamental structure of the model can also change the results. Examples of structural changes include the addition of new pollutants, or reformulation to achieve a finer geographical resolution. Structural changes can also involve the addition of new feedback effects between the atmospheric and terrestrial systems. The uncertainties associated with structural changes are much more difficult to quantify. They require one to compare model results before and after the change in model structure. But changing the structure in a satisfying manner is not easy (otherwise, it would have been done long ago). So, what general conclusions might be drawn about structural uncertainty?

Ultimately, statements about structural uncertainty come down to the scientists' intuition on whether the omitted structure will act in a stabilizing or a destabilizing manner. In some cases, adding new relationships to a model will close negative feedback loops that can act to stabilize the simulated system. In these cases, the new structure could lead to a narrower band of uncertainty. However, the customary process of model development is to first include most of the pervasive, well-understood processes. These turn out to be controlled by the negative feedback loops in the system. The less understood feedback loops are often left to future work (when more evidence about their role becomes available.) These omitted feedbacks are often the positive feedback loops that act to destabilize the system. This general tendency means that the true range of uncertainty will be considerably larger than parametric intervals like those in figure 23.4.

Positive feedback effects are difficult to simulate, but their destabilizing effects are important to consider when scientists and policymakers think about the uncertainty of the system. This is why scientists often conclude their quantitative analysis of uncertainty with a qualitative assessment of the possible surprises from relationships that have not yet been simulated in a model. An example is the concluding remarks about the analysis of parametric uncertainty published by Webster et al. (2003, 317):

> As with all investigations of complex and only partially understood systems, the results presented here must be treated with appropriate caution.
>
> Current knowledge of the stability of the great ice sheets, stability of thermohaline circulation, ecosystem transition dynamics . . . is limited. Therefore abrupt-changes or "surprises" not currently evident from model studies, including our uncertainty studies summarized here, may occur.

Feedbacks in the Climate System

The climate system is a complex web of feedback connections. Selected loops are shown here to characterize the general practice in climate modeling. Figure 23.5 begins with two loops that operate through increased photosynthetic activity in a world with elevated CO$_2$ and temperature. Trees and other green plants may respond to warmer air temperature with accelerated growth. Green plant photosynthesis would lead to greater CO$_2$ absorption, less CO$_2$ in the atmosphere, and a reduction in the temperature. Plant growth could also be acceler-

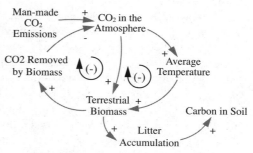

Figure 23.5. Negative feedbacks influencing terrestrial biomass.

ated by higher CO_2 concentration. Higher plant growth would then lead to more CO_2 removal and less CO_2 remaining in the atmosphere.

These feedbacks act to negate the external disturbance from anthropogenic emissions. As drawn here, they also act to increase the amount of carbon stored in the soil. The processes of photosynthesis and growth are among the best-understood processes (Field et al. 2007). Consequently, they are well represented in the climate models. Indeed, "in many modeling studies, these are the only mechanisms through which ecosystem responses to climate change feed back to climate." Field et al. (2007) consider this to be "a highly simplified and potentially misleading view," one that leads to the general result that "increasing CO_2 leads to a current carbon sink that persists through the century." As drawn in figure 23.5, the carbon sink takes the form of carbon in the soil.

Figure 23.6 expands the view of plant growth to include soil carbon decomposition. The new diagram shows a small negative loop involving removal of carbon from the soil and a positive loop involving the pathway from litter to soil to the atmosphere. This is the outer loop in the diagram. The inner loop generates positive feedback with the potential for run-away behavior. To illustrate, imagine there is an increase in atmospheric CO_2, which leads to an increase in temperature. This can accelerate the decomposition of soil carbon, adding more CO_2 to the atmosphere, increasing the temperature further, and leading to still faster soil decomposition.

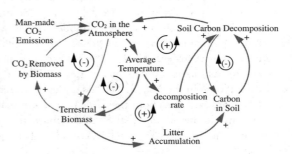

Figure 23.6. Positive feedbacks involving soil carbon decomposition.

Closing the new loops in figure 23.6 requires scientists to combine models of the biomass and soils into a single model. Scientists refer to this approach as a *coupled model*. With this terminology, sorting out the relative power of the soil carbon feedbacks requires detailed analysis with "fully coupled models" (Kump 2002; Govindasamy et al. 2005). Field et al. (2007, 7) describe the results of a modeling comparison project in which models are operated as coupled or uncoupled. The important result is that the coupled mode of operation always delivers a smaller amount of carbon storage: "The storage is always smaller when the model is run in a coupled rather than in an uncoupled mode because coupling creates a positive feedback in which decreased carbon sinks lead to increased temperatures, which further decrease carbon sinks." These and other analyses show the possibility for soil carbon to change from a net sink to a net source of carbon to the atmosphere. The possible reversal of the net flow from the atmosphere to the soils is a major source of uncertainty, one that is not easily resolved without further research.

An equally important source of uncertainty involves the water vapor in the atmosphere. Water vapor is the most important of the greenhouse gases, accounting for a larger share of the greenhouse effect than trace gases such as CO_2 and methane. A warmer world will bring increased evaporation, creating more water vapor in the atmosphere. This will lead to greater retention of the outgoing long-wave (LW) radiation, still higher temperatures, more evaporation, and still more water vapor in the air. This is the positive feedback loop in figure 23.7. It could lead to runaway behavior were it not for the negative feedback loop.

The negative loop involves increased formation of clouds that could occur with increased water vapor in the atmosphere. The cloud cover acts to reflect some of the incoming short-wave (SW) radiation. With greater cloud cover, we would see less radiation reaching the surface of the earth and less outgoing long-wave radiation subject to the greenhouse effect. This

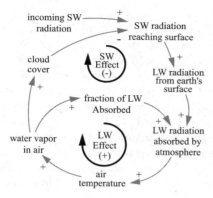

Figure 23.7. Water vapor and cloud feedbacks.

negative loop is sometimes called the *short-wave effect*. The relative balance of these two effects is a major source of uncertainty. Indeed, it may well be the "largest source of uncertainty in current model predictions of climate sensitivity" (Soden and Held 2006, 3359).

Positive Feedbacks and the Termination of the Ice Ages

The geologic record offers clues to help scientists sort out the relative strengths of positive and negative feedbacks in the climate system. Scientists have long searched for an explanation of the ice ages and the pattern of rapid warming that occurs at the terminus of each ice age (Weart 2003). The warming periods are often associated with changes in solar intensity due to minute changes in the Earth's orbit. (The changes involve the inclination of the spin axis, the eccentricity of the orbit, and the season of closest approach.) The effect of these changes is somewhat greater solar intensity every 100,000 years. This timing coincides approximately with the termination of the ice ages.

But the solar changes are minute compared with the radiative forcing needed to explain the warming. Scientists have come to view the 100,000-year solar cycle as the Earth's "pacemaker." It sends a minute signal, and the Earth responds with natural processes that amplify the small signal into the large changes seen at the termination of the ice ages. Weart (2003) explains the surprisingly rapid warming:

> *Swings in temperature that were believed in the 1950s to take tens of thousands of years, and in the 1980s to take hundreds of years, were now found to take only decades.*

Positive feedbacks are the key to understanding how the global system could amplify the minute signals from the pacemaker. A possible source of positive feedback is the release of methane that is sequestered in the permafrost or in swamps and bogs. If a minute change in solar radiation leads to a small increase in temperature, it would be slightly warmer in the permafrost regions. A small shrinkage of the permafrost could then release the methane embedded in the clathrate sediments into the atmosphere. Methane is a powerful greenhouse gas, so an increased concentration in the atmosphere could lead to further increases in temperature and still further shrinkage of the permafrost. A similar effect could involve methane in swamps and bogs. If a minute change in the solar cycle causes a small temperature increase, there could be accelerated decomposition of dead organic matter in bogs and swamps. This would release methane into the atmosphere, leading to still further temperature increases.

An alternative explanation of the rapid warmings is given by Hansen et al. (2007). He examines the geologic record stretching back 400,000 years. The record shows an asymmetric pattern with rapid warmings, usually followed by a slower descent into a colder climate.

Hansen focused on the termination of the last four ice ages. He explained the external changes in the solar cycles and looked to natural processes to amplify the small incoming signal. He discusses a variety of positive feedbacks, including the water vapor effect shown in figure 23.7. However, he finds the best explanation to be ice sheet disintegration in a process called *albedo flip* that happens when snow and ice become wet. Imagine a small rise in temperature from the external changes in the solar cycle. The small change in temperature leads to some melting of ice. The ice has a high albedo (high reflectivity), and some of the reflective power is lost when the ice melts. When sea ice melts, we would have more area covered by the darker ocean water. This leads to more absorption of solar energy, still higher temperatures, and more melting of the ice. Hansen believes that ice sheet collapse provides the best explanation of the termination of the ice ages, and he concludes that the earth's climate is remarkably sensitive to external changes. As a result, "our climate has the potential for large, rapid fluctuations. Indeed, the Earth and the creatures struggling to exist on the planet, have been repeatedly whipsawed between climate states."

Most scientists view rapid climate change as clearly evident in the past record and as a serious possibility in the future. Nevertheless, such changes are not necessarily explainable with the main global circulation modeling. Weart (2003) draws on a National Academy study (NAS 2002) to caution that "the abrupt changes of the past are not fully explained yet and climate models typically underestimate the size, speed and extent of those changes. Hence, . . . climate surprises are to be expected." Hansen et al. (2007) reaches a similar conclusion. He observes that a "global warming of approximately 3°C is predicted by practically all climate models" if greenhouse gases continue to grow in a business-as-usual scenario. The models typically project a sea level rise of only a fraction of a meter. Hansen challenges this estimate because he believes the IPCC models are assuming an "inertia for ice sheets that is incompatible with the palaeoclimate data." He believes that the IPCC projections are understating the sea level rise because "the existing ice sheet models are missing realistic (if any) representation of the physics of ice streams and ice quakes, processes that are needed to obtain realistic nonlinear behavior."

Climate Models

This chapter concludes with a discussion of the different types of computer models that have been developed by scientists around the world. The review sets the stage for identifying the appropriate role for system dynamics modeling. The review draws on the classification scheme by Claussen et al. (2000), in which climate models are described as simple, comprehensive, or intermediate.

The *simple models* are sometimes called *box models* since they represent the storage in the system by highly aggregated stocks. The models generally produce zonal averages or global averages. The parameters are usually selected to match the results from more complicated models, and the parameters can be altered for purposes of sensitivity analysis. Also, the models can be initialized in a steady state without the computational cost of the more complex models. The simple models can be simulated faster on the computer, and the results are easier to interpret. This makes them valuable in conducting extensive sensitivity studies and in scenario analysis. A primer on simple climate models is provided by the IPCC (1997). A system dynamics example of a simple climate model is the climate sector of the coupled climate-economic model by Fiddaman (2002).

Comprehensive models aim to represent all the important processes and to simulate them in a highly detailed manner. Comprehensive models are sometimes called *general circulation models*, and they are usually maintained by large research centers. They can be used to describe circulation in the atmosphere or the ocean. They can be coupled to simulate both the ocean and atmospheric circulation in a simultaneous, interacting fashion. Claussen et al. (2000) describe these coupled models as the "most comprehensive" of the models. Comprehensive models are particularly useful when a high spatial resolution is required. However, a disadvantage is that only a limited number of multidecadal experiments can be performed even when using the most powerful computers. *Intermediate models* help scientists bridge the gap between the simple and the comprehensive models. Claussen et al. (2000) describe eleven models of intermediate complexity. They aim to "preserve the geographic integrity of the Earth system" while still providing the opportunity for multiple simulations to "explore the parameter space with some completeness. Thus, they are more suitable for assessing uncertainty." An example of an intermediate model is the MIT model used by Webster et al. (2003).

The Role of System Dynamics Modeling

Figure 23.8 depicts the three dimensions of climate modeling—detail, processes, and integration. The surfaces are positioned in a three-dimensional space to portray the goals of the model developers. The comprehensive models are at the base of the diagram; they aim for a strong command of detail and processes, but they do so knowing that they sacrifice the ability to integrate across boundaries and to provide for multiple simulations. The diagram is organized with the comprehensive models at the base because they provide a foundation for the other models.

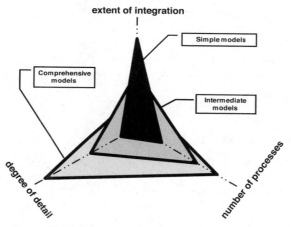

Figure 23.8. Classification of climate models.

The simple models aim for high integration across the terrestrial, soils, atmospheric, and oceanic systems. Their contribution is ease of simulation and opportunity for multiple sensitivity studies. The simple models strive for height on the integration dimension, but the diagram leaves the impression that they do so with a relatively weak foundation along the detail and processes dimensions. This design is intentional. The simple models limit the number of processes and the degree of detail because their parameter values can be supported by the comprehensive models. The intermediate models occupy the middle position in the diagram. Compared with the comprehensive models, they achieve more integration and are more suitable for sensitivity studies.

System dynamics modeling can contribute to improved understanding of global warming and climate change. Indeed, the climate modeling community and the system dynamics community have a shared language and way of thinking about complex systems. Both communities are conversant with the concepts of stocks, flows, positive feedback, negative feedback, homeostasis, and span of control. And both groups view computer modeling as essential to checking our thinking about complex systems. But the system dynamics groups must address the question faced by all climate modeling teams—are we aiming for a simple, an intermediate, or a comprehensive model?

I believe that system dynamics groups will be most productive if their models are designed to fit in the simple category. The goal should be to deliver improved integration across the different sectors of the climate system. By providing greater coupling of the terrestrial, atmospheric, and oceanic systems, simple models can deliver insights that are beyond the reach of the other classes of models. The insights could be especially valuable if the models aim for better understanding of the positive feedbacks that could destabilize the climate system. Improved simulation of the positive feedbacks is crucial if we are to understand the span of control of the climatic system.

Final Thought on Dangerous Anthropogenic Interference

Our current emissions of CO_2 and other greenhouse gases is driving atmospheric CO_2 to unprecedented levels. At some point, the disturbance could push the system beyond a tipping point and trigger runaway behavior owing to positive feedbacks. We are, in effect, conducting a dangerous experiment with the global system. This danger has been formally recognized by the 190 nations that have signed the United Nations Framework Convention on Climate Change. The framework convention aims to stabilize greenhouse gas concentrations at a level that would prevent *dangerous anthropogenic interference* with the climate system (UNFCCC 1992).

Massive reductions in the emissions of CO_2 are required if we are to limit the amount of CO_2 in the atmosphere. Achieving those reductions will require a combination of regulatory policies and placing a price on carbon emissions. These policies will transform the production and use of energy, especially in the electric power sector. Indeed, the electric power sector is expected to lead the way in reducing emissions, as explained by Ford (2008) and in appendix F.

Acknowledgment

- Figure 23.1 is adapted from a web schematic derived from multiple sources: the Center for Climate Research, Institute for Environmental Studies, University of Wisconsin at Madison; Okanagan University College in Canada, Department of Geography; *World Watch*, November–December 1998; and *Nature*.

Chapter 24

Concluding Perspective

You have reached the final chapter of the book, and you see the world differently than when you began. You can now visualize the stocks, flows, and feedbacks in an environmental system. And you have learned how to build computer models to test your theory of the dynamic behavior. At this point, you are probably wondering what to expect when you put the new knowledge to use. This chapter provides perspective on what to expect in your first application of modeling to environmental problems.

Modeling Is a Challenge

The first thing to remember is that modeling is a challenging and difficult endeavor. A textbook can describe the modeling process and can illustrate it with examples. If you have followed the lessons in this text, you may have the impression that modeling is easy. If this is your impression, brace yourself for a shock the first time you develop a model on your own. The process of modeling is much harder than it looks. You may find yourself getting off to a confusing start and becoming bogged down with little prospect for insight and learning.

Don't be discouraged. This happens to almost everyone in their first application. The problems that arise in the first few applications will be challenging and exasperating, far more difficult to solve than the pitfalls described in chapter 17. The pitfalls are minor problems, easily overcome; your first experiences will feel more like quagmires with no way out. Learn from these early experiences, and you will be able to avoid the quagmires.

Reminders on the Modeling Process

The most common cause of a modeling quagmire is lack of clarity in the purpose. This can happen if you do not invest the time to become familiar with the clients' views of the problem. It's crucial to define the nature of the problem early in the project; drawing a reference mode is the most important step. Some projects bog down from confusion over whether the project is aiming for a forecasting model or a model for learning. If you feel the team is yearning for a forecasting model, point them to other methods and wish them luck. If they are interested in gaining a shared understanding, proceed with system dynamics, with your eye focused on the reference mode. Some projects become trapped in a difficult situation if the reference mode is drawn with both short-term and long-term dynamics. The better approach is to draw separate patterns and develop several models. You'll learn more from developing a portfolio of models rather than a single, all-encompassing model.

The modeling process can also bog down if you forget the advice to start simple and expand the model in an incremental fashion, adding structure as needed to explain the dynamic problem. You should leave plenty of time to complete the steps of modeling several times, with active involvement by all members of the team. Remember the slogan to simulate early and often, the key lesson from the group modeling projects described in chapter 13. And remember to validate (i.e., to build confidence in) the model as you go along.

Don't be afraid to aim for a big-picture perspective in your first application. Your goal should be to close the key feedback loops in the system, regardless of where they may lead. The information and material flows will often cross disciplinary boundaries, so you should be willing to follow along by expanding the model boundary. The end result will be models that combine elements and ideas from several disciplines. The greatest insights on serious problems are often found with interdisciplinary models.

This advice is shared by experienced environmental modelers and confirmed from my own experiences working on a combination of energy and environmental problems. Similar advice is given by experts with experience in widely different systems. For example, Nicolson et al. (2002) describe ten heuristics (rules of thumb) for interdisciplinary modeling of ecosystems. They view interdisciplinary modeling as "part science and part craft" where "there are no general, infallible rules." They are certainly right about the lack of infallible rules, but there is a surprising consensus about pragmatic suggestions. Indeed, Nicolson et al.'s ten suggestions align quite closely with the list of twelve principles for effective use of system dynamics in business systems by Sterman (2000, 79). And both Nicolson's and Sterman's suggestions align with the advice in chapter 13 of this book.

It's encouraging to receive similar advice from a variety of experts, and you will find your own way to fit the general suggestions to your specific situation. As you make your way, brace yourself for major differences in perspective among the participants in the modeling process.

An Inward or Outward Perspective?

You've learned that insights on dynamic behavior are best found by looking inside the system at the feedbacks and delays. System dynamics facilitates the development of an endogenous theory of the dynamic behavior, and you expect the boundary diagram to confirm that you are hitting the bull's-eye. But brace yourself for discussions with individuals inclined to look outside the system for the cause of the problem. This tendency is particularly strong in business systems, where the source of the difficult behavior may lie in the decision making of the top managers and executives. One example is commercial real estate (chapter 19), where the boom and bust cycle arises from the decision making of real estate developers. However, most developers are not inclined to look inside the industry for an explanation of the cyclical behavior. More common is denial that cycles exist or blaming the problematical behavior on external factors that are beyond one's control. This point of view gets in the way of improved understanding and limits the options for improving system behavior. Also, this point of view translates into an appetite for forecasting models whose purpose is to predict the changes in the external factors that are blamed for the problem.

You know that dynamic behavior is shaped by a combination of external and internal forces. Exogenous factors certainly influence the dynamics, but the endogenous forces are often crucial to our understanding. An example is global climate change, described in chapter 23. You read about small changes in the sun's energy that can act as the Earth's pacemaker (Weart 2003). These small changes control the timing of the 100,000-year cycle of cooling and warming, but their signal is minute compared with the temperature swings at the terminus of the ice ages. We must look inside the climate system for an explanation of the rapid warming. It is to be found in positive feedbacks that can act like *vicious circles* to create the rapid warming evident in the geologic record. Understanding these vicious circles is the key to our assessment of the threat of global warming and global change.

Vicious Circles and the Boomtown Story

Your knowledge of systems will help you identify vicious circles in your first environmental application. You will know that it's important to make a good effort to estimate the relevant parameters and to simulate the strength of the vicious circle. But prepare yourself—others will not necessarily share your interest in the vicious circles.

I discovered this lack of interest in my first application of system dynamics. The topic was the boomtown problem in small towns in the energy-rich regions of the Rocky Mountain West. The towns were overwhelmed with huge population surges as thousands of workers moved in to work on an energy construction project. Figure 24.1 shows a vicious circle involving the link between productivity of these workers and the availability of housing. This diagram depicts the housing portion of the general shortfall of private and public facilities during the boom. The vicious circle could lead to major increases in the work force, cost overruns for the project developer, and aggravated conditions in the town. This vicious circle was important and widely known among the dozens of groups conducting boomtown studies during the 1970s and 1980s, as explained on the book's website, the BWeb.

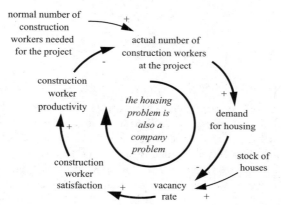

Figure 24.1. A vicious circle in a construction boomtown.

But the surprising result was the unwillingness of almost all the modeling teams to confront this problem. Their preference was to specify the number of construction workers as an exogenous input and to concentrate their calculations on the increased requirement for public and private facilities. Population and economic impacts were the center of their attention, and the calculations proceeded in a linear fashion: more workers means more population, which means fewer facilities per capita. These calculations were implemented with various degrees of detail, but there was no attempt to close the loop in figure 24.1. The information on lack of adequate facilities was seldom fed back to influence the productivity of the construction workforce.

The main reason for this reluctance was unfamiliarity with the feedback-oriented methods that you have learned in this book. Lacking a method to close the loop, most groups found reasons to argue that it was not important to do so. The standard argument was that the link to productivity was highly uncertain, and that adding an uncertain relationship would be confusing rather than illuminating. The many models missed the opportunity to add insight on the severity of the boomtown problem. They also missed the opportunity to represent the benefits to the construction company of lowering cost overruns by contributing to solutions of the boomtown infrastructure problem.

The exception to this tendency was a small collection of system dynamics models (BWeb). These simulated the feedback structure of the boomtown system and generated a fundamentally different perspective on the problem. My experiences from the boomtown situation were then reinforced by experiences in modeling of financial problems in the electric power industry. Once again, a vicious circle was at the center of the story.

Background on the Death Spiral

Electric power is a massive industry with major impacts on the environment. It is also a capital-intensive industry with a rich tradition of long-range planning. System dynamics has been used extensively in this industry, and practitioners have accumulated an impressive record of implementation (Ford 1997). The focus of this story takes place in the 1970s, a period of extreme financial difficulties in constructing power plants to keep pace with rapidly growing demand. Power plants had become larger and larger, and they were taking much longer to build. Their costs were much higher than in the 1950s and 1960s. Utilities found themselves with declining internal funds to pay for construction, and they faced unusually high costs of capital from Wall Street. The financial problems were painfully evident from the headlines in the business press:

> *Utilities: Weak Point in the Energy Future*
>
> *Utilities Need Help—Now!*
>
> *Con Edison: Archetype of the Ailing Utility*
>
> *Electric Industry Cutback Could Result in Blackouts*

Some utility companies cut back on the construction of new power stations, and some experts feared that the long, successful history of keeping the lights on was about to be broken. The utility executives asked the regulators for approval of higher electric rates, and regulators responded by approving many rate increases. But the regulators were not sure that meeting all the requests would solve the problem. They wondered whether higher rates would depress the sale of electricity and possibly reduce utility revenues. If this were to happen, the utility might return for yet another rate increase. Indeed, this is just what happened, and news headlines spoke of a vicious circle involving higher rates:

> *The Vicious Circle That Utilities Can't Seem to Break: New Plants Are Forcing Rate Increases—Further Cutting the Growth in Demand*
>
> *The Electricity Curve Ball: Declining Demand and Increasing Rates*

Figure 24.2 portrays the vicious circle along with other feedback loops at work in the system. The vicious circle is labeled the *death spiral,* a term commonly used to describe the possibility for runaway behavior in a capital-intensive regulated industry. The indicated price in the diagram stands for the price of electricity that regulators would normally allow to generate the allowed revenues. The actual price follows the indicated price after a delay for regulatory review. An increase in the actual price would lead to lower electricity consumption and yet another request for a price increase. Both utility executives and their regulators wondered whether this loop would lead to a spiral of declining consumption and increasing rates.

Figure 24.2 shows how the death spiral fits within the larger system. The outer loop shows negative feedback: an increase in the actual price would lead to a decline in electricity consumption, reduced capacity initiations, reduced capacity, a lower rate base, a reduced revenue target, and a subsequent reduction in the price. The third loop is the construction loop, a goal-oriented loop to bring installed capacity into balance with required capacity. This third loop includes only one delay, but the delay was quite long for utilities building large coal or nuclear power plants.

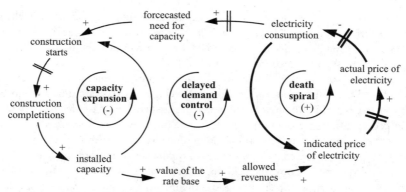

Figure 24.2. Key feedback loops in the utility system.

Utility Efforts to Simulate the Death Spiral

The death spiral was on the minds of utility managers, regulators, and even the headline writers in the newspapers. Large utilities had decades of experience in the use of models of all kinds—operational, financial, accounting, forecasting, and capacity planning. A large utility might well have 30 models to deal with different aspects of the company and its environment. Each of the models might be staffed by 10 or more employees, so the important lesson from this period is what 300 or more modelers could do to help top management better understand the death spiral.

The common utility approach was to link a series of departmental models together, as shown in figure 24.3. (This diagram is limited to 3 models, but you should imagine 30 or more models linked in a similar fashion.) The diagram begins with a set of electric rates that are fed into a model of electricity demand. The output of the demand model is a forecast of electric load for each of 20 years in the future. These results are then fed to a capacity expansion model, whose output is a plan for new power plant construction. This plan is the input for a third model to calculate the utility revenue requirements and electric rates. The electric

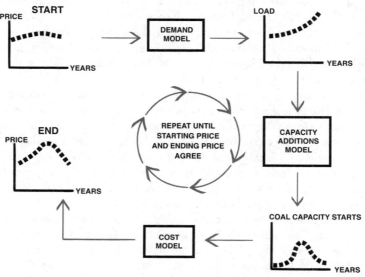

Figure 24.3. Linking existing models together in a large electric utility.

rates at the end of the process are then compared with the rates used at the start. If the two sets of rates were different, the analysts might adjust the input rates and repeat the entire sequence of calculations. Through artful manipulation of the starting rates, the modeling team might obtain a consistent set of projections.

This approach was popular with utilities because it allowed them to take advantage of existing models. The existing models were often developed in separate departments and had grown to be quite complex in order to serve each department's need for detail. They were often implemented in different programming languages, and they sometimes resided on different computers. The principal drawback was the long time interval needed to complete a single calculation. Completing many calculations to find internal consistency took even longer.

In practical terms, the approach seldom resulted in a consistent set of projections. The more common approach was to simply ignore the inconsistencies that arose from the lack of information feedback within the system. However, ignoring the inconsistencies is equivalent to ignoring the price feedback that creates the death spiral. Connecting existing models together was not the way to simulate the key feedbacks in the system.

Utilities Construct Single Models of the Corporation

Some utilities reached this conclusion and turned to a fresh approach. The idea was to build a single model with simplified representations of the calculations from the 30 or more models. The single models were designed with one programming language and to reside on one computer. They were successful in reducing the time interval required to obtain a single projection. But the vast majority of these models failed to simulate the information feedback from the projected prices to the growth in electricity demand. This failure was documented in a modeling forum with 12 corporate models and again in a separate workshop with a collection of 13 models (BWeb). A total of 25 models were represented in these discussions; only two simulated the price feedback needed to study the death spiral. These exceptional models were the only models to use system dynamics.

Insights from the System Dynamics Models

Further information on the system dynamics models is given on the BWeb. The supporting information explains that the focus on information feedback led to dramatically different conclusions about utility policy. One of the conclusions was that utilities could lower their financial vulnerability by shifting investments from long-lead-time technologies (e.g., nuclear plants or large coal plants) to short-lead-time technologies (e.g., small coal plants or geothermal stations). We also learned that the debilitating effects of the death spiral would be greatly reduced if the utilities could slow the growth in demand.

Utility managers and regulators gradually came around to these conclusions, and the 1980s were distinctive for the shift from large power stations with long lead times to smaller, shorter-lead-time resources. The move to smaller scale was evident in the cancellation of nuclear construction and a shift to smaller coal plant construction. The 1980s were also known for the increased emphasis on conservation. Utilities no longer viewed rapid growth in demand as desirable. If their customers could be encouraged to use electricity more efficiently, the pace of demand growth could be slowed, and utilities could reduce the risks of carrying long-lead-time construction projects to completion.

Figure 24.4 dramatizes the shift to conservation with an image of the Northwest region wrapped in a blanket of insulation. Northwest utilities went from home to home to encourage their customers to take advantage of cost-effective measures like insulation and better lighting. Utilities provided audits, loans, and direct financial incentives to encourage investment in ef-

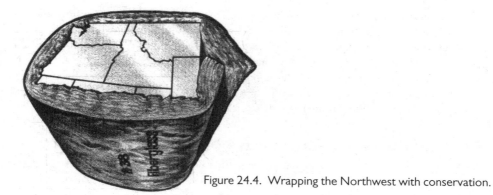

Figure 24.4. Wrapping the Northwest with conservation.

ficiency. These programs would have been inconceivable in previous decades. Indeed, it is probably hard for many business leaders to appreciate why a private company would encourage its customers to use less of its product. But, to their credit, utility leaders saw the wisdom in encouraging efficiency. They had learned that it made good business sense to help customers plug the leaks in their houses. Helping them plug the leaks was much less risky than investing in long-lead-time power plants.

General Lessons

There are important lessons that could transfer from the boomtown and utility situations to your own situation. The first conclusion applies if you are working with an organization with many separate models. Stitching the existing models together will not lead to a simulation of the key feedbacks in the system. The many models in the different departments of a large company or agency are typically developed to deal with different problems. These models may involve different time frames, different geographic areas, different methodologies, and different software. Linking their input and outputs together does not portray the major feedbacks in the system. If you want to simulate the feedbacks, you need to start fresh with a new model. A second, general conclusion is that the new model should be a system dynamics model. Other modeling methods do not focus on the feedbacks in the system. They were designed for other purposes and are not useful when you need to close the loops in the system.

System dynamics models are especially helpful when we need to understand the vicious circles in the system. But you must have the courage to include highly uncertain parameters in your model. Nicolson et al.'s (2002, 378) advice is relevant here—be willing to "guess at the unknown." The corresponding advice from chapter 13 is to proceed with your estimate if you know "it is better than zero." You should expect that highly uncertain parameters will be involved because the vicious circles are often inactive when the system operates within its span of control (figure 10.5). Think of the vicious circles as sitting in the wings, waiting to take effect when the homeostatic forces are overpowered. When they do take effect, there can be rapid change and perhaps runaway behavior. This leaves us with little recorded data to support parameter estimation, so it's important that you draw from the full range of information sources shown in figure 13.3.

The final lesson learned from the utility situation is the most important lesson of all—system dynamics modeling has the potential to dramatically change the thinking of the participants. In the utilities example, system dynamics simulations contributed to the transformation in thinking during the 1980s. The industry had prospered for decades thinking that rapid growth in demand and construction of ever larger power plants were the keys to their success.

But the 1980s were characterized by a transition to smaller power plants and a serious effort to slow the growth in demand through utility-funded conservation programs. System dynamics modeling was part of this transformation. The modeling efforts helped regulators and managers see the benefits of investing in conservation programs and of building power plants with shorter lead times. The new strategy turned out to be beneficial to the utilities' financial performance, and it also helped lower the customers' monthly bills. The transformation was good for both stockholders and customers, and it was good for the environment as well.

The Benefits of Interdisciplinary Modeling

The boomtown and utility examples demonstrate the value of interdisciplinary modeling. The boomtown models succeeded in incorporating impacts in the housing, municipal, and retail sectors of the local economy. They also included demographics and the feedback to construction productivity. An interdisciplinary approach was needed to follow the closed chain of cause and effect around the loops, like the loop shown in figure 24.1. An interdisciplinary approach was also important in the utility modeling of the death spiral. The models dealt with operations, accounting, finance, regulatory behavior, consumer behavior, and capacity expansion planning. The key loops worked their way through all these sectors. Their simulation required interactions with experts from multiple disciplines.

Examples of interdisciplinary modeling appear in several chapters of this book as well. The Mono Lake case is the first example. It began with the hydrology model and then included the population biology of the brine shrimp. The combination of the two models (figure 5.13) goes beyond a hydrological model with elevation as a proxy for wider impacts. The brine shrimp appear in the expanded model, so export can be tested with the brine shrimp population as an explicit goal.

The salmon model is a second example. The initial model (figure 15.1) is population biology, but the student expansion (figure 15.15) drew on fluvial geomorphology. (The student was a fluvial geomorphologist with practical experience in river restoration.) Exercise 15.16 asks you to combine the models and simulate a $7 million restoration project followed by salmon harvesting. If you did the exercise properly, you discovered the dramatic result that surprised the original student—that the salmon harvesting can pay back the $7 million in less time than it took to restore the river. This surprising result arose from experimentation with a combination of models from different disciplines. The geomorphologist observed that such a model would seldom be constructed, owing to the separation of experts into their different disciplines.

Chapter 23, on CO_2 accumulation, explains the benefit of interdisciplinary modeling to build our understanding of climate change. The climate system is the ultimate challenge to interdisciplinary research and modeling, as the positive feedbacks in figure 23.6 illustrate: their simulation requires expertise in the atmospheric, terrestrial biomass, and soil systems. The vicious circle involving increased soil carbon decomposition still remains to be resolved with further research and the use of fully coupled models that close the loops in the system.

The Challenges of Interdisciplinary Modeling

Interdisciplinary modeling requires time and effort by participants with expertise in the different disciplines. Nicolson et al. (2002, 378) describe the best participants as those with a deep grasp of their own disciplines and the "willingness to simplify their field and to guess at the unknown factors." They advise participants to approach interdisciplinary modeling with a sense of humility. Participants may be world-class experts in their field, but "they are all likely to be amateurs when it comes to the system as a whole." Cockerill et al. (2007, 31) remind us that participants in cooperative environmental modeling projects will "still be on a steep learning curve and all lessons are valuable." The communication challenges are substantial and difficult to overestimate (Nicolson et al. 2002; Cockerill et al. 2007; Saito et al. 2007).

The difficulties of interdisciplinary modeling can be traced to our educational systems. Knowledge has been broken down into disciplines to permit specialization and deep understanding within each discipline. The disciplinary boundaries are useful and often inevitable. But the reductionist perspective leaves us poorly informed on other disciplines, and we are often inclined to linger within the safety of our own discipline. But the serious problems of the environment are interdisciplinary. For example, the National Research Council (NRC 2004) describes aquatic systems problems as requiring contributions from the physical, chemical, biological, and social sciences. A similar mix of disciplines is required to address problems in the nitrogen cycle, one of the grand challenges identified by the National Academy of Engineering (NAE 2008). However, communication among experts from the different disciplines is difficult, and there are few programs to educate and train students on interdisciplinary approaches (Saito et al. 2007).

System dynamics can aid in the interdisciplinary training and education of students by encouraging contributions from different disciplines in a single model. The clarity of the stock and flows and the emphasis on feedback control provide a common language that can be understood by scientists from many disciplines. And system dynamics software aids in the formulation and testing of models in an iterative fashion. The approach stresses clarity and transparency, and is ideally suited for cooperative modeling involving participation by experts from multiple disciplines and by stakeholders. These and other reasons have led system dynamics to become a common platform for cooperative modeling of environmental systems (Stave 2002; Tidwell et al. 2004; Van den Belt 2004; Videira 2005; Beall 2007; Cockerill et al. 2007).

Environmental Problems in the 21st Century

Environmental problems are numerous, substantial, and interdisciplinary. They challenge our understanding and our willingness to become involved. System dynamics can add to our general understanding, and it can help groups and individuals who are committed to problem solving. This book demonstrates the system dynamics method in examples ranging from the local to the global. Local challenges include increasing the water level in Mono Lake and increasing the salmon population in the Tucannon River. The global challenges include reducing the DDT in the oceans and reducing the CO_2 in the atmosphere. The CO_2 accumulation has led to global warming and other changes on a global scale.

Global warming has emerged as the dominant environmental challenge of the 21st century. The goal in the first half of the century should be to cut emissions sufficiently to stabilize atmospheric CO_2 and other greenhouse gases. Stabilization is needed if we are to reduce the likelihood of dangerous anthropogenic interference with the climate system. Hopefully, the next 50 years will witness the nations of the world adopting stringent policies to reduce the emissions of CO_2. The most likely policies will be a combination of direct regulations and putting a price on emissions. The mitigation of climate change will lead to a transformation of the way energy is produced and consumed around the world. The transformation will be especially pronounced in the electric power sector.

We can hope for policy changes, but they are highly uncertain. However, one thing is certain—climate changes are coming, owing to the momentum in the global carbon cycle. The 21st century will bear witness to rising temperatures, rising sea levels, and a wide variety of global changes. All of the biosphere will be affected, challenging us to adapt to the new conditions.

Adaptation and mitigation are the dual challenges of climate change. They stand out among the many environmental problems of the new century. They will challenge you to get involved, and you will be confronted with dynamics that are difficult to understand. May you find system dynamics helpful in building understanding. And may your efforts lead to a better world.

Acknowledgment

- The drawing in figure 24.4 is used courtesy of the Bonneville Power Administration.

APPENDIXES

REVIEW AND ADVANCED METHODS

Appendix A

Review of Units

We must pay attention to the units of measurement if we expect to build useful models. We start by selecting units that will make the meaning of the variables clear to the people who will use the model. Then we check the equations to make sure that the variables are combined in a consistent manner. This appendix provides an opportunity to refresh your understanding of units. It then offers pragmatic advice on the use of units in system dynamics models.

Review

You have probably already learned the simple rules for combining units. One of the first rules is that variables to be added or subtracted must be expressed in the same units. Addition or subtraction leaves the units of measurement unchanged. But this is not the case with multiplication or division. To illustrate, consider a rectangular solid with dimensions measured in centimeters (cm):

> length = 10 cm
> width = 5 cm
> height = 4 cm

To find the volume, we multiply 10*4*5 to get 200. You know that we multiply the units to get the units for volume: cm*cm*cm, which gives cm^3, or cubic centimeters. If you use division to find an attribute of the solid, you would divide the units as well as the numbers. For example, we might define the relative length as 10 cm divided by 5 cm, which is 2. In other words, the length is twice the width. When dividing cm by cm, the units cancel out, and we are left with a number that does not have any units. Such a number might be called a *unitless number*, a *normalized number*, or a *dimensionless number*. This book uses the term *dimensionless* for a number that has no units.

The most famous dimensionless number can be found by measuring the circumference of a circle and dividing it by the diameter. Suppose we measured a circumference of 31.4159 cm and then divide by a diameter of 10 cm. We would get 3.14159, a number with no units. This is the famous number π (pi). It gives a fundamental property of all circles. The circle can be big or small, and it can be measured in any country. It doesn't matter what units are used; they will cancel out when we take the ratio of the circumference to the diameter.

Dimensionless Variables

Dimensionless variables deserve special attention because they often reveal the essential features of a system. What is true for circles is true for systems in general: we should expect the fundamental properties to reveal themselves in dimensionless form. Hastings (1997, 95) argues that it is no coincidence that a key indicator turns out to be dimensionless:

> One important concept that arises is that of nondimensional parameters: parameters without units. We could measure biomass of a population in kilograms, pounds, tons, yielding different values for a carrying capacity, since carrying capacity and the population have the same units. But the stability of the model and of the population obviously cannot change as we change the units. Similarly, the units of intrinsic rate of increase are inverse time units, and the stability of the population cannot depend on whether time is measured in weeks, years or centuries. Our conclusion is that stability can only depend on nondimensional groups of parameters, combinations for which all the units cancel.

Most dimensionless variables in this book are formed by dividing two variables measured in the same units. The units cancel out, and we are left with a dimensionless number. However, it's sometimes useful to include a dimensionless index in a model. An example is the "sound index" in exercise 3.10. It takes on the values 1, 2, or 3. These numbers have no units because they are simply our way of numerically summarizing the interview results from Joe's description of filling the gas tank.

The Importance of Units in Modeling

The importance of units was appreciated at the very outset of the field of system dynamics. Forrester (1961, 64) observed that dimensional analysis plays an important part in guiding equation writing in engineering and in the physical sciences. He argued that the same standards should apply in the emerging field of industrial dynamics:

> In writing equations, meticulous attention must be given to the correctness of the dimensions (units of measure) associated with each term. . . . Inconsistency in dimensions often divulges an incorrect equation formulation. The units of measure of all variables and constants should be precisely stated and should be checked for consistency in each equation that is written. Carelessness on this point can lead to much needless confusion.

This good advice applies to all models, not just models of industrial systems. When modeling biological or ecological systems, we should heed the advice of Nisbet and Gurney (1982, 21):

> Any equations representing a real biological system should be valid irrespective of the units in which we measure the quantities involved in the system. This will only be the case if both sides of the equation have the same dimensions and if, on either side of the equation, quantities to be added or subtracted have the same dimensions.

They emphasize that this simple advice is of considerable practical importance. Regularly checking the dimensions in the course of a long calculation provides a "rapid check on algebraic accuracy." They note that a dimensionally correct equation need not be correct, but that "a dimensionally incorrect equation is invariably nonsense!" If you develop a regular habit of checking the units, you will find that it improves your progress toward a clear and consistent model.

Units for the Stocks and Flows

The advice in chapter 3 on stocks and flows is good way to start. Each flow must be measured in the units of the stock divided by the units that you have selected for time. If the stock is measured in tons, and time is measured in years, all flows in and out of the stock will be measured in tons/yr. If you find yourself connecting a flow that is not measured in the correct units, take the opportunity to rethink the model structure. Chances are, you will come up with a better combination of stocks and flows.

Units for the Rest of the Model

Once the stocks and flows are in the proper units, you should proceed to each of the converters to make sure they are consistent as well. A good way to check the entire model is to include the units in an equilibrium diagram. This diagram is convenient because you can check the units and the numbers at the same time. As an example, look at the equilibrium diagram for the third model of Mono Lake (figure 6.1). The units for the stocks and flows can be checked with a quick glance at the diagram. The units for several of the converters are easily verified as well. Quick checking is possible because the model uses "friendly algebra," which involves a simple combination of add, subtract, multiply, or divide. But the specific gravity is the exception. This is a dimensionless number to show the weight of Mono Lake's water relative to fresh water. You cannot check the units in your head, but you can write out the equation (see chapter 5) and verify that the units make sense.

Three of the converters in figure 6.1 use graphical functions (~), and their units may not make sense. Take the elevation as an example. It is measured in feet above sea level; water in lake is measured in thousand acre-feet (KAF). How can we get a result in feet by an algebraic formula with an input measured in KAF? The answer is we can't. The graphical function (~) does not use algebra. Rather, it shows a nonlinear relationship between one variable and another, regardless of their units. Unit consistency is not required when we use graphical functions.

Automatic or Manual Checking?

Writing out the units by hand on an equilibrium diagram is a useful exercise for several reasons. We check for consistency by simply looking at the variable names and thinking through the equations. If we have used friendly algebra, the units will easy to check. On the other hand, some experts prefer to use the software tools to check the units automatically. Both Stella and Vensim allow you to enter the units in a separate window. If you take the time to enter all the units, the software can check for consistency and report back errors. This automatic checking comes in handy for large models and if you happen to write complicated equations. Either approach is useful; the main thing is to check the units. I recommend the manual method, especially for first-time modelers.

Concluding Advice

Another pragmatic piece of advice is to select units that are familiar within your organization. If you are a newcomer, you may encounter unusual and unfamiliar units. Take the time to become familiar with them. For example, if you have not worked with watersheds, you might be surprised by the unit *acre-foot*, used to measure volume of water. An acre-foot is the volume formed by covering 1 acre of area to a depth of 1 foot. If watershed planners talk in acre-feet, your model should talk in acre-feet.

After learning the common units, learn whether you are dealing with unusually large numbers. If river flows can range from 1,000 to 50,000 acre-feet/yr, you should specify your units in thousands of acre-feet/yr. This was the practice in the Mono Lake model. We used KAF to stand for 1,000 acre-feet, and you can see the use of this unit throughout the model (figure 6.1). Following this advice will help you avoid the errors from entering the wrong number of zeros, as in the error in entering the births in figure 2.3. This practice is followed throughout this book, and the units and their abbreviations are explained each time they are used.

The final word of advice is to be sure that each and every variable has a unit. If you cannot specify a unit of measurement, it does not make sense to proceed with formulating equations. Your inability to specify the unit is a sign that you need to firm up your understanding of the system.

Exercises

Exercise A.1. Leaf area index

The leaf area index (LAI) is useful for explaining how plants obtain their energy from photosynthesis (Pratt 1995, 8). It describes the relationship between leaves and the amount of sunlight they intercept as light passes through the various layers of leaves on a plant. The LAI is defined as the ratio of the leaf area per unit of ground area. What are the units for LAI in the United States, where we measure area in square feet? What are the units for LAI in Germany, where the area is measured in square meters?

Exercise A.2. Verify that *Reynold's number* is dimensionless

A famous number in the study of fluids is the *Reynolds number*, named after Osborne Reynolds. Reynolds was interested in the conditions that would cause the flow of a fluid to change from laminar flow to turbulent flow. He summarized his experiments by reporting results for different values of R, a combination of the velocity (V), diameter (d), density (p), and viscosity (u):

$R = Vdp/u$

Since R teaches us about a fundamental property of fluids, we suspect that it must be dimensionless. Prove that R is dimensionless if:

V = mean velocity of the fluid in the pipe (meters/second)
d = diameter of the pipe (meters)
p = density of the fluid (kilograms per cubic meter)
u = viscosity of the fluid (kilograms per meter-second)

Exercise A.3. Value of the conversion factor

Figure A.1 shows a model with the water in the reservoir measured in millions of acre-feet (MAF). The flows are in MAF per year. If there is a flow of 10 MAF/yr, we would see 10 million acre-feet passing by the measuring point in 1 year. However, river flows are often measured in thousands of cubic feet passing by the measuring point in 1 second (TCFS). The extra variable in the model is called *TCFS outflow*, so we have two different ways to show the outflow. The equation for the new variable is:

TCFS outflow = conversion factor * outflow

There are 43,560 cubic feet in an acre-foot, and there are 8,760 hours in a year. What is the numerical value of the conversion factor?

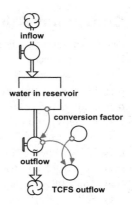

Figure A.1. Conversion factor.

Exercise A.4. Find the errors

Find an error in each of these equations based on the units shown in table A.1.

equation #1: $A = C + F*G + T$

equation #2: $A = B*e^{ET}$

equation #3: $A = C*e^F$

equation #4: $A = B*T + F*G + D$

equation #5: $D = (A/B) + B*T$

equation #6: $G = (A + C)/(F + D)$

equation #7: $B = A*T$

equation #8: $B = (A/F) + (B*T/A)$

equation #9: $X = A*D/C$

Table A.1. Units.

Variable	Units
A	insects
B	insects/yr
C	insects
D	dimensionless
E	1/yr
F	acres
G	insects/acre
H	years
X	insects/yr

Exercise A.5. Find the bad variable

One of the variables in figure A.2 does not make sense if the variables have the units shown in table A.1. Which variable is the problem?

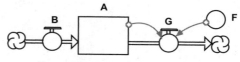

Figure A.2. Which variable is wrong?

Exercise A.6. Verify error free

The variables in figure A.3 have the units in table A.1. The model does not have any obvious errors in units. To confirm this is true, write an equation for the flow B and the flow X.

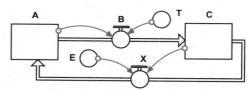

Figure A.3. Write equations for the flows.

Exercise A.7. The units for utility?

The discrete choice model in figure 16.2 uses the following market share equation for a multinomial logit model:

$$MSv = \exp Uv \bigg/ \sum_{i=1}^{5} \exp Ui$$

MS_v is a dimensionless number to represent the market share of vehicle v. The U_v stands for the utility of vehicle v. The term *utility* is roughly the same as satisfaction. The *exp* in the equation stands for the exponential function. What are the units of U if the units of this equation are to be consistent?

Exercise A.8. Utility measured in utils

Some experts in discrete choice theory discuss utility as a measurable quantity with units of *utils*. Does this unit make sense given your answer to the question in exercise A.7? (To learn more about this puzzling situation, look to the economics literature on cardinal utility versus ordinal utility.) If we agree that U is measured in utils, what are the units for coef_6 in the Stella equations in table 16.3?

Further Reading

- Harte (1988) provides an extensive list of useful numbers and their units for the environmental scientist.
- The "critical primer" by Riggs (1963) in *The Mathematical Approach to Physiological Problems* includes numerous exercises on units.
- Jacobsen and Bronson (1987) describe the challenges of selecting variables (and their units) for system dynamics models of sociological systems.

Appendix B

Review of Exponential Growth

This book assumes that you have learned introductory algebra and how to interpret charts and tables. These are the basic prerequisites for using the book. If you've had a course in calculus, you'll find that your knowledge of differentiation and integration is helpful. You'll recognize that stocks act to integrate the effects of the flows over time. And if you've studied differential equations, you'll recognize that a system dynamics model may be viewed as a coupled set of first-order differential equations. These insights will be helpful, but they are certainly not essential to build and test useful models. Indeed, many students are learning system dynamics in K–12 classrooms with the same software used in this book (Draper and Swanson 1990; Fisher 2007).

This appendix will help refresh your memory of exponential growth and exponential decay, topics normally covered in high school. These topics merit review because of the tendency for environmental systems to grow in an exponential fashion. And when they are in decline, the natural pattern is usually exponential decay. A review of these fundamental patterns is a good way to prepare for modeling of environmental systems. If you can work through the implications of exponential growth in your head, you will be in a much stronger position to judge the plausibility of model simulations.

Exponential Growth

Figure B.1 shows an example of how the exponential function appears when we are describing systems that grow exponentially over time. The equation would be:

$$Y(t) = e^{rt}$$

where t is time and r is the growth rate. The graph shows the special case where r is 0.069/yr and time is in years. When t is zero, we have $Y(t) = 1$. By the time t reaches 10 years, the product of r and t is 0.69, and $Y(t)$ has grown to 2. When t reaches 20 years, the product of r and t is 1.39, and $Y(t)$ has grown to 4. The graph in figure B.1 shows that an exponentially growing function will double in size in a fixed time interval. This time interval is called the *doubling time*. For a growth rate of 0.069/yr, the doubling time turns out to be 10 years. The product of the growth rate and the doubling time must

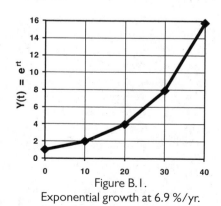

Figure B.1.
Exponential growth at 6.9 %/yr.

be 0.69. (You can check this by raising *e* to the power 0.69 with your hand calculator. The answer will be 2.)

The purpose of this appendix is to provide an easily remembered rule for estimating the effect of exponential growth. For this purpose, we round the 0.69 to 0.7. Growth rates are often expressed in %/yr, so we would say:

growth rate (%/yr) * doubling time (years) = ~ 70

Let's call this the *rule of 70*. Table B.1 shows some examples of growth rates and doubling times that obey this rule. The first four examples involve time in years. Notice that the product of the growth rate and the doubling time is always 70. If we know the growth rate, we can estimate the doubling time. However, in some cases we know the doubling time, but not the growth rate. When this is the case, we use the "rule of 70" in reverse. For example, if we know the system doubles in 5 years, we know that the growth rate is 14%/yr. If it doubles in 10 years, the growth rate must be 7%/yr. We should keep such examples in mind whenever we are simulating an exponentially growing system. Once we know the doubling time, we can guess the size of the system in the future. The guessed values can provide a good check on what we see in the simulation model.

Table B.1. Growth rates and doubling times.

Growth rate	Doubling time
3.5%/yr	20 years
7%/yr	10 years
14%/yr	5 years
28%/yr	2.5 years
3.5%/day	20 days
7%/day	10 days
14%/day	5 days
1%/century	70 centuries
7%/century	10 centuries
14%/century	5 centuries

Exponential Decline

Figure B.2 shows an example of how the exponential function appears when we are describing systems that decline exponentially over time. The equation would be:

$$Y(t) = e^{-rt} = 1/e^{rt}$$

where *t* is time and *r* is the rate of decay. The graph shows the special case where *r* is 0.069/yr and time is in years. When *t* is zero, we have $Y(t) = 1$. By the time *t* reaches 10 years, the product of *r* and *t* is 0.69, so $Y(t)$ has declined to 1/2. When *t* reaches 20 years, the product of *r* and *t* is 1.39, and $Y(t)$ has declined to 1/4. The graph shows that the system is cut in half in a fixed time interval. This interval is called the *half-life*. We round the 0.69 to 0.7 and express the decay rate in %/yr. The rule for the half-life can be written as:

decay rate (%/yr) * half-life (years) = ~ 70

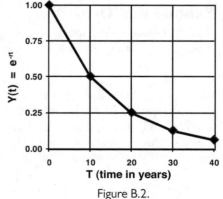

Figure B.2.
Exponential decay at 6.9%/yr.

So the "rule of 70" can help us describe exponential decay. Table B.2 shows examples that obey the rule with exponential decay. The first four examples involve time in years. Notice that the product of the decay rate and the doubling time is always 70. If we know the decay rate, we can estimate the half-life. However, in some cases we know the half-life, but not the decay rate. If this is the case, we use the "rule of 70" in reverse. For example, if we know the system is cut in half in 5 years, we know the decay rate is 14%/yr. If it takes 10 years for the system to be cut in half, the decay rate must be 7%/yr. We can keep such examples in mind whenever we are simulating a system in exponential decay. Once we know the half-life, we can guess the size of the system in the future. The guessed values can provide a good check on what we should see in the simulation model.

Exercises

Table B.2. Decay rates and half-lives.

Decay rate	Half-life
3.5%/yr	20 years
7%/yr	10 years
14%/yr	5 years
28%/yr	2.5 years
3.5%/day	20 days
7%/day	10 days
14%/day	5 days
1%/century	70 centuries
7%/century	10 centuries
14%/century	5 centuries

Exercise B.1. Guessing China's population

The book *Land of 500 Million: A Geography of China* was published in 1955. Suppose the opening paragraph says China's population had been growing in an exponential fashion at 3.5%/yr. And suppose this rate of growth were to have continued to the year 1995. What is your guess for the population in 1995?

Exercise B.2. Toxic decay

A train wreck in 1997 leads to a spill of 480 tons of the chemical XYZ, which becomes absorbed in the soil. Experts are called to the scene. They announce that the chemical will degrade with a half-life of 20 years, and they say there will be no danger once the amount of chemical in the soil falls below 30 tons. In what year will the soil be safe?

Exercise B.3. Doubling times for large animals

Pratt (1995, 58) summarizes the range of biotic potentials for four broad classes of animals. The *biotic potential* is defined as the rate of exponential growth expected when the animal is free to grow without constraints from the environment. For large mammals, the biotic potential can range from a low of 2%/yr to a high of 50%/yr. What is the corresponding range of doubling times?

Exercise B.4. Doubling time in days

Pratt reports that biotic potentials for small mammals can be as high as 800%/yr. The 800%/yr growth rate suggests that years is not an appropriate time for simulating this population. Change the growth rate to %/day and give the doubling time in days.

Exercise B.5. Reach $1 million

You open a savings account on the day your child is born. The bank guarantees that the balance in the account will grow exponentially at 10%/yr forever. Your goal is to have $1 million in the account by the time your child reaches 70 years of age, and you plan no further deposits. How much do you deposit in the account?

Exercise B.6. Advice for the king

D. H. Meadows et al. (1972, 29) describe a Persian legend about a clever courtier who presented a beautiful chessboard to his king. The courtier asked the king to show his favor by simply giving the courtier 1 grain of rice for the first square on the board. He would then follow that with 2 grains for the second square, 4 grains for the third, and so on. Suppose the king turns to you for advice. He tells you that the kingdom's entire storage of 1 million tons is available. You count 64 squares on a chessboard, and you know there are 5,760 grains in a pound and 2,000 pounds in a ton. Does the king have enough rice to accept the chessboard?

Exercise B.7. Rice weevil optimum

Odum (1971, 182) describes the doubling time for the rice weevil at just under 1 week, provided the temperature is near the optimum value. Assume there are no limits on growth and the weevil population doubles in 1 week. How long will it take for the weevils to increase by a millionfold?

Exercise B.8. Rice weevils feel the heat

Odum reports a doubling time of 5.78 weeks if the temperature is 4.5°C above the optimum. How long will it take for the weevils to increase by a millionfold under these conditions?

Exercise B.9. World population growth

Odum (1971, 182) notes a doubling time of around 35 years for the world's human population during the 1960s. What is the corresponding rate of exponential growth?

Exercise B.10. Exaggeration or truth?

Suppose someone from the 1960s looked at population trends over the previous three or four decades and saw the same data as Odum. That person might say that "the population has increased as much in the last 35 years as in all previously recorded time." This sounds like an extraordinary statement. Is it an exaggeration or the truth?

Exercise B.11. Kaibab growth

Chapter 21 describes a herd of 4,000 deer whose biotic potential is around 50%/yr in the absence of predators. The deer herd begins to grow when predators are removed from the region. If the herd continues to grow at 50%/yr, how long will it take for the population to exceed 120,000?

Exercise B.12. Interval between pig festivals

The book's website, the BWeb, describes the Tsembaga people in the New Guinea Highlands. The Tsembaga clan maintain a pig herd that grows exponentially at 14%/yr. When the pigs become too numerous for the women to care for, the Tsembaga launch a pig festival that results in the slaughter of 75% of the pig population. The remaining pig population begins to grow again at the same rate as before, and there are approximately the same number of women as before. How long will it take before the pigs once again become too numerous to care for?

Exercise B.13. Save the lily pond

There is a French children's riddle (D. H. Meadows et al. 1972, 29) about a pond with a water lily in it. The lily doubles in size every day. In 30 days, it will cover the entire pond, killing the other creatures in it. The owner wants to avoid that, but he sees no hurry. He will wait till the lily plant covers half the pond before taking action. On what day will he take action?

Exercise B.14. Taking action for the pond

Suppose the owner of the lily pond takes big action—he doubles the size of the pond. How much time will elapse before he needs to take more action to save the pond?

Appendix C

Software Choices and Individual-Based Models

This appendix describes a variety of software programs for system dynamics modeling. It begins with spreadsheet programs, since these are probably the most widely recognized software. The appendix demonstrates that spreadsheets can be constructed to simulate behavior over time. But this is not their best use; spreadsheets are better used to support dynamic modeling in programs like Stella or Vensim.

Most current applications of system dynamics are implemented in Stella, Vensim, or Powersim. Many useful applications from earlier decades were implemented in Dynamo, the first software to implement the ideas of Forrester (1961). This appendix shows the visual similarity of Stella, Vensim, Powersim, and Dynamo. It then describes related software such as Simile, Simulink, and GoldSim. These programs also provide icon-based methods for numerical simulation of system dynamics models.

The appendix then changes the topic from system dynamics modeling to individual-based modeling. This new type of model represents each and every individual in the population. If we know the rules to describe individual behavior, the models could lead to insights on the emergent behavior of the entire population. These results can then be used to support the aggregate relationships that appear in system dynamics models.

Spreadsheets: The Most Familiar Software

Spreadsheet programs are in wide use. If you are looking for software that is widely known, spreadsheets might be the answer. So it is natural to ask whether we can build system dynamics models in a spreadsheet? And if we can, should we do so?

Let's use the sales model from chapter 7 for illustration. It may be implemented in a spreadsheet if we make the effort to deal with time. Time is the most important variable in dynamic models. Stella and Vensim make it easy to define and control the passage of time, but we have to deal with time ourselves in a spreadsheet. Table C.1 illustrates with the sales model. It shows that time occupies an entire row and is advanced by one DT (the step size in the simulation) as we move from column to column. DT is set to 0.25 year, so we need four columns for the first year. (The other 19 years of the simulation are outside our view.)

The spreadsheet begins with the constants; these may be located at the top, because their values will not change over time. We assign a row to time, and a new row for each of the stocks. We have only one stock in the sales model (the size of the sales force). We set the initial

Table C.1. Introductory columns in a spreadsheet model of the sales force.

Sales Model from Chaper 7

Start with the constants; then define time to fill a row. The stocks are next.

Then define the converters (called auxiliaries in Vensim). The flows are last.

There are five constants

average annual salary	$25,000				
exit rate (fraction/year)	0.20				
fraction of revenues to sales	0.50				
hiring fraction (fraction/year)	1.00				
widget price	$100				

Time (in years with DT = 0.25 years)	0.00	0.25	0.50	0.75	1.00

Size of Sales Force (persons)	50.00	53.25	56.71	60.40	64.32	...

There are five converters

effectiveness (widgets per person per day)*	2.00	2.00	2.00	2.00	2.00	...
widget sales (widgets/year)	36,500	38,873	41,399	44,090	46,956	...
annual revenues ($/yr)	$3,650,000	$3,887,250	$4,139,921	$4,409,016	$4,695,602	...
sales department budget ($/yr)	$1,825,000	$1,943,625	$2,069,961	$2,204,508	$2,347,801	...
budgeted number of sales persons	73.0	77.7	82.8	88.2	93.9	...

There are two flows

new hires (persons/yr)	23.0	24.5	26.1	27.8	29.6	...
departures (persons/yr)	10.0	10.7	11.3	12.1	12.9	...

The flows are used to update the size of the sales force in the next column.

*The effectiveness is found in a vertical lookup function (located off screen).

value of this stock to 50 in the first column. The subsequent values will be found by the action of the flows located at the bottom of the spreadsheet. The next step is to assign a separate row to each of the converters. Their values are found by formulas based on the discussion in chapter 7. For example, the number of initial widget sales is 36,500. This make sense when 50 people are selling 2 widgets per day for 365 days per year.

Nonlinear relationships are somewhat more complicated in spreadsheets. A nonlinear relationship is needed to find the effectiveness of the average salesperson. Table C.1 shows the effectiveness at 2.00 in the first five columns. These values are found in a vertical lookup function (see the book's website, the BWeb). The flows appear at the bottom of the sheet. The simulation starts with 23 new hires per year and 10 departures per year. The effect of these flows over the 0.25-year interval is a net addition of 3.25 persons, so there are 53.25 persons in the next column. (The numbers are a continuous approximation to the actual number of workers.) The remaining columns are a repetition of previous columns. The hard work is to set up the initial values and formulas. If you build this sheet, you will find that it gives the results shown in chapter 7.

The Appeal of Spreadsheets

This example demonstrates that we can build stock-and-flow models in a spreadsheet. But should we do so? Spreadsheets allow us to see each and every number, and we can confirm the algebraic manipulations to arrive at each number. If you want to display the numbers in a clear and orderly fashion, spreadsheets are hard to beat. Spreadsheets are also popular because of their widespread use. If your colleagues think in terms of spreadsheets, you may find that the best chance to contribute within your organization is through spreadsheets. Also, the widespread popularity of spreadsheets has spawned a market for supplemental software to aid in the analysis of uncertainty (e.g., @RISK) or to search for an optimum solution (e.g., What'sBest!). Other specialized software supports a wide variety of applications, especially in management science and operations research (LeBlanc and Grossman 2008). Many of the specialized add-ins have been available for more than two decades, providing support for decision analysis, risk analysis, project management, expert systems, and forecasting (Bodily 1986).

Clearly, spreadsheets are an intriguing option for modeling. Indeed, Nicolson et al. (2002, 382) suggest that spreadsheets could be a good choice for interdisciplinary modeling, especially compared with "traditional programming languages such as FORTRAN, BASIC or C++." I believe spreadsheets could be our best choice if we are asked to perform a complex calculation for a single point in time. But this book focuses on behavior that changes over time. Dynamic behavior is more easily and clearly simulated with programs like Stella and Vensim.

The better role for spreadsheets is in support of Stella or Vensim. For example, spreadsheets can provide a convenient way to store, display, and check model inputs that are imported to Stella or Vensim. Spreadsheets also provide a convenient way to display or analyze results from dynamic models. Examples are explained later in the book:

- Appendix D describes Vensim results exported to a special spreadsheet to search for the most important inputs to a model.
- Appendix F shows Vensim results exported to a spreadsheet for display of the hour-by-hour operations of the electric power system in the western United States.
- Appendix G shows Vensim inputs from a spreadsheet with shaded display of elevations in a catchment.

Stella, Vensim, and Powersim

Stella, Vensim, and Powersim are the most widely used programs to implement the system dynamics ideas explained in this book. The programs start with the stocks and flows, the building blocks of system dynamics modeling. The models may be viewed as a collection of first-order differential equations, with a separate equation for each stock in the model. The equations in realistic models are almost always nonlinear, so it makes sense to solve the equations through numerical simulation. The three programs are icon based, so they promote the development of models with visual clarity. The programs are visually similar, as illustrated with the sales company models. Figures C.1 and C.2 show the Stella and Vensim models. By now, the icon conventions are familiar to you. You can see the similarity at a glance. And if you are inclined to strive for even more similarity, you can change the size and shape of the icons (BWeb). The Powersim version of the sales model is shown in figure C.3. The software is grounded in the system dynamics philosophy first published by Forrester (1961). And, like Stella and Vensim, Powersim enforces the philosophy by the connections that are permitted in the model.

Figure C.3 illustrates the visual similarity of Powersim with a flow diagram for the sales model. The stock is the size of the sales force; the flows are new hires and departures. The flows may link one stock to another, or they may link the stocks to clouds, as shown in this diagram. The diamonds are used for the exit rate, the widget price, and other constants in the model.

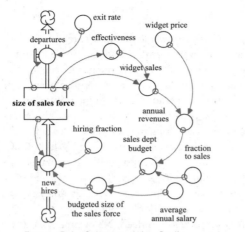

Figure C.1. Sales model in Stella.

Figure C.2. Sales model in Vensim.

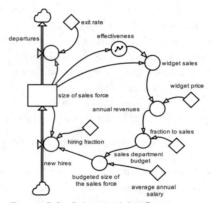

Figure C.3. Sales model in Powersim.

The circles are used for widget sales, annual revenues, and other variables that change over time. An extra symbol is inserted inside the circle for effectiveness. This reminds us that the effectiveness of the average salesperson is determined as a graph function based on the size of the sales force.

I have not used Powersim sufficiently in my own work to include a detailed description in this book. But my colleagues who use Powersim can testify to its usefulness in dynamic modeling. Powersim is valued for its core features facilitating system dynamics modeling. It is also valued for its capability to simulate in multiple dimensions, to support hierarchical modeling, and for its ease of interface design (BWeb).

Dynamo

Dynamo is the original program developed for implementing system dynamics models. It is described by Forrester (1961) and by Richardson and Pugh (1985). A wide range of applications were implemented during the 1960s–1980s, including several important applications to environmental and resource problems. The most widely known application is undoubtedly the World3 model used in *The Limits to Growth* (D. H. Meadows et al. 1972). The work from the 1960s–1980s provides good examples of modeling practice. A knowledge of Dynamo will help you appreciate and learn from this previous work.

Figure C.4 shows a diagram drawn in preparation for writing the Dynamo equations for the sales model. This diagram was drawn to help you see the structure of the model and to get organized to write the equations. Dynamo allows up to seven characters in a variable name, so a short name appears alongside the long name. In this example, *SSF* represents the size of the sales force. *SSF* would be called a *stock* in this book, but the custom with Dynamo is to call it a *level*. The levels grow or decline based on the *rates*, so *NH* and *D* are the rates. The short lines represent variables like *ER* (exit rate) or *WP* (widget price). These remain constant over

Figure C.4. Sales model Dynamo diagram.

time, so they are called *constants* in the Dynamo equations. The circled variables are called *auxiliaries*. They may change during the simulation; their role is to help explain the rates (i.e., they are viewed as auxiliary to the rates). An example is *WS*, the widget sales, one of the auxiliaries that helps us understand *NH*, the new-hires rate. The diagram shows lines above and below the auxiliary *E* (effectiveness). These lines alert us to the nonlinear relationship for the effectiveness.

This brief explanation is sufficient for you to see the similarity between Dynamo and the icon-based programs used in this book. The BWeb provides more information, including the Dynamo equations for the sales model. It also provides exercises to learn from some of the models published in *Toward Global Equilibrium* (D. H. Meadows and D. L. Meadows 1973). This book of readings deals with a wide range of environmental and resource problems, including exploration for natural gas, eutrophication of lakes, and cycles in the human and pig populations in the New Guinea Highlands.

Simile

A variety of icon-based programs can be put to good use in dynamic modeling. The programs were developed for general-purpose modeling, so they can be adapted to build and test models in the fashion explained in this book. One example is Simile, the software developed by Simulistics (BWeb). It is described by the developers as "simulation software with a system dynamics heart." Simile grew out of applications to earth, environmental, and life sciences. It provides capabilities similar to those of Stella and Vensim, including icons with a close resemblance to the stocks and flows shown in this book. Figure C.5 illustrates Simile with the familiar model of the sales company. The

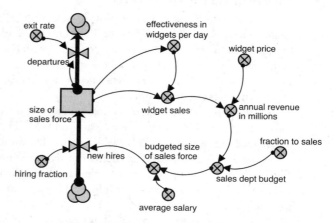

Figure C.5. Sales model in Simile.

visual similarity is clear, but the terminology is slightly different. (Simile uses the term *compartment* in the same way as Stella uses the term *stock*.)

Figure C.6 further illustrates the close similarity between Simile and Stella. This diagram is taken from the library of models at the Simulistics website (BWeb). It corresponds to the second model of Mono Lake shown in chapter 5.

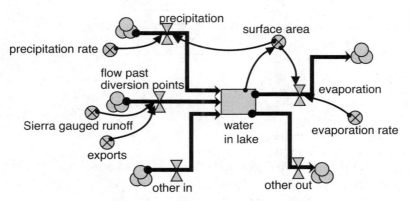

Figure C.6. The second model of Mono Lake in Simile.

I have only used Simile sufficiently to appreciate the developers' slogan that the software was developed with a "system dynamics heart." My initial tests were also sufficient to see that the developers are striving to serve multiple purposes, many of which go beyond the goals of the system dynamics approach first explained by Forrester (1961). The greater generality brings greater challenges in learning, but it provides opportunities for more complex simulations. My environmental modeling colleagues are particularly interested in exploring Simile's capability for combining spatial modeling with dynamic modeling. These explorations are in progress, so the current publications are limited (Boer 2004; Raper et al. 2005; Voinov 2008). Perhaps more will be available later (BWeb). Meanwhile, spatially explicit modeling with Stella or Vensim is described in appendix G.

Simulink and GoldSim

A variety of icon-based programs provide capability for dynamic simulations. Simulink and Goldsim are discussed here. These programs make use of icons to represent stocks and flows, but the visual correspondence is not as strong as with Simile. Readers who are already familiar with Simulink and Goldsim will see the correspondence and be able to develop models similar to those shown in this book. Readers who are not familiar with these programs will make more progress using Stella or Vensim.

Simulink is the dynamic modeling component of MATLAB, a multipurpose modeling software developed by The MathWorks (BWeb). MATLAB has become the de facto standard for engineering calculations in academic circles. The Simulink component uses a combination of electrical circuit icons and mathematical symbols to represent the structure of the model. The mathematically sophisticated engineer will see the correspondence between a Simulink model and the models shown in this book. However, my experiments with Simulink lead me to conclude that dynamic modeling is easier and more productive with Stella or Vensim. In my view, the key feature of Simulink is its position within MATLAB. MATLAB is valued in engineering circles for its ease of use, its versatility, and its very large library of functions. MATLAB code can also be compiled into stand-alone DLL (dynamic link library). This feature permits a productive link with Vensim's DLL, as explained in appendix E.

GoldSim is a dynamic modeling software developed by the GoldSim Technology Group (BWeb). It was designed from the ground up as a general-purpose, probabilistic simulation framework. The initial applications were in civil and environmental engineering, with special emphasis on stochastic simulation. Subsequent applications span a range of systems where statistical analysis of stochastic simulations was needed. Readers familiar with GoldSim will know how to implement some of the models shown in this book. Figure C.7 shows a GoldSim diagram of the second model of Mono Lake. The stock variable is the water volume, the amount of water stored in Mono Lake. It changes over time as the software integrates the effect of the total inflows and the total outflows. The diagram illustrates the mix of mathematical and visual symbols used in GoldSim models.

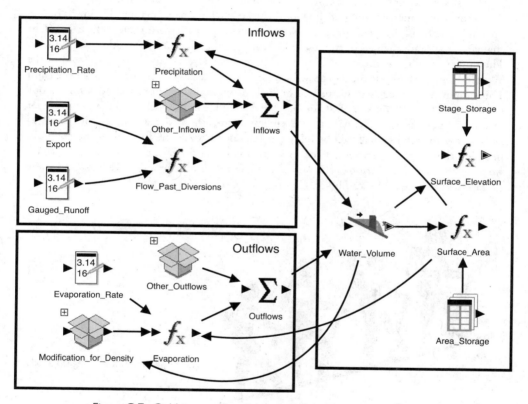

Figure C.7. Goldsim version of the second model of Mono Lake.

GoldSim, Simulink, Simile, Powersim, Vensim, and Stella are among the programs that provide icon-based support for dynamic modeling. These programs are most frequently used to represent the average behavior of the individuals in the system. For example, the sales model simulates the average behavior of the members of the sales force. The average effectiveness is multiplied by the size of the sales force to give total sales, total revenues, and the ability of the company to grow. The model delivers insight on the company's dynamic behavior without representing each individual in the sales force.

Individual-Based Modeling

Now imagine the challenge of building a model of the behavior of each and every person in a large company. There could be hundreds or thousands of employees. Such a model would seem to impose overwhelming computational requirements. But advances in computer power have spawned the development of models that represent each individual in the system. Such models have been used extensively in environmental systems (Grimm and Railsback 2005). A careful review of their use in ecology is provided by Grimm (1999; Grimm et al. 2005). Individual-based models have been used in business systems as well. The individuals may be individual customers or individual firms. These participants are viewed as the agents of behavior, and the models are called *agent-based models*. Grimm and Railsback (2005) provide a recent review of the effectiveness of such models in ecosystems, financial systems, and urban systems. North and Macal (2007) provide a detailed text on the development of agent-based modeling in business and governmental systems.

The simple example in figure C.8 is on the book's website (BWeb). It illustrates the type of insight that may emerge from modeling individual behavior. There are four individuals hiking up a hill. The first hiker is assumed to walk at a natural pace (yards/minute). His pace might change as the slope changes during the hike, but it is not affected by the other three hikers. The other hikers are assumed to adjust their pace to achieve a suitable separation from the hiker in front of them. (If the gap grows too large, they accelerate the pace to close the gap.) This simple model may be implemented in Stella or Vensim with a separate stock for the distance walked by each individual. If you build and simulate this model (with the slowest hiker in the lead position), a clear pattern will emerge within the first few minutes of the simulation—the four individuals will be hiking at the same pace as the first hiker, and they will maintain the desired separations. Their progress over the entire hike can then be calculated if we know the pace of the first hiker. If we want to speed their total progress, we must find a way to improve the pace of the lead hiker (Goldratt and Cox 1986).

Figure C.8. Hikers walking up a hill.

This simple exercise is easily implemented in Stella because of the small number of individuals and clear rules for how they adjust their behavior. The model has a simple stock-and-flow structure, with stocks representing distance and flows representing pace. Three of the hikers exhibit goal-oriented behavior (i.e., maintaining a suitable separation from the hiker in front of them), so the model includes three negative feedback loops to achieve the desired separations. The example is easily implemented in Stella because of the relative simplicity of the spatial relationships between the individuals. Each hiker is assumed to follow the same trail up the hill, so their progress can be represented by the distanced covered on the trail. In other words, we are simulating the individuals' behavior along a single dimension (the distance covered along the trail.)

Spatial Complexity

The hikers example illustrates an important point about the spatial complexity of the landscape. The hikers may be traveling across a complicated landscape, but that complexity does not necessarily mean that the model of the hikers has to be spatially complex. The key assumption is that the four hikers stay on the trail. This eliminates the need for the model to represent the individuals as they make their way through a three-dimensional landscape.

Let's turn now to situations with more spatial complexity. Think of the V pattern in the sky when geese are migrating. Ask yourself how the geese maintain the V pattern? And how do they select the lead individual? What is their average speed in migration? These are the sort of questions that could be addressed with an individual-based model. If we are willing to adopt some simple assumptions on how each individual controls its relative position in three-dimensional space, the model could help us anticipate the time and effort required to complete the migration.

A model of the geese migration is much more complex than the hikers' model because the geese are moving in three dimensions. A still more complicated situation is presented by the interaction of deer and cougars, the predator and prey populations described in chapter 20. The deer travel across the complicated landscape looking for water and forage while hoping to evade detection by the cougars. The cougars travel across the same landscape looking for opportunities for predation. The spatial interactions are complex indeed! Such interactions can be represented by individual-based models if we are willing to adopt assumptions on the individuals' behavior (BWeb).

The complexity of individual-based modeling has led to the development of software to support spatially explicit simulation of hundreds or thousands of independent agents. One example is NetLogo, a multiagent programming language developed at Northwestern University (BWeb). NetLogo is free software with an extensive library of models from a variety of domains (e.g., biology, chemistry, economics, physics, and psychology). NetLogo also includes a "system dynamics modeler environment" to enable stock-and-flow models with relatively close visual similarity to the stock-and flow models in this book.

Other programs for support of individual-based modeling are described by North and Macal (2007). They describe "desktop software" such as spreadsheets, Repast Py, NetLogo, StarLogo, Mathematica, and MATLAB . They view spreadsheets as the simplest approach, one that can lead to insightful modeling. "Programs such as Repast Py and NetLogo avoid the spreadsheet limitations on the diversity of agents and restrictions on their behavior. They also describe large-scale software such as Swarm and Repast. They view Repast as the leading free and open-source toolkit for support of large-scale, agent-based modeling.

Individual-Based Modeling and System Dynamics

Individual-based models are sometimes described as *bottom-up models*, since the emergent behavior arises from the assumptions adopted for each individual in the population. System dynamics models, on the other hand, are sometimes characterized as *top-down models*, since they represent the average behavior of large groups of individuals. Grimm uses the terms *top-town* or *state variable modeling* in ecology. His "state variable" category is not necessarily synonymous with the system dynamics approach explained in this book. But it is sufficiently close for Grimm's conclusions to apply to the benefits of using individual-based modeling as a complement to system dynamics. Grimm (1999, 139) believes that the bottom-up and top-down approaches "are not exclusive alternatives but rather complementary approaches that are mutually dependent."

I believe that Grimm is right to point out the complementary nature of the two approaches. Indeed, the findings from individual-based models may provide valuable support for the aggregate relationships in system dynamics models. A model of salmon survival in the Tuolumne River in California will illustrate. The model was developed by Jager et al. (1997) to represent the day-by-day struggle for survival by thousands of juvenile salmon. The model simulates the competition in a river with space for around 40,000 redds. The potential redd sites (and the potential feeding locations) are dependent on the flow, which is subject to external disturbances and decisions on reservoir releases. The Tuolumne model was designed to help scientists understand the impact of reservoir operations on salmon survival. Such a model might be put to other uses as well. If the model were simulated with a wide range of values for egg deposition, for example, it could provide useful information to support a system dynamics model.

To illustrate, think of the salmon model in chapter 15. It simulates juvenile survival as part of the life cycle of salmon population of the Tucannon River in Washington state. Juvenile survival is represented by the nonlinear curve in figure 15.2. The general shape corresponds to the Beverton-Holt curve, which has proved useful in estimating fish survival. The curve parameters were estimated by Bjornn (1987) based on observations in the Tucannon watershed. Support for these parameters could be provided by an individual-based model similar to Jager et al.'s (1997) model for the Tuolumne River salmon. Their model represented the juveniles competing for suitable feeding locations. The individuals that survive from one day to another become larger and are then in an improved position to command better feeding locations for the next day of the simulation. After 365 days of simulated competition, the model estimates the number of juveniles that outmigrate as smolts. If one were to build a similar model for the Tucannon, it could be used to show the number of outmigrating smolts from simulations with a wide range of values for the emergent fry. The results of these simulations could then inform the shape of the juvenile survival curve in figure 15.2.

Further Reading

- Voinov (2008) describes a wide range of modeling software, including Stella, Vensim, and Powersim. These are classified as "modeling systems," which are said to be "completely prepackaged and do not allow any additions to the methods provided." He also includes other examples of modeling systems such as Madonna and ModelMaker. He classifies GoldSim and Simulink as extendable modeling systems — "modeling packages that allow specific code to be added by the user if the existing methods are not sufficient for their purposes." Voinov's software spectrum extends from computer languages (e.g., FORTRAN or C) at one extreme to individual models (e.g., SimCity) at the other. He provides a detailed description of software for spatial modeling (see appendix G). The extendability of system dynamics software (e.g., Vensim's external functions) is discussed in appendix E.
- Schmickl and Crailsheim (2006) describe an individual-based model of an artificial population of predators and prey. The predators are assumed to adopt a hunting behavior that focuses on an apparently weak individual in the prey population. The individuals in the prey population avoid predation by speed of escape (which may be treated as an evolved characteristic). The model was designed to contribute to the field of artificial life (Adami 1998).
- Rahmandad and Sterman (2008) ask, When is it better to use agent-based models, and when should system dynamics models be used? (The system dynamics models are ex-

pressed in the form of differential equations.) The question is addressed with models of the spread of an epidemic. Agent-based models are valued for simulating heterogeneity among individuals. They can also be used to represent the complex network of interactions between groups of individuals. System dynamics models are valued for their ability to simulate behavioral feedbacks and the ease of conducting sensitivity analysis of the behavioral assumptions. The system dynamics approach makes more sense if the impact of feedback effects are expected to be larger than the impact of network structure and heterogeneity among individuals.

- Osgood (2009) summarizes the trade-offs between individual-based and aggregate models. The individual-based models are valued "when studying the impact of interventions in systems where populations exhibit high heterogeneity, small size or clustering, and complex and dynamic network structures." But these models "frequently require significantly greater time to understand and analyze than do their aggregate cousins." He then explains a technique for dimensional analysis and scale modeling to reduce the long simulation times for individual-based models of large populations.

Appendix D

Sensitivity Analysis and Uncertainty

This appendix describes a systematic and comprehensive approach to sensitivity analysis. It begins with background on the informal method of sensitivity analysis normally conducted in the early stages of a modeling project. It then describes sampling methods that make comprehensive analysis possible. The analysis can lead to tolerance intervals, a common measure of the uncertainty in the simulation results. The analysis requires the important inputs to be independent of one another. The appendix describes statistical screening to find the important inputs and then suggests an iterative method to arrive at tolerance intervals. The appendix concludes with a brief discussion of modeling and policy making in an uncertain world.

Background

Sensitivity analysis is a key step in the modeling process, especially when the model contains highly uncertain parameters. Some parameters may be based on expert judgment; others may be based on our intuition. The estimates may be highly uncertain, but the uncertainty should not be used as an excuse to leave the parameters out of the model. It makes better sense to face up to the uncertainty and take the time to conduct sensitivity analysis. Analysis of system dynamics models has often revealed that the uncertainties in many parameters do not make much difference in the results. On the other hand, some parameter values may be crucial—small changes in their estimates can lead to major changes in the results. These are the parameters that warrant closer study.

The case studies in this book use an informal style of sensitivity analysis that makes sense in your early iterations through the model building process. The sensitivity results of the Kaibab deer model (chapter 21) illustrates what often happens in well-structured models. The tests reveal that the overshoot pattern appears again and again. The model's fundamental pattern is said to be *robust*—that is, it continues to appear regardless of changes in uncertain parameters. This was an important finding, but you might be wondering if the Kaibab model tests were sufficient. Perhaps there are surprise results waiting to be discovered if only we take the time to conduct more tests. It would seem that a more comprehensive sensitivity analysis should be part of the modeling process.

Is Comprehensive Analysis Possible?

Suppose you are testing a relatively small model with 15 uncertain inputs. You might select low, medium, and high values for each of the parameters. Then you might design a collection of simulations with all possible combinations. The number of simulations would be 3 values for the first parameter, multiplied by 3 values for the second parameter, multiplied by 3 values for the third parameter, etc. The total number is 3 to the 15th power. Type this into your calculator, and you may be surprised by what you see—we would need over 14 million simulations!

The idea of conducting so many simulations leads some researchers to conclude that comprehensive testing is not possible. In describing a model of fluid loss in a patient, Bush et al. (1985, 22) argue that "minimum tests for sensitivity include a high and low value for each important factor around a baseline case. The possible combinations of parameter values grow exponentially with the number of variables being studied as well as with the number of levels taken of each variable." They concluded that "a comprehensive analysis quickly becomes intractable." This view illustrates a common misperception. Many practitioners seem to believe that comprehensive sensitivity analysis is simply not feasible in models that have grown to include more than just a few parameters. This is an unfortunate misperception that is based on an inefficient sampling method.

Sampling Methods

Sampling method refers to how we select a collection of experiments with the model. I believe that a *control panel approach* is useful in the early stages of a modeling project. The modeler should take the time to develop a user-friendly control panel and simply experiment with the model. An example is the sales model control panel shown in figure 14.1. Each of the six parameters is assigned to a slider, with the width of the slider corresponding to the range of uncertainty. The team can experiment with changes in any or all of the inputs, and Vensim responds instantly with new results. This can be a fun and illuminating way to learn about the importance of the inputs. It should be a routine part of the modeling process in the early stages of a project. A more thorough approach is needed in the later stages of a project. But how many simulations should be conducted? And how should we specify the inputs for each of the simulations?

Suppose we decide to conduct 40 simulations, with changes in 100 inputs. We need 40 sets of values for 100 parameters. This selection could be based on random sampling, stratified sampling, and importance sampling (McKay, Conover, and Beckman 1979; Morgan and Henrion 1990; Frey 1997). Formal comparisons have revealed that Latin hypercube sampling (LHS) is a highly efficient way to test a model (McKay, Conover, and Beckman 1979; Reilly et al. 1987). Several software packages (e.g., @RISK) became available in the 1980s to facilitate sensitivity analysis with LHS. For system dynamics models, the best approach was Hypersens, a customized package by Backus and Amlin (1985). It was put to good use in a study of the uncertainty in the Northwest electric system (Ford 1990). A comprehensive analysis required only 40 simulations, and the results provided valuable insights to electricity planners.

But how did we know that the sample of 40 simulations was sufficient? This question was addressed in a pragmatic manner by doubling the sample size. A new analysis with 80 simulations showed the same uncertainty intervals and confirmed that the sample of 40 runs was sufficient for the study. Customized software was required for the Northwest electricity study, but that is no longer the case. The PLE+ version of Vensim provides LHS as part of the sensitivity simulation. This feature is illustrated with the sales model in chapter 7.

Intervals of Uncertainty in the Sales Model

Table D.1 shows the six input parameters that were assigned sliders in the control panel in figure 14.1. Five of the six parameters are highly uncertain; their range is plus or minus 50%. The widget price is known with greater certainty; its range is plus or minus 2%. The book's website, BWeb, shows Vensim's setup window for assigning ranges of uncertainty with a sample size of 50 runs. Vensim allows for a variety of statistical distributions, but this illustration uses the uniform random distribution for all inputs. We ask Vensim to store the results of the 50 runs in the .vsc file, and we specify a file with a list of variables to be saved. For this illustration, we save the size of the sales force and the values for each of the uncertain variables. Vensim will then ask for the name of the .vdf file to store the results of the sensitivity analysis.

Table D.1. Uncertain parameters in the sales model.

Base Case Values	Ranges of Uncertainty
Initial sales force = 50 persons	Uniform (25,75)
Saturation size = 1,000 persons	Uniform (500,1500)
Average salary = $25,000/yr	Uniform (20000,30000)
Widget price = $100	Uniform (98,102)
Exit rate = 0.2/yr	Uniform (0.15,0.25)
Maximum effectiveness = 2 widgets/day	Uniform (1.5,2.5)

The first 16 years of results in the 50 simulations are shown in two versions of Vensim's "sensitivity graph." Figure D.1 shows all 50 runs as individual traces, an option available by right-clicking on the sensitivity graph. Some traces show the company growing to around 1,200 persons within the first 5 years; other traces show that the company would be unable to grow. Figure D.2 shows a different perspective by graphing the percentile intervals with 50% in light gray, 75% in dark gray, and 90% in black. The mean result is shown by the black curve.

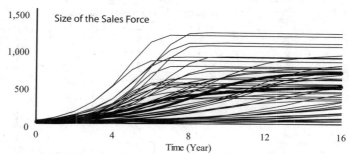

Figure D.1. Size of the sales force in a sensitivity test with 50 simulations.

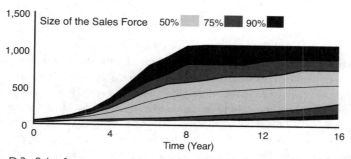

Figure D.2. Sales force percentile intervals with 50%, 75%, and 90% coverage.

The BWeb explains how percentile intervals may be converted to tolerance intervals. The translation rests on the assumption that the inputs may be varied independently of one another. This assumption poses a problem for most system dynamics models, which typically contain exogenous inputs that are interdependent. These interdependencies should be removed if we are to trust the estimated size of the tolerance intervals. But a large model with many inputs could contain dozens of interdependencies. To remove them all could be quite time consuming. It makes more sense to identify the key inputs and to consider whether they are independent of one another.

Estimating Tolerance Intervals

Figure D.3 shows an iterative approach to tolerance interval estimation. We begin by setting the range of uncertainties for inputs to the model. Past experience suggests that some individuals will be reluctant to assign ranges of uncertainty. If investigators disagree on the appropriate range of uncertainty, simply repeat the simulations with a new range of uncertainty. You may well find that the size of the uncertain intervals on many inputs does not make an appreciable difference in the tolerance interval for the output. Experience also suggests that we should guard against overconfidence in parameter estimates. Stainforth et al. (2005) warn that experts tend to underestimate the range of possible values. Sterman (2000, 884) gives a similar warning; he recommends that we "test over a range at least twice as wide as statistical and judgmental considerations suggest."

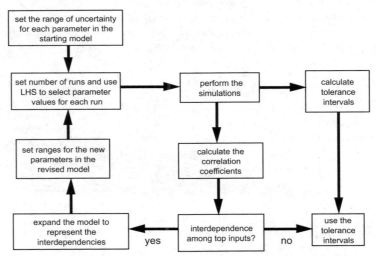

Figure D.3. Iterative approach to obtaining tolerance intervals.

The next step is to decide on the number of runs and to assign values to each of the parameters. We started with 50 simulations, and experience suggests that this should be sufficient. But you can verify by repeating the analysis with 100 runs. With many system dynamics models, 50 or 100 simulations can be executed in a few seconds. The final step is to calculate the tolerance intervals. This is easily done with Vensim's sensitivity graph. It shows color-coded intervals, with the term *confidence bounds* in the dialog box for creating the graph. A *confidence bound* should be thought of as a *percentile interval*, and the BWeb explains its correspondence to the more commonly used *tolerance interval*. If we have independence among the key inputs, the tolerance interval provides a good estimate of the range of uncertainty on the output of the model.

Statistical Screening for the Important Inputs

The next step in figure D.3 is to find the key inputs to the model. This is best done by statistical screening, as described by Ford and Flynn (2005) and illustrated with the sales model. We begin by exporting the 50 simulations to a spreadsheet template designed to receive the values assigned to 6 uncertain inputs with results for the key output saved in 20 time periods. The results of the Vensim sensitivity analysis are stored in a .vdf file, which may be exported to a tab file (BWeb). The contents of the tab file will be difficult to read, but they are ready to be imported into the template. In this illustration, we open CC Template 6x50x20.xls since we have 6 inputs, 50 runs, and 20 years' worth of results. The results of the sensitivity analysis will appear in the boxed cells, and a graph of the correlation coefficients will appear on the second worksheet.

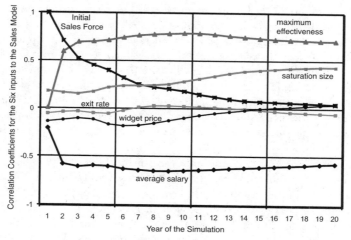

Figure D.4. Correlation coefficients for the six inputs to the sales model.

Figure D.4 shows the correlation coefficient (CC) for each input for each year of the 20-year simulation. The highest CC appears in very first year, where the initial value assigned to the stock stands out. A CC of 1.0 is expected, since the initial size of the sales force is totally explained by the initial value of the stock. This parameter is the most important input for the first two years of the simulation. By the third year, however, the influential inputs appear to be the maximum effectiveness and the average salary. The CC for maximum effectiveness is around 0.7; the CC for the average salary is almost as important, with a negative correlation of around 0.6. It makes sense that these parameters emerge as most influential, since they control the gain around the positive feedback loop highlighted in figure 9.18. A higher maximum effectiveness makes this loop stronger; a higher salary makes it weaker. Figure D.4 shows that these inputs remain dominant for the remainder of the simulation. The positive correlation for the maximum effectiveness makes sense because a higher effectiveness leads to a larger company. The negative correlation for the average salary makes sense because a larger salary means that you must budget for a smaller workforce.

Now, let's ask whether the top inputs are sufficiently independent of one another that we can use the tolerance intervals. In this example, we must ask if the maximum effectiveness can be specified independently of the average salary. Most readers would answer that it cannot. Companies that pay higher salaries are probably going to attract personnel with a higher maximum effectiveness. So we would expect these inputs to be positively correlated.

Now what is to be done about the positive correlation? Frey (1997, 5) describes a variety of approaches, including explicit modeling of the dependence. I recommend this approach, as noted in the lower-left corner of figure D.3. In the sales example, the model should be expanded to describe how sales effectiveness would increase if the company committed to higher salaries. The new model would include additional parameters with their own ranges of uncertainty. We would specify the new ranges and begin the next iteration. After an iteration or two, we should have intervals that can be interpreted as tolerance intervals.

Deterministic and Stochastic Uncertainty

The uncertainty intervals shown in this appendix arise from the uncertainty in the model inputs. We do not know their values, but we expect the values to remain constant over time. The resulting uncertainty in the output might be called *deterministic uncertainty*, since each of the simulations is a deterministic simulation. This book uses the term *stochastic uncertainty* for the variability in model results associated with randomness in the inputs. The average values of these inputs may be well known, but their value may vary from year to year in a fashion that we cannot explain. These are called *random inputs*, and you have learned in chapter 14 how to include randomness to produce stochastic simulations. In some cases, the stochastic simulations can be surprising, as in Figures 15.6 and 21.20. At this point, you're probably wondering whether system dynamics models can represent the combined effect of deterministic and stochastic uncertainty. This question is addressed with illustrative examples on the BWeb.

The terms *deterministic* and *stochastic* are used in this book, but you should be aware that other authors assign different labels. Frey (1997) describes quantitative analysis of variability and uncertainty in energy and environmental systems. The terms *variability* and *uncertainty* are his choices to make distinctions between many sources of uncertainty. But a variety of terms are used, depending on the background of the investigator. Frey notes that "variability is sometimes referred to as aleatory uncertainty, stochastic variability, and interindividual variability." On the other hand, "uncertainty has been referred to as fundamental or epistemic uncertainty."

Pragmatic Benefits of Sensitivity Analysis

Sensitivity analysis is a key step in the modeling process. I recommend informal sensitivity testing from a user-friendly control panel in the early stages of a project. Comprehensive testing should be performed later in the project. A collection of 50 simulations may be all you need to study the range of behavior with wide variations in the uncertain inputs. Pragmatic benefits await the modeler who takes the time to conduct this analysis. Two are mentioned here: *error detection* and *policy interpretation*.

Most real applications of system dynamics involve a team of individuals working under time pressure. There are bound to be poorly formulated equations in such projects. The key to productive modeling is to discover the mistakes through a continual process of checking and discussion. Comprehensive sensitivity analysis can help with the checking. For example, you might test the model with 50 simulations and ask for a graph with traces of all the simulations.

Then study the traces, looking for a simulation that stands out as clearly erroneous. You should congratulate yourself if there are no erroneous results—this is a sign that your feedback structure is well designed to deal with the wide range of conditions. If you do find a problematic simulation, look at the assigned values of the inputs. These will allow you to recreate that simulation for closer study. It's often useful to set a pause interval and study the problematic simulation looking for the first variable to exhibit highly questionable behavior. At this point you will be closing in on the formulation error in the model.

The policy implications of sensitivity analysis usually require two sets of simulations. The first set teaches us about the range of uncertainty. We may then repeat the sensitivity analysis with a change in a key policy variable. Comparing the tolerance intervals can help managers who are planning for and operating in an uncertain world (Ford 1990). The policy implications of sensitivity analysis are particularly striking when the model is prone to create distinctive clusters of results. The BWeb illustrates with a sage-grouse model and a project management model. The sage-grouse model simulates the bird population in central Washington based on land use decisions. One cluster of simulations shows the bird population growing to safe size; the other cluster shows the population falling inexorably to zero. This pattern teaches us about the turning point in the sage-grouse system. A strikingly similar result is shown for a project management model to simulate scheduling, work, and rework. The sensitivity testing reveals a cluster of completed projects and a cluster of projects that become permanently trapped under a burden of rework. Once again, the analysis teaches us about the tipping point in the system.

Final Thoughts on Uncertainty and Policy

This appendix has explained how system dynamics can be used to show tolerance intervals on key outputs. The intervals can teach us about the dynamics of uncertainty. In some cases, the uncertainty in the key output grows over time. An example is the first eight years of figure D.2. Growth in uncertainty intervals is usually attributed to the cumulative effects of the uncertain gain around a positive feedback loop. But the uncertainty does not necessarily grow forever, as we can see in the second half of figure D.2. And in some systems, the uncertainty interval can become more and more narrow over time. The narrowing of uncertainty is usually caused by the action of negative feedback loops that negate the effect of the changes in the uncertain inputs.

System dynamics simulations can also help teach us about the policy implications of uncertainty. Indeed, this may be where system dynamics delivers its greatest benefit. But we must first determine the organization's policy for dealing with uncertainty. The policy may sometimes take the form of a simple, easily remembered rule. For example, the policy for electric resource planning in the Northwest (Ford 1990) was summarized as "build for the medium, option to the high." The ability to simulate this policy in the system dynamics model turned out to be far more important than the precise characterization of the uncertainty in the future electricity demands.

In highly uncertain environmental systems, regulatory policies are sometimes summarized as general principles. A well-known example is the *precautionary principle*. It calls on regulators to act in anticipation of potential harm to the environment. Another example is the *polluter pays principle*, long advocated by environmental economists. But the implementation of "polluter pays" is often hampered by large uncertainties about the possible damages to the environment. How can economic incentives be implemented when the damages are highly uncertain? One approach is to require polluters to pay for bonding insurance against the uncertainty in future damages. Costanza and Cornwell (1992) call this the *precautionary polluter pays principle*.

These three examples draw our attention away from the timing and size of the uncertain impacts. The new focus is on policies to deal with the uncertainty. Such policies deserve our attention early in the modeling process, and the organization's preferred policy should be included in the model. Policy-oriented modeling is the best way for system dynamics to deliver insights in an uncertain world.

Appendix E

Incorporating Other Methods in a System Dynamics Model

The challenge in system dynamics modeling is to formulate the best combination of stocks, flows, and feedbacks to explain a dynamic problem. It is often easiest to start with the stocks, add the flows, and then add the feedback relationships to explain each flow. Stella, Vensim, and Powersim facilitate the development of such models through visual clarity, ease of model development, and ease of simulation. All three programs are grounded in the system dynamics philosophy first published by Forrester (1961). Moreover, they each enforce the philosophy by the connections and the calculations that are permitted in a model. Their focus on stocks, flows, and feedbacks makes the programs easier than general-purpose programs. The ease of understanding is one of the greatest strengths of the system dynamics approach, especially in projects with individuals from widely different backgrounds.

What If System Dynamics Doesn't Fit?

But what can be done when a portion of the system does not appear to be easily described by a combination of stocks, flows, and feedbacks? If you are new to modeling, an example may not come readily to mind. After all, you are nearing the end of a book with examples ranging from sales companies to salmon populations. So far, it seems that stocks, flows, and feedbacks are sufficient to cover many systems.

An oil refinery is a good example to illustrate the limits of the system dynamics approach. Refineries are complicated facilities that convert crude oil into a mix of fuels such as gasoline and jet fuel. Their owners strive for an optimal mix of fuels, given the current (or expected) fuel prices. The optimal mix is hard to find since there are so many possible combinations, and the refinery must obey a long list of physical and chemical rules. The operation cannot be found by the standard tools of system dynamics, but it is well suited to optimization methods such as linear or nonlinear programming. Mathematical programming can deliver important insights to the refinery owners, and many companies rely on such models to guide their operations.

Now suppose we have the luxury of two models—an optimization model of the short-term refinery operation and a long-term model similar to the vehicles model in chapter 16. How might we take advantage of the short-term operational model in the longer-term vehicles model? The answer to this question depends on our ability to condense the operational results into rules of thumb for the mix of output fuels. Such rules could then be inserted into a system dynamics model of the long-term changes in the larger system. In other words, we would maintain the two models, each implemented with its own software. The detailed model of short-term operation yields approximate decision rules that are distilled into the aggregate relationships in the system dynamics model. You've seen this two-step approach before. It was recommended in appendix C where the results of an individual-based model of salmon survival could be used to support the aggregate relationships in the salmon model in chapter 15. And it will be recommended again in appendix G, where detailed spatial modeling results are used to develop approximate relationships in a nonspatial model.

What If the Detailed Model Cannot Be Summarized?

This appendix deals with the situation where the two-step approach breaks down. This can happen when the team responsible for the detailed model of short-term operations cannot arrive at summary rules of thumb. Perhaps they have examined hundreds of cases, and there is no way to summarize the results. They may conclude that each and every situation is different, and that the only way to determine the short-term operation is to rerun the model for every possible situation.

This difficult situation begs the question of whether the short-term model could be incorporated directly into a system dynamics model. The answer is yes, provided the team has the programming skills needed to deal with external functions. This approach is explained briefly with a Vensim model of the electric power system in the western United States.

The Western Power System

A Vensim model of the western power system was developed to look at the long-term impact of carbon policies in the Western Electricity Coordinating Council (WECC). The WECC is a vast, highly interconnected market that ranges from Southern California to British Columbia, and from New Mexico to Alberta. The model is comprised of around 50 views, so it is much larger than the models in this book (all of which would fit on 1 view). The model was used to show the rapid reduction in CO_2 emissions that could be achieved with the carbon market proposed in the Climate Stewardship Act of 2003 (Ford 2008).

Over 90% of the model is developed with the system dynamics approach. The challenging 10% deals with power flows across the transmission system. This WECC was represented as seven areas interconnected by 10 transmission lines. Most of the tie lines extend over long distances, and their available transmission capacity (ATC) is limited by stability concerns. (Their ATCs can differ depending on the direction of power flow.) Three of the lines are more likely to be limited by thermal constraints; their ATC will be the same in both directions. These constraints and the subsequent power flows are not amenable to standard system dynamics representation. The wholesale market prices in each of the seven areas are also beyond the reach of standard system dynamics methods. These prices can vary greatly over the course of a 24-hour day owing to the large swings in the demand for power. But the prices and power flows can be simulated with optimal power flow (OPF) methods familiar to the power systems engineer. For our purposes, it made sense to incorporate a compact, reduced version of the direct current optimum power flow calculation (DC OPF). (The term DC simply means that the calculation ignores the reactive power.)

Vensim's External Functions

The Vensim software provides a library of standard functions such as MAX, MIN, and EXP. These are typically used to characterize the exogenous inputs to a model (see chapter 14). The DSS (decision support system) version of Vensim also allows the user to define external functions that can perform arbitrary computations during each time step of the simulation. The external functions receive values from the rest of the model at each step in the simulation The external function uses those values to perform a calculation (presumably one that could not be accommodated with standard stocks and flows). The results of the calculation are returned to the rest of the model, and the simulation proceeds to the next time step.

The WECC model represents each month by a typical day. The prices vary over the 24 hours, reaching a peak in the afternoon during the hour of peak demand. Time was measured in months, and the model simulated the seasonal variations in demand and generation over the 12 months of a year. A typical simulation ran for 240 months to cover the 20-year interval for analysis of carbon policies. Meanwhile, the prices and electricity generations for each hour of the typical days were tracked through the use of time-related subscript (hour of day: hr1, hr2 . . . hr24). Other subscripts were used to represent different areas in the WECC, different types of generating technologies, and different transmission lines.

An example of an external function is FIND PRICE, shown in table E.1. This equation gives the wholesale price of electricity in each hour of the day in each area of the WECC. The price depends on the variable cost of operating the thermal generating units in each area. They range from a low to a high, and the equation lists the first elements of the two-dimensional array ($A1$ for the first area; Nu for nuclear, the first of the market-run generating technologies). The next variable is the thermal generating capacity available after maintenance. This is also a two-dimensional array to tell the external function about the eight types of capacity in each area. The loads to market are next; they represent the demand for power after accounting for the must-run generation in the WECC. The final entry is the price cap, the administrative limit imposed on the wholesale price of electricity.

Table E.1. Example of an external function in the WECC model.

Prices[Hour,Area] = FIND PRICE (VarCostLow[A1,Nu], VarCostHigh[A1,Nu], Thermal Gen Cap

Available to Run[A1,Nu], Loads to Market[Hour,A1],Price Cap)

Vensim calls the external function during each time step of the simulation. The gateway for these external functions is provided by a dynamic link library (DLL). The library is usually created in a C/C++ developing environment, but other languages may also be used. This DLL can contain any number of functions, and it can also call other DLLs. (Calling other DLLs is used when it is not possible to directly compile required functions within the environment used to create the gateway DLL.)

A variety of tools can be used to implement the engineering calculations. In recent times, MATLAB has become the de facto standard in academic circles because of its ease of use, versatility, and large library of functions. MATLAB code can also be compiled into a stand-alone DLL, which can then be linked to Vensim's DLL. For computational efficiency, the code can be first translated into C and then compiled.

The Appeal and Challenge of External Functions

The external functions were very useful in the WECC model. They enabled us to enforce algebraic constraints and perform complex mathematical calculations using the appropriate engineering techniques for the transmission system. This approach may make sense if you confront a system that is not amenable to the standard system dynamics approach. But you should brace yourself—programming with dynamic link libraries will require considerable time and effort.

This brief description is sufficient to alert readers to the possibility of stretching the system dynamics approach beyond the standard tools of stocks, flows, and feedbacks. Modelers proficient in the use of dynamic data links and C language programming will find that complex calculation may be called at each step of a Vensim simulation. This feature will allow system dynamics methods and other methods to operate in unison over simulated time.

Further Reading

- Further details on optimal power flow method and the programming calculations are provided by Dimitrovski, Tomsovic, and Ford (2007).
- Details on external functions appear in the manuals provided at the Help command in the DSS version of the Vensim software.

Appendix F

Short-Run and Long-Run
Dynamics in a Single Model

The models in this book simulate dynamics across widely different time scales, as shown in table 1.1. The majority of the models operate in days, months, or years, but the principles of modeling apply across all time scales. You are free to select the unit of time that best fits your dynamic problem. This appendix deals with the difficult situation in which you feel compelled to combine short-run and long-run dynamics within the same model. Perhaps you are interested in short-run changes best simulated in days, but your time horizon stretches 20 years into the future. At this point, you may be tempted to simulate the daily dynamics in a model that simulates time in years. You should resist this temptation; you'll make more progress by developing two models. The first model deals with the short run, the second with the long run, and the results of the short-run model can support the relationships in the long-run model.

The general advice is to avoid simulating both short-run and long-run dynamics within a single model. This appendix deals with exceptions to this general rule. It begins with previous models in the book.

Previous Examples

The first example is the flowers model in figure 14.10. The original flowers model was simulated in years, and the flowered area reached equilibrium within 15 years (figure 7.5). The new model simulates time in months in order to include seasonal variations. DT (delta time, the increment of time between steps in the simulation) is 1 month, and the 180-month simulation requires only 180 steps. The simulation is rapid, and the seasonal variations add realism to the results.

A second example is shown in figure 5.13. The Mono Lake hydrology model is changed from years to months to allow it to be merged with a brine shrimp population model that operates in months. A 50-year simulation would take 600 steps (if DT is 1 month). This is a useful merger because it focuses on water export, the key policy variable in the Mono Lake case. A 600-step simulation will appear rapidly on the computer, so we can run the model in a rapid, interactive fashion.

The third example is more difficult. It involves two models of the salmon population. The model in figure 15.1 (salmon life cycle) operates in months, whereas the model for figure 15.16 (smolt migration) operates in days. Suppose we were to merge these two models, as discussed in exercise 15.19. If DT were 1 day, a 20-year simulation would require over 7,000 steps. Those who attempt exercise 15.19 usually conclude that the models are best operated separately. For example, the results of a smolts simulation may be used to set the smolt migration loss fraction in the longer-term model.

One Thousand Steps Should Be Enough

The number of steps required to complete the simulation is an indicator of the extent to which we are mixing time dynamics across dramatically different time scales. The Mono Lake brine shrimp model merger made sense, but imagine the situation if brine shrimp dynamics required time to be in hours. If DT were 1 hour, a single year would require 8,760 steps. A 50-year simulation would require 438,000 steps. Such a model would not deliver rapid simulations, and we would lose the opportunity to conduct multiple simulations in an interactive manner.

You should think twice if you find yourself building a model that takes more than 1,000 steps. It will probably make more sense to construct two models. You may find yourself violating the 1,000-steps rule of thumb if you include a high-turnover stock in the model. Figure 17.6 shows an example of a high-turnover stock in a model of the global hydrologic cycle. An accurate simulation of the water stored in the atmosphere required a DT of 1/64 year, so we would need over 6,000 steps to complete a 100-year simulation. You know that high-turnover stocks can be eliminated without eliminating their important impacts on the system. The key is to represent their impacts with an algebraic equation rather than integrating the flows acting on a stock variable.

Let's turn now to the difficult situation where rapidly changing dynamics must be included in a long-term model. Your goal should be to obtain rapid simulations over a long time horizon and still see the short-run dynamics that are important to you and your client. This unusual situation requires an innovative approach. This appendix describes two approaches that have proved useful in simulating policies in the power industry.

An Abbreviated Calendar in a Model of the California Electricity Crisis

The first approach was used to simulate the California electricity crisis conditions of 2000 and 2001 (Ford 2002). A long time horizon was required to show the simulated pattern of power plant construction with the new market rules. (The construction appears in waves of boom and bust similar to the real estate construction in chapter 19.) A model of the construction patterns would normally be in years. However, the debate over the California crisis was dominated by discussions of price spikes that often appeared during the peak hours of the day. The price spikes were crucial to the revenues of the generating companies, and many believed that the hourly prices should be simulated within a model of the long-term trends in construction.

The situation called for the seemingly impossible—simulating price spikes that could break out during difficult hours of the day while still maintaining rapid simulations over a 12-year time horizon. Clearly, we were not going to comply with the 1,000-steps rule of thumb. If DT were 1 hour, a 12-year simulation would require over 100,000 steps. But the situation was even worse, since the wholesale market prices could increase by tenfold within a single hour. Simulating these price spikes required setting DT to 1/16 hour, and a 12-year simulation would take over 1.7 million steps!

The choices were daunting. We might increase the size of DT, but the model would no longer simulate the price spikes that the client viewed as a key part of the system. We might shorten the time horizon, but that would prevent us from seeing the pattern of boom and bust in construction and a return of price spikes in the future. Clearly, we could not maintain rapid simulations if we simulated all 8,760 hours in the year. The innovative approach was to simulate a typical 24-hour day for winter, spring, summer, and fall. With this abbreviated calendar, an entire year was compressed into 96 simulated hours. The simulated electricity prices were then compared with the prices in the wholesale market. Most generators were assumed to bid their units into the wholesale market at variable costs. However, some generators would submit bids well above variable costs (a form of strategic behavior known as economic withholding). Without strategic behavior, the simulated prices reflected competitive conditions. These results were reasonably close to the counterfactual prices published by the California Independent System Operator. With strategic behavior, the simulated prices were reasonably close to the actual prices in the wholesale market. These historical checks confirmed that an abbreviated calendar would provide useful simulations of the short-run price spikes within a long-run model. The simulations were realistic, but somewhat slower than normal. (The DT was set to 1/16 hour; a 12-year simulation required around 18,000 steps; and the simulation took about 1 minute on the computer.) This is a slow response compared with models in this book, but it was sufficiently fast to allow for testing of policy proposals at the California Energy Commission (see the book's website, the BWeb).

The abbreviated calendar was innovative and useful at the time, but the model was encumbered by many unit conversions to deal with expenses, revenues, profits, and other variables normally expressed on an annual basis. Time delays and forecasting horizons were also more complicated, with 96 hours in a simulated year. Keeping track of the stocks of generating capacity, financial assets, and liabilities was tedious as well. Although the electricity model served its original purpose, it was not readily amenable to further expansion. A new approach was needed if we were to improve the scope and speed of the simulations.

A Monthly Model of the Western Power System

The second approach was used to simulate the carbon dioxide emissions (CO_2) from the western power system (Ford 2008). The long time horizon is evident from figure F.1. The simulations begin in the year 2005 and end in 2025. This graph shows emissions in the Western Electricity Coordinating Council (WECC) in a scenario with no significant policies to discourage the release of CO_2 to the atmosphere. There are major variations in emissions during the course of a year. Emissions are low in the spring, when there is high runoff in the rivers, large hydroelectric generation, low electric loads, and when many of the thermal generating stations are taken down for planned maintenance. The emissions are much higher in the summer, when runoff is low, hydroelectric generation is down, and thermal plants are called back into operation to meet the peak loads. Figure F.1 shows emissions from coal plants, gas-fired

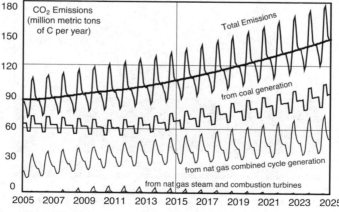

Figure F.1. CO_2 emissions (MMTC/yr) in a business-as-usual scenario.

combined-cycle (CC) plants, older gas-fired steam plants, plus a small amount from the gas-fired combustion turbines (CTs). The coal plants create the majority of the emissions, accounting for two-thirds of the emissions at the end of the simulation.

The CO_2 model simulates time in months, with a full calendar of 12 months per year. Time is in months, but figure F.1 shows results from 2005 to 2025. This convenient display makes use of Vensim's TIME BASE function (BWeb). The WECC model operates with DT set to 1 month, so the 20-year simulation requires only 240 steps to complete. These simulations appear on the screen with the same rapid response that you are accustomed to seeing with the examples in this book. A typical simulation requires 3 seconds. This rapid response allows for interactive experimentation that is essential to learning.

Hourly Operations within the Monthly Model

A 3-second simulation is surprisingly rapid for a model that is representing hourly operations within a 20-year simulation of the western power system. The hourly operations are important to the way power plants are scheduled. The daily schedule is simulated endogenously, as explained in appendix E. The user specifies a 24-hour time profile for power demands in the seven areas of the WECC for a typical day in each month of the year. The user also specifies time profiles for the daily operation of must-run generators. The must-run generation is subtracted from the demands to obtain the demands imposed on the market. An external function is then used to find the electricity prices in each hour of the typical day for each month. The key to fast operation was the assignment of a time-related subscript (hour of day: hr1, hr2 . . . hr24). Electricity prices tend to be low during the off-peak hours in the middle of the night, but they are much higher during the on-peak hours in the middle of the afternoon. Power plants with low variable costs run all day and earn net revenues regardless of the hourly variations in price. But other plants are more expensive to operate. They may operate only for a few hours when electricity prices are sufficiently high.

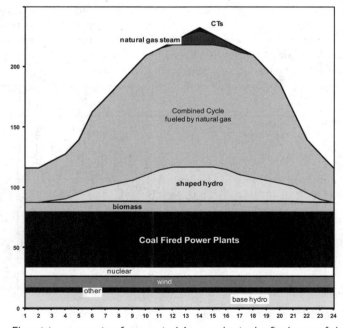

Figure F.2 . Electricity generation for a typical August day in the final year of the simulation.

The daily schedule is depicted in figure F.2. This example is taken from a typical August day at the end of the simulation. The peak demand appears around 2 p.m., at over 230 gigawatts (GW). The demand for power is lowest at 2 a.m. Figure F.2 shows that nuclear, coal, and biomass generators would be running for the entire day, providing most of the power needed for minimum loads. Some gas-fired CCs are running at this hour as well, but many more CCs are needed as the demand for power rises during the course of the day. By the noon hour, all of the available gas CCs are in operation. At this point, the market prices must rise sufficiently high to bring the more expensive gas steam units into operation. And by 2 p.m., the price rises high enough to bring the combustion turbines into operation. These units operate for only a few hours, and they contribute little to the region's CO_2 emissions.

The daily schedule in figure F.2 is a business-as-usual situation. The focus of the WECC model was the change in power plant construction and operations if the nation puts a price on carbon emissions. (The model was used to simulate the carbon prices that would emerge with the Climate Stewardship Act of 2003.) When generators must pay for CO_2 allowances, the daily schedule will be altered substantially, and CO_2 emissions could be reduced by as much as 75% by 2025. Further details on these results are provided by Ford (2008). Details on the model structure are provided on the BWeb.

Advice on Short-Run Dynamics

The best advice on short-run and long-run dynamics is to avoid simulating them within a single model. You will make more progress by developing a separate model for the short-run dynamics in your system. The results of the short-run model may then be used to inform the parameters in the long-run model.

The innovative approach used in the WECC model of CO_2 emissions may prove useful if you face an exceptional case where both the short-run and long-run dynamics should be simulated within the same model. The key to this particular model was the adoption of a time-related subscript to represent the hourly operations of the power system. Perhaps you will think of a time-related subscript to represent the short-term dynamics in your own system.

Appendix G

Spatial Dynamics and Spatial Displays

This appendix describes how system dynamics software may be used to simulate the flow and accumulation of nitrogen in a complex landscape. The model illustrates the challenge of including spatial complexity within a dynamic model. The nitrogen example raises a question for landscape systems in general—should we attempt to simulate both spatial and dynamic complexity within the same model? The answer to this question depends on the purpose of the model and the programming skills of the modeling team. Some background on geographic information systems (GIS) is needed if we are to appreciate the programming demands of spatial/temporal modeling.

Geographic Information Systems

A geographic information system (GIS) is a system for capturing, storing, checking, manipulating, analyzing, and displaying spatial data. Many landscape modeling projects make use of a GIS (Clarke, Parks, and Crane 2002; Costanza and Voinov 2003; Maguire, Batty, and Goodchild 2005; Voinov 2008). System dynamics models have been developed to operate with a GIS. The models include spatial detail and the necessary software to exchange information with the GIS. Voinov (2008) describes two programs that he believes are most suitable to provide the exchange of information. The first program is Simile, the software described briefly in appendix C. The developers of Simile included a wide variety of features that were intentionally omitted from Stella. Voinov (2008, 167) warns us to expect extra time to master the advanced features. Once they are mastered, however, "Simile is ready for a much more advanced user to build quite sophisticated models." Voinov illustrates the Simile approach to spatial modeling with a model of rabbit grazing patterns. The highlight of the application is the dramatic maps that reveal the simulated effects of the grazing.

The second program described by Voinov (2008) is the Spatial Modeling Environment (SME) pioneered by Maxwell and Costanza (1997). SME was designed to encourage collaboration among modelers and to overcome a tendency of some complex models to be "idiosyncratic and nearly incomprehensible" (BenDor and Metcalf 2006). Voinov (2008, 180) explains that

> SME is not quite a modeling system, since it does not require a language or formalism of its own. It can take the equations from your Stella model and translate them into an intermediate Modular Modeling Language (MML), which is then translated into C++ code. At the same time, SME will link your model to spatial data. . . .
>
> These days, the simplest way to generate these maps is to use ArcInfo or ArcGIS, the monopolist on the GIS market

Both SME and Simile provide the software for simulations that deal with spatial complexity and dynamic complexity within the same model. Voinov (2008, 193) favors the use of the SME, especially for teams with expertise in C++. He concludes that the SME approach is

> *not as easy as putting together a model in Stella, or even Simile. However, for somebody comfortable with C++, it may not be a big deal and actually turn out to be simpler than learning the new formalism required for Simile. Once we are in the programming language mode, we have all the power we need to create any complex model. So, in a way, SME may be treated as a nice interface between Stella and C++ power modeling.*

A few system dynamics models have been developed in conjunction with a GIS. An exemplary example is the analysis of the infestation of Midwest forests by the emerald ash beetle (BenDor and Metcalf 2006). Stella was used for the population dynamics, with the spatial results exchanged with a GIS via SME. The beetle application illustrates the insights to be found if a team invests the time and effort to develop a spatial/temporal model. Another important lesson from the beetle infestation study is that the construction of spatial/temporal models requires a team with substantial programming resources that are beyond the reach of many groups. Some groups lack the time and the resources to develop and test a spatial/temporal model. Other groups may wish to retain close stakeholder involvement in the modeling process and choose not to enter the "programming language mode" required for spatial/temporal modeling.

I believe the more productive approach for such groups is to deal with spatial complexity in a stand-alone GIS. The GIS would be maintained separately from the group's dynamic model of the larger system. The GIS may provide support for the aggregate (nonspatial) parameters in the dynamic model. If policy changes lead to new spatial arrangements, the GIS may then be used to obtain new parameters for the dynamic model. This approach will be easier to implement and easier to explain, and the clarity of the modeling can enhance the prospects for stakeholder involvement. An excellent example of stand-alone GIS support is the stakeholder-driven modeling of the sage-grouse population in central Washington (Beall and Zeoli 2008).

But the map from a GIS gives a static view of the landscape. What should we do if the landscape is changing over time? Perhaps the only answer is to work with SME and invest the time to develop a single model with both spatial complexity and dynamic complexity. An alternative approach is to develop two models to deal with the two forms of complexity. One model can focus exclusively on the spatial complexity. The second model can deal with the big-picture interactions of the environmental system with human decision making. The second model could stress the stocks, flows, and feedbacks while aiming for visual clarity. In other words, it would follow the advice given throughout this book. Meanwhile, the spatial model can be constructed with Stella or Vensim and then used to support the approximate relationships in the big-picture model.

A Spatial Model of Nitrogen Accumulation

This two-step approach is demonstrated by simulating nitrogen flows in a landscape. The landscape elevations are stored in a spreadsheet, since spreadsheets are familiar, and they provide a convenient way to store and display spatial information. The demonstration uses an adaptation of Huggett's (1993) model of nitrogen flows in a catchment. Huggett assumed that nitrogen would flow from cell to cell based on the downward slope between the cells. For this illustration, we begin with 1 kilogram of nitrogen stored in each acre of a 320-acre catchment. The highest elevations are 100 meters in the northwest and southwest corners of the catchment. The low elevation is 40 meters, located in the middle of the catchment. Figure G.1 shows elevations in a spreadsheet "surface chart" shaded in 5-meter intervals (e.g., 95–100 meters in black, 90–95 meters in gray, and 85–90 meters in white, etc.). The nitrogen flows

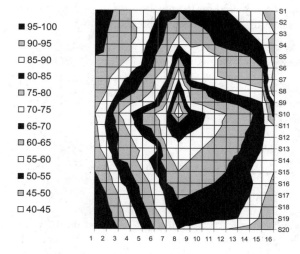

- ■ 95-100
- □ 90-95
- □ 85-90
- ■ 80-85
- □ 75-80
- □ 70-75
- ■ 65-70
- □ 60-65
- □ 55-60
- ■ 50-55
- □ 45-50
- □ 40-45

Figure G.1. Excel surface chart of elevations for a catchment.

downhill, following complex routes over time, with the entire 320 kilograms eventually located at the low point, the catchment.

Figure G.2 shows a Stella model of *NS*, the nitrogen stored in each of the cells. The three-dimensional icons remind us that all the variables are array variables. The model begins with *NS* initialized at 1 kilogram. Two interconnected flows simulate the flows in the east-west directions. The flows to the east depend on the nitrogen stored in each cell and the slope measured in the eastern direction. The slopes to the east are calculated by comparing the elevations of the cells in the neighboring column. The flows to the east represent the nitrogen that flows from left to right in the spreadsheet. This flow connects to a cloud in Figure G.2. Clouds normally stand for sinks that are outside the boundary of analysis, so the diagram creates the impression that the eastward flow is removing nitrogen from the catchment. This is a false impression, since the nitrogen is merely moving into the next cell to the east. The cell-by-cell flows are captured mathematically by connecting the flows to the east to the flows from the west. A similar approach is used for flows in the south-north directions.

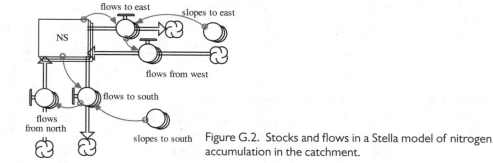

Figure G.2. Stocks and flows in a Stella model of nitrogen accumulation in the catchment.

Figure G.2 looks like a simple model, but the implementation requires extra time to master the equation syntax for Stella's two-dimensional array variables. Details are on the book's website (BWeb). Figure G.3 shows a Vensim version of the same model. *NS* stands for the nitrogen stored in each of the 320 cells in the 16-by-20 grid. Flows to the east depend on the NS and the slopes measured to the east (abbreviated as *Se*). These flows are then connected directly to the flows from the west. A similar approach is used for the north-south flows. The flows to the south depend on NS and the slopes measured to the south (abbreviated as *Ss*).

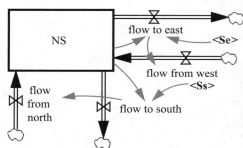

Figure G.3. Stocks and flows in a Vensim model of nitrogen accumulation in the catchment.

The model makes use of Vensim's subscripts feature. A subscript row takes on the values r1, r2 . . . r20. A subscript col (for column) takes on the values c1, c2 . . . c16. These are the full range of values. Vensim allows subranges, and this model is easier to construct with a subrange of col defined as "all but first col." A subrange of row called "all but first row" is also useful. Table G.1 shows selected equations. They begin with the slopes in the eastern direction and in the southern direction. The next equation calculates the flow to the east by checking whether the slope in the eastern direction is positive or negative. The flow from west is then connected to the flow to east. A similar approach is used for flows in the southern and northern directions. Table G.1 concludes with the equation for NS, the amount of nitrogen stored in each cell of the catchment.

Table G.1. Selected equations for the Vensim model in figure G.3.

Se[row,all but last col] = (elevations[row,all but last col]-elevations[row,next col])/width
Ss[all but last row,col] = (elevations[all but last row,col]-elevations[next row,col])/width
flow to east[row,all but last col] = IF THEN ELSE(Se[row,all but last col]>0,
    NS[row,all but last col]*Se[row,all but last col] ,NS[row,next col]*Se[row,all but last col])
flow from west[row,all but first col] = flow to east[row,previous col]
flow to south[all but last row,col] = IF THEN ELSE(Ss[all but last row,col]>0,
    NS[all but last row,col]*Ss[all but last row,col] ,NS[next row,col])*Ss[all but last row,col]
flow from north[all but first row,col] = flow to south[previous row,col]
NS[row,col] = INTEG (–flow to east[row,col]+flow from west[row,col]
    –flow to south[row,col]+flow from north[row,col])

Writing Code for the Nitrogen Model

The models in figures G.2 and G.3 appear to be quite simple. After all, they include only one stock and four flows. The flows are controlled by size of the stock and the slopes between the cells. When we look just at stocks and flows, it appears that these are two of the simplest models in the book. But the challenge in building the nitrogen model is not in connecting the stocks and flows; the challenge is in mastering the programming syntax for defining locations in two-dimensional arrays. If you construct these models to match the results on the BWeb, you will discover that the bulk of your time is devoted to writing code. This may remind you of Joe's question from chapter 2:

When do I get to write code?

The answer was that Joe should not expect to be writing code when using system dynamics. The normal challenge in system dynamics is to formulate the best combination of stocks, flows, and feedbacks to explain a dynamic problem. But the challenge in the nitrogen model is entirely different—most of our time is devoted to mastering software syntax for defining relative positions in a two-dimensional array. We also need to import the elevations from a spreadsheet. Once this is done, the spatial model can teach us about the flows through a complex landscape.

Learning from the Spatial Model

Let's use the spatial model to learn about the levels of nitrogen that will eventually appear in the low point of the catchment. The movements are all in the downhill direction, so the entire 320 kilograms eventually accumulates in the low point. If you simulate the BWeb model, you'll see that around 75% of the nitrogen will be in the low point by the 50th day. Almost all the nitrogen is in the low point by the 100th day.

Now suppose we are concerned about excessive nitrogen concentration in hot spots on the landscape, and that the primary purpose of the model is to learn about the nitrogen in the lowest cell. Let's also assume that the dynamics of nitrogen in the lowest cell is important to a larger model dealing with the application and regulation of nitrogen in the region. The larger model might include a variety of stocks to deal with the economic activities, the costs and magnitudes of nitrogen use, and the challenges of measuring and regulating the application of nitrogen. One might then expand the larger model to include the spatial details and complicated syntax associated with the array variables in the spatial model.

But this appendix takes a different approach. The larger model would be easier to build (and easier to understand) if we find a way to approximate the key result without getting bogged down in the spatial details. This is a two-step approach. The spatial model in figure G.3 teaches us about the accumulation in the lowest cell. We then find a way to approximate this result, and the approximation is inserted into the larger model.

Figure G.4 shows an extra stock and flow to experiment with an approximation. The total storage is found by SUM(NS[row!,Col!]), which is Vensim's function to add up the contents of every cell. Since no nitrogen enters or leaves the system, the total storage will be constant at 320 kilograms. Figure G.4 assumes that the approximate value will gradually approach 320 kilograms. Experiments with the spatial model show a pattern of nitrogen accumulation that resembles a first-order approach. This suggests we can define the flow:

add to approximation = (total storage – Approximate N in Lowest Cell)/adjustment time for approximation

and we experiment with the value of the adjustment time. Such experimentation is easy and quick in a control panel with a slider assigned to the adjustment time (BWeb). A few experiments will reveal that a 34-day adjustment time gives a reasonable fit, as shown in figure G.5. The thick line is the actual N stored in the lowest cell. It grows to 320 kilograms as the flows from all the other cells gradually find their way to the low point in the catchment. The thin line is the approximate value. It is somewhat high in the first 50 days and somewhat low for

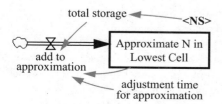

Figure G.4. An approximation to the nitrogen in the lowest cell of the catchment.

the next 100 days. By the 150th day, all the nitrogen has reached the low point, and the approximation is nearly exact. The close fit in figure G.5 may be quite sufficient for a wide variety of uses of the larger model. This means that the larger model need only include the small structure shown in figure G.4. This is an important benefit, as it allows the team engaged in the larger model development to avoid the coding challenges associated with relative positions in two-dimensional arrays.

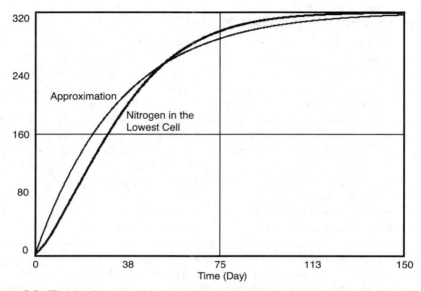

Figure G.5. The thin line provides a good approximation to the nitrogen in the lowest cell.

Advice on the Two-Models Approach

The simple stock-and-flow structure in figure G.4 and the close fit in figure G.5 may leave you with the impression that the two-models approach is easy. But the nitrogen example was intentionally simple so you could see how the use of two models allows a model of a larger system to benefit from the results of a spatial model. Following the two-models approach will not necessarily be easy for a complex landscape of interest to you. But it will probably be easier than the alternative—the inclusion of both spatial complexity and dynamic complexity in a single model.

The use of multiple models is good practice for a wide variety of systems, not just for landscape systems. Energy companies often carry a portfolio of models dealing with the challenges of forecasting, planning, and operations. The long-term, big-picture models frequently benefit from approximations obtained from the operational models (Ford 1994, 1997). Another example involves the combination of system dynamics and individual-based models. Appendix C explains how multiple simulations of an individual-based model can inform the nonlinear relationships in the system dynamics models of the entire population.

Another illustration of the two-models approach is the modeling of the Snake River flows in southern Idaho. The "big-picture" model was designed for interactive simulation by a group of stakeholders, so clarity of structure and speed of simulation were important. The model combined surface-water flows and groundwater flows along with the many uses of the water (irrigated agriculture, electricity generation, and minimum flows for wildlife). The aquifer discharge at two points on the Snake was a crucial part of the system. But the mathematical approach in groundwater modeling is entirely different than in system dynamics. Mathematically speaking, system dynamics models are a collection of first-order differential equations. But groundwater models are typically based on a collection of partial differential equations (with the solution found by numeric methods). Groundwater models require considerable programming skills as well as detailed data on both the surface water and the groundwater characteristics. They can also take a while for the computer to find the numerical solution. Such a model would be entirely inappropriate to include within the system dynamics model of the Snake River (Ford 1996). So we turned to the two-models approach. Fortunately, the groundwater modelers had exercised their model under a wide variety of assumptions on groundwater pumping. They then summarized the many simulations with time patterns for aquifer discharge. The aquifer flows turned out to be quite similar to the first-order shape shown in figure G.5. This allowed the system dynamics model to represent aquifer discharge as the contents of the aquifer divided by an average residence time. (The value of the residence time was selected to match the results from the hydrologic model.) The system dynamics model provided realistic results on aquifer volume and discharge and allowed for the rapid simulations needed for active discussion and learning among the stakeholders. You can experience the learning advantages of rapid, interactive simulation of a complex river system by visiting the Idagon (see figure 13.4) on the BWeb.

Spatial Displays

The highlights of landscape modeling systems are the maps that reveal the changes in landscape over time (Costanza and Voinov 2003). A GIS display can reveal important relationships in a single glance. An example is the land bridge to Negit Island in the Mono Basin GIS (BWeb). Spatial information can also be displayed in a simple spreadsheet. The elevation surface chart at the start of this appendix (figure G.1) illustrates with one of the standard graphing options in Excel. Spreadsheets may also be used to display spatial information by simply shading the elevation cells through conditional formatting. Both Stella and Vensim allow for dynamic links with spreadsheets. This allows for spatial results to be displayed in two dimensions with widely available and widely recognized software. Spatial displays are also made easy by Stella's spatial map, as illustrated in figure G.6.

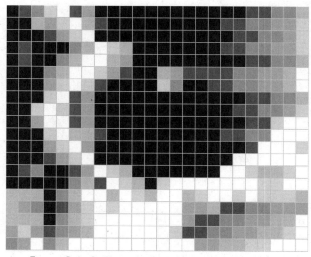

Figure G.6. Stella map of a hypothetical landscape.

The spatial map is a stand-alone application that allows one to transfer array data from the model to colored values in a two-dimensional display. The x and y axes in the map correspond to the bounds of a two-dimensional array in Stella. You assign different colors to the values in each cell, and you elect whether the color transitions are smooth or abrupt. Stella responds with a two-dimensional rendering of the data, as illustrated in figure G.6. This is a Stella map of a hypothetical landscape with the highest elevations in black and the lowest elevations in white. This example is a static map, a map that makes sense if the elevations are not changing over time. A dynamic map would be useful if the elevations were changing over time (e.g., owing to the simulated effects of erosion). With dynamic mapping, Stella refreshes the spatial map as the simulation proceeds through time (BWeb). This new feature puts dynamic mapping within closer reach of groups that lack the programming skills to undertake spatial/temporal modeling with programs like SME or Simile.

References

Adami, C. 1998. *Introduction to artificial life*. New York: Springer-Verlag.

Alee, W., A. Emerson, O. Park, T. Park, and K. Schmidt. 1949. *Principles of animal ecology*. Philadelphia: W. B. Saunders.

Anderson, A., and J. Anderson. 1973. System simulation to identify environmental research needs: Mercury contamination. In *Toward global equilibrium*, ed. D. H. Meadows and D. L. Meadows. Waltham, MA: Pegasus Communications.

Andrewartha, H. 1961. *Introduction to the study of animal populations*. Chicago: University of Chicago Press.

Armstrong, D. 1987. *Rocky mountain mammals*. 2nd ed. Boulder: University Press of Colorado.

Army Corps of Engineers, Bonneville Power Administration, and the Bureau of Reclamation. 1992. *Columbia River salmon flow measures options analysis/EIS*. Walla Walla, WA: U.S. Department of the Army.

Backus, G., and J. Amlin. 1985. Combined multidimensional simulation language, database manager and sensitivity/confidence analysis package for system dynamics modeling. In *Proceedings of the 1985 International Conference of the System Dynamics Society, Keystone, Colorado*. Albany, NY: System Dynamics Society.

Beall, A. 2007.Participatory environmental modeling and system dynamics: Integrating natural resource science and social concerns. PhD diss., Washington State University, Pullman.

Beall, A., and L. Zeoli. 2008. Participatory modeling of endangered wildlife systems: Simulating sage-grouse and land use in central Washington. *Ecological Economics* 68 (1–2): 24–33.

BenDor, T., and S. Metcalf. 2006. The spatial dynamics of invasive species spread. *System Dynamics Review* 22 (1): 27–50.

Bjornn, T. 1987. *A model for predicting production and yield of Tucannon River salmon and steelhead stocks in relation to land management practices*. Technical Report 98-1. Idaho Fish and Wildlife Research Unit, University of Idaho, Moscow, ID 83843.

Bodily, S. 1986. Spreadsheet modeling as a stepping stone. *Interfaces* 16 (5): 34–52.

Boer, E. 2004. Simile: Finally bridging the gap between temporal and spatial models? Thesis report GIRS-2004-38. Wageningen University, Wageningen, The Netherlands.

Botkin, D. 1990. *Discordant harmonies*. New York: Oxford University Press.

Botkin, D., and E. Keller. 1998. *Environmental science*. New York: Wiley.

Boyce, S. 1991. *Models for managers in natural resources*. Published by the author, 180 College View Terrace, Brevard, NC 28712.

[BPA] Bonneville Power Administration. 2008. Fact sheet: Federal agencies announce agreements to benefit Columbia River Basin fish. Portland, OR: Bonneville Power Administration.

Brennan, M., and M. Shelley. 1999. A model of the uptake, translocation and accumulation of lead (Pb) by maize for the purpose of phytoextraction. *Ecological Engineering* 12:271–97.

Bunch, D., M. Bradley, T. Golob, R. Kitamura, and G. Occhiuzzo. 1992. *Demand for clean fuel personal vehicles in California: A discrete choice stated preference survey*. Irvine: University of California Institute of Transportation Studies.

Bush, J., A. Schneider, T. Wachtel, and J. Brimm. 1985. Fluid therapy in acute large area burns: A system dynamics model. *System Dynamics Review* 13 (4): 271–88.

Cannon, W. 1932. *The wisdom of the body*. New York: Norton.

Capra, F. 1996. *The web of life*. New York: Doubleday.

Carson, R. 1962. *Silent spring*. Boston: Houghton Mifflin.

Caughley, G. 1970. Eruption of ungulate populations, with emphasis on Himalayan thar in New Zealand. *Ecology* 51 (1): 53–72.

Cavana, R., and A. Ford. 2004. Environmental and resource systems: Editors' introduction. *System Dynamics Review* 20 (2): 89–98.

Clark, C. 1985. *Bioeconomic modelling and fisheries management*. New York: Wiley Interscience.

Clarke, K., B. Parks, and M. Crane, eds. 2002. *Geographic information systems and environmental modeling*. Saddle River, NJ: Prentice Hall.

Claussen, M., L. Mysak, A. Weaver, M. Crucifix, T. Fichefet, M. Loutre, S. Weber, et al. 2000. Earth system models of intermediate complexity: Closing the gap in the spectrum of climate system models. *Climate Dynamics* 18:579–86.

Clutton-Brock, T., and S. Albon. 1992. Trial and error in the highlands. *Nature* 358 (July 2): 11.

Cockerill, K., H. Passell, and V. Tidwell. 2006. Cooperative modeling: Building bridges between science and the public. *Journal of the American Water Resources Association* 42 (2): 457–71.

Cockerill, K., V. Tidwell, H. Passell, and L. Malczynski. 2007. Cooperative modeling lessons for environmental management. *Environmental Practice* 9 (1): 28–41.

Corburn, J. 2005. *Street science: Community knowledge and environmental health justice*. Cambridge, MA: MIT Press.

Costanza, R., and L. Cornwell. 1992. The 4P approach to dealing with scientific uncertainty. *Environment* 34 (9): 12–20.

Costanza, R., and M. Ruth. 1998. Using dynamic modeling to scope environmental problems and build consensus. *Environmental Management* 22 (2): 183–95.

Costanza, R., and A. Voinov, eds. 2003. *Landscape simulation modeling*. New York: Springer-Verlag.

Coyle, G. 1977. *Management system dynamics*. New York: Wiley.

———. 1996. *System dynamics modeling: A practical approach*. New York. Chapman & Hall/CRC.

Dangerfield, B., Y. Fang, and C. Roberts. 2001. Model-based scenarios for the epidemiology of HIV/AIDS: The consequences of highly active antiretroviral therapy. *System Dynamics Review* 17 (2): 119–50.

Dawkins, R.1986. *The extended phenotype*. New York: Oxford University Press.

Deaton, M., and J. Winebrake. 2000. *Dynamic modeling of environmental systems*. New York: Springer-Verlag.

Dimitrovski, A., K. Tomsovic, and A. Ford. 2007. Comprehensive long term modeling of the dynamics of investment and network planning in electric power systems. *Journal of Critical Infrastructures* 3 (1–2): 235–64.

DiPasquale, D., and W. Wheaton. 1996. *Urban economics and real estate markets*. Saddle River, NJ: Prentice Hall.

Doolittle, W. 1981. Is nature really motherly? *CoEvolution Quarterly* (Spring): 58–63.

Draper, F., and M. Swanson. 1990. Learner-directed systems education. *System Dynamics Review* 6 (2): 209–12.

Dudley, R. 2008. A basis for understanding fishery management dynamics. *System Dynamics Review* 24 (1): 1–29.

Edwards, W. 1977. How to use multi-attribute utility measurement for social decision making. *IEEE Transactions on Systems, Man and Cybernetics*. SMC-7 (May): 326–40.

EIR. 1993. Draft environmental impact report for the review of Mono Basin water rights to the City of Los Angeles. Prepared for the California State Water Resources Control Board, Division of Water Rights, Sacramento, CA, by Jones & Stokes Associates (JSA 90-171). May.

Elton, C. 1933. The Canadian snowshoe rabbit enquiry, 1931–1932. *Canadian Field Naturalist* 47:63–86.

Fiddaman, T. 2002. Exploring policy options with a behavioral climate-economy model. *System Dynamics Review* 18(2): 243–67.

Field, C., D. Lobell, H. Peters, and N. Chiariello. 2007. Feedbacks of terrestrial ecosystems to climate change. *Annual Review of Environment and Resources* 32:1–29.

Fisher, D. 2007. *Modeling dynamic systems: Lessons for a first course*. 2nd ed. Lebanon, NH: iSee Systems.

Fisher, T., and R. Hinrichsen. 2005. *Preliminary abundance-based trend results for Columbia Basin salmon and steelhead ESUs*. Portland, OR: Division of Environment, Fish and Wildlife, Bonneville Power Administration.

Ford, A. 1978. Breaking the stalemate: An analysis of boom town mitigation policies. *Journal of Interdisciplinary Modeling and Simulation* (January).

———. 1990. Estimating the impact of efficiency standards on the uncertainty of the Northwest electric system. *Operations Research* 38 (4): 580–97.

———. 1994. Electric vehicles and the electric utility company. *Energy Policy* 22 (7): 555–70.

———. 1996. Testing the Snake River explorer. *System Dynamics Review* 12 (4): 305–29.

———. 1997. System dynamics and the electric power industry. *System Dynamics Review* 13 (1): 57–85.

———. 2002. Boom and bust in power plant construction: Lessons from the California electricity crisis. *Journal of Industry, Competition and Trade* 2 (1–2): 59–74.

———. 2008. Simulation scenarios for rapid reduction in carbon dioxide emissions in the western electricity system. *Energy Policy* 36:44–55.

Ford, A., and H. Flynn. 2005. Statistical screening of system dynamics models. *System Dynamics Review* 21:273–303.

Forrester, J. 1961. *Industrial dynamics.* Waltham, MA. Pegasus Communications.

———. 1968. *Urban dynamics.* Waltham, MA. Pegasus Communications.

———. 1973. Counterintuitive behavior of social systems. In *Toward global equilibrium*, ed. D. H. Meadows and D. L. Meadows. Waltham, MA: Pegasus Communications.

———. 1980. Information sources for modeling the national economy. *Journal of the American Statistical Association* 75 (371): 555–56.

Forrester, J., and P. Senge. 1980. Tests for building confidence in system dynamics models. *System Dynamics: TIMS Studies in the Management Sciences*, vol. 14, 201–28. Amsterdam: North Holland Press.

Frey, H. 1997. Quantitative analysis of variability and uncertainty in energy and environmental systems. In *Uncertainty modeling and analysis in civil engineering*, ed. B. Ayyub, 383–85. Boca Raton, FL: CRC Press.

Frissell, C., W. Liss, C. Warren, and M. Hurley. 1986. A hierarchical framework for stream habitat classification. *Environmental Management* 10:199–214.

Gallaher, E. 1996. Biological system dynamics. *Simulation* 66 (4): 243–57.

Gardiner, P., and W. Edwards. 1975. Public values: Multiattribute utility measurement for social decision making. In *Human judgment and decision processes*, ed. M. F. Kaplan and S. Schwartz, 1–37. New York: Academic Press.

Gardiner, P., and A. Ford. 1980. Which policy run is best, and who says so? In *System Dynamics: TIMS Studies in the Management Sciences*, vol. 14, 241–57. Amsterdam: North Holland Press.

Gill, R., L. Anderson, H. Polley, H. Johnson, and R. Jackson. 2006. Potential nitrogen constraints on soil carbon sequestration under low and elevated atmospheric CO_2. *Ecology* 87 (1): 2006: 41–52.

Gleick, J. 1988. *Chaos: Making a new science.* New York: Penguin.

Goldratt, E., and J. Cox. 1986. *The goal.* Croton-on-Hudson, NY: North River Press.

Gordon, D. 1991. *Steering a new course: Transportation, energy and the environment.* Washington, DC: Island Press.

Govindasamy, B., S. Thompson, A. Mirin, M. Wickett, K. Caldeira, and C. Delire. 2005. Increase of carbon cycle feedback with climate sensitivity. *Tellus* 57B:153–63.

Grant, W. 1986. *Systems analysis and simulation in wildlife and fisheries sciences.* New York: Wiley.

Grant, W., E. Pedersen, and S. Marin. 1997. *Ecology and natural resource management: Systems analysis and simulation.* New York: Wiley.

Greenberger, M., M. Crenson, and B. Crissey. 1976. *Models in the policy process.* New York: Russell Sage.

Gregory, R. 1998. *Eye and brain: The psychology of seeing.* New York. Oxford University Press.

Grimm, V. 1999. Ten years of individual-based modeling in ecology. *Ecological Modelling* 115:129–48.

Grimm, V., and S. Railsback. 2005. *Individual-based modeling and ecology.* Princeton, NJ: Princeton University Press.

Grimm, V., E. Revilla, U. Berger, F. Jeltsch, W. Mooij, S. Railsback, H. Thulke, J. Weiner, T. Wiegand, and D. DeAngelis. 2005. Pattern-oriented modeling of agent-based complex systems: Lessons from ecology. *Science* 310 (Nov. 11): 987–91.

Guyton, A., and J. Hall. 1996. *Textbook of medical physiology.* 9th ed. Philadelphia: W. B. Saunders.

Hannon, B., and M. Ruth. 1997. *Modeling dynamic biological systems.* New York: Springer-Verlag.

Hansen, J., M. Sato, P. Kharecha, G. Russell, D. W. Lea, and M. Siddall. 2007. Climate change and trace gases. *Philosophical Transactions of the Royal Society* 365 (May): 1925–54.

Hardin, G. 1966. *Biology: Its principles and implications.* 2nd ed. San Francisco: W. H. Freeman.

———. 1968. The tragedy of the commons. *Science* 162 (3859): 1243–48.

Harrison, J. 1992. Tackling the Tucannon. *Northwest Energy News* (March). Portland, OR: Northwest Power Planning Council.

Hart, J. 1996. *Storm over Mono: The Mono Lake battle and the California water future.* Berkeley and Los Angeles: University of California Press.

Harte, J. 1988. *Consider a spherical cow: A course in environmental problem solving.* Sausalito, CA: University Science Books.

Hastings, A. 1997. *Population biology: Concepts and models.* New York: Springer-Verlag.

Holling, C. 1978. *Adaptive environmental assessment and management.* Caldwell, NJ: Blackburn Press.

Houghton, J. 2004. *Global warming: The complete briefing.* 3rd ed. Cambridge, UK: Cambridge University Press.

House, P., and J. McLeod. 1977. *Large scale models for policy evaluation.* New York: Wiley.

Hoyt, H. 1933. *One hundred years of land values in Chicago.* University of Chicago Press.

Huffaker, C. 1958. Experimental studies on predation. *Hilgardia* 27:343–83.

Huggett, R. 1993. *Modelling the human impact on nature.* New York: Oxford University Press.

[IPCC] Intergovernmental Panel on Climate Change. 1997. *An introduction to simple climate models used in the IPCC second assessment report.* ISBN 92-9169-101-1.

———. 2001. *Climate change 2001: The scientific basis—summary for policymakers.* www.ipcc.ch.

———. 2007. *Climate Change 2007: The physical science basis—summary for policymakers.* www.ipcc.ch.

Iudicello, S., M. Weber, and R. Wieland. 1999. *Fish, markets and fishermen: The economics of overfishing.* Washington, DC: Island Press.

Jacobsen, C., and R. Bronson. 1987. Defining sociological concepts as variables for system dynamics modeling. *System Dynamics Review* 3 (1): 1–7.

Jager, H., H. Caldwell, M. Sale, M. Bevelhimer, C. Coutant, and W. Van Winkle. 1997. Modelling the linkages between flow management and salmon recruitment in rivers. *Ecological Modelling* 103:171–91.

Joseph, L. 1990. *Gaia: The growth of an idea.* New York: St. Martin's Press.

Kalman, R. 1960. A new approach to linear filtering and prediction problems. *Journal of Basic Engineering* D82:35–45.

Keeney, R., and H. Raiffa. 1976. *Decisions with multiple objectives: Preferences and value trade-offs.* New York: Wiley.

Kirchner, J. 1989. The Gaia hypothesis: Can it be tested? *Review of Geophysics* 27 (2): 223–35.

Kirkwood, C. 1992. An overview of methods for applied decision analysis. *Interfaces* 22 (6): 28–39.

Kitching, R. 1983. *Systems ecology: An introduction to ecological modelling.* St. Lucia, Queensland, Australia: University of Queensland Press.

Kormondy, E. 1969. *Concepts of ecology.* Englewood Cliffs, NJ: Prentice Hall.

Kump, L. 2002. Reducing uncertainty about carbon dioxide as a climate driver. *Nature* 419 (September 12): 188–90.

Lack, D. 1954. *The natural regulation of animal numbers.* New York: Oxford University Press.

LeBlanc, L., and T. Grossman. 2008. Introduction: The use of spreadsheet software in the application of management science and operations research. *Interfaces* 38 (4): 225–27.

Lee, K. 1993. *Compass and gyroscope: Integrating science and politics for the environment.* Washington, DC: Island Press.

Leopold, A. 1943. Deer irruptions. *Wisconsin Conservation Bulletin* 8 (8).

———. 1949. *A Sand County almanac and sketches here and there.* New York: Oxford University Press.

Levine, L. 1993. Gaia: Goddess and idea. *BioSystems* 31:85–92.

Levins, R. 1966. The strategy of model building in population biology. *American Scientist* 54 (4): 421–31.

Lotka, A. 1925. *Elements of physical biology.* Baltimore: Williams & Wilkins.

Louviere, J., J. Swait, and D. Hensher. 2000. *Stated choice methods: Analysis and applications.* Cambridge, UK: Cambridge University Press.

Lovelock, J. 1990. Hands up for the Gaia Hypothesis. *Nature* 344: 100–102.

———. 1991. *Healing Gaia.* New York: Harmony Books.

———. 1995. *Gaia, a new look at life on Earth.* New York: Oxford University Press.

———. 1998. *The ages of Gaia: A biography of our living earth.* New York: Norton.

Lovelock, J., and L. Kump. 1994. Failure of climate regulation in a geophysiological model. *Nature* 369 (June): 732–34.

Lovelock, J., and L. Margulis. 1974. Atmospheric homeostasis: The Gaia hypothesis. *Tellus* 26:1–10.

Maguire, D., M. Batty, and M. Goodchild. 2005. *GIS, spatial analysis and modeling.* Redlands, CA: Esri Press.

Margulis, L., and J. Lovelock. 1989. Gaia and geognosy. In *Global ecology*, ed. M. Rambler, L. Margulis, and R. Fester. Boston: Academic Press.

Mass, N., and P. Senge. 1980. Alternative tests for selecting model variables. In *Elements of the system dynamics method*, ed. J. Randers. Waltham, MA: Pegasus Communications.

Matson, P., and A. Berryman. 1992. Special feature: Predator-prey theory. *Ecology* 73 (5): 1529.

Maxwell, T., and R. Costanza. 1997. A language for modular spatio-temporal simulation. *Ecological modelling* 103:105–13.

McKay, M., W. Conover, and R. Beckman. 1979. A comparison of three methods for selecting values of input variables in the analysis of output from a computer code. *Technometrics* 21:239–45.

Meadows, D. H. 2009. *Thinking in systems: A primer.* White River Jct., VT: Chelsea Green.

Meadows, D. H., and D. L. Meadows, eds. 1973. *Toward global equilibrium.* Waltham, MA: Pegasus Communications.

Meadows, D. H., D. L. Meadows, J. Randers, and W. Behrens. 1972. *The limits to growth.* New York: Universe Books.

Meadows, D. H., and J. M. Robinson. 1985. *The electronic oracle: Computer models and social decisions.* New York: Wiley.

Meadows, D. L., T. Fiddaman, and D. Shannon. 1993. *Fish Banks, Ltd.* Durham, NH: Laboratory for Interactive Learning, University of New Hampshire.

Morecroft, J. 1988. System dynamics and microworlds for policymakers. *European Journal of Operational Research* 35:301–20.

———. 2007. *Strategic modelling and business dynamics: A feedback systems approach.* West Sussex, UK: Wiley.

Morgan, M. G., and M. Henrion. 1990. *Uncertainty: A guide to dealing with quantitative risk and policy analysis.* New York: Cambridge University Press.

Moxnes, E. 2000. Not only the tragedy of the commons: Misperceptions of feedback and policies for sustainable development. *System Dynamics Review* 16 (4): 325–48.

Murray, J. 2002. *Mathematical biology: An introduction.* 3rd ed. Berlin: Springer-Verlag.

[NAE] National Academy of Engineering. 2008. Grand Challenges for Engineering: Manage the nitrogen cycle. www.engineeringchallenges.org/cms/8996/9132.aspx.

[NAS] National Academy of Sciences. 2002. *Abrupt climate change: Inevitable surprises.* Washington, DC: National Academy Press.

Nicolson, C., A. Starfield, G. Kovinas, and J. Kruse. 2002. Ten heuristics for interdisciplinary modeling projects. *Ecosystems* 5:376–84.

[NOAA] National Oceanic and Atmospheric Administration. 2008. Final Columbia-Snake Biological Opinions. National Marine Fisheries Service, Northwest Regional Office. www.nwr.noaa.gov/Salmon-Hydropower/Columbia-Snake-Basin/final-BOs.cfm.

Nisbet, R. M. and W. S. C. Garney. 1982. *Modelling fluctuating populations.* New York: John Wiley.

North, M., and C. Macal. 2007. *Managing business complexity: Discovering strategic solutions with agent-based modeling and simulation.* New York: Oxford University Press.

[NPCC] Northwest Power and Conservation Council. 2003. *Twenty years of progress: Columbia River Basin fish and wildlife program.* Portland, OR: NPCC.

———. 2005. *Pocket guide to the Columbia River Basin.* 2005 ed. Portland, OR: NPCC.

[NRC] National Research Council. 2004. *Confronting the nation's water problems: The role of research.* Washington, DC: National Academy Press.

Odum, E. 1971. *Fundamentals of ecology.* 3rd ed. Philadelphia: Saunders.

Odum, H., and E. Odum. 2000. *Modeling for all scales.* San Diego: Academic Press.

Osgood, N. 2009. Lightening the performance burden of individual-based models through dimensional analysis and scale modeling, *System Dynamics Review* 25 (2): 101–34.

Picardi, A. 1975. A systems analysis of pastoralism in the West Africa Sahel. ScD. thesis, Massachusetts Institute of Technology, Department of Civil Engineering, Cambridge, MA.

Pimentel, D. 1968. Population regulation and genetic feedback. *Science* 159 (March 29): 1432–37.

Pratt, C. 1995. *Ecology.* Springhouse, PA: Springhouse Corp.

Rahmandad, H., and J. Sterman. 2008. Heterogeneity and network structure in the dynamics of diffusion: Comparing agent-based and differential equation models. *Management Science* 54 (5): 998–1014.

Randers, J. 1973. DDT movement in the global environment. In *Toward global equilibrium*, ed. D. H. Meadows and D. L. Meadows. Waltham, MA: Pegasus Communications.

Raper, J., H. Miller, S. Guhathakurta, R. Muetzelfeldt, and T. Cheng. 2005. Time as well: An introduction. In *Representing GIS*, ed. P. Fisher and D. Unwin, 195–98. New York: Wiley.

Rasmussen, D. 1941. Biotic communities of the Kaibab Plateau, Arizona. *Ecological monographs* 3:229–75.

Reilly, J., J. Edmonds, R. Gardner, and A. Brenkert. 1987. Uncertainty analysis of the IEA/ORNU carbon dioxide emissions model. *Energy Journal* 18 (3).

[Richardson, G. 1986. Problems with causal loop diagrams. *System Dynamics Review* 2 (2): 158–70.

———. 1991. *Feedback thought in social science and systems theory*. Philadelphia: University of Pennsylvania Press.

Richardson, G., and A. Pugh. 1981. *Introduction to system dynamics modeling with Dynamo*. Waltham, MA: Pegasus Communications.

[Ricklefs, R. 1990. *Ecology*. 3rd ed. New York: W. H. Freeman.

Riggs, D. 1963. *The mathematical approach to physiological problems*. Baltimore: Williams & Wilkins.

Ritchie-Dunham, J., and J. Galvan. 1999. Evaluating epidemic intervention policies with systems thinking: A case study of dengue fever in Mexico. *System Dynamics Review* 15 (2): 119–38.

Robinson, J. 1980. Managerial sketches of the steps of modeling. In *Elements of the System Dynamics Method*, ed. J. Randers. Waltham, MA: Pegasus Communications.

Rouwette, E., J. Vennix, and T. van Mullekom. 2002. Group model building effectiveness: A review of assessment studies. *System Dynamics Review* 18 (1): 5–45.

Royal Society. 2005. Ocean acidification due to increasing atmospheric carbon dioxide. Policy document 12/05. www.royalsoc.ac.uk.

Russo, J. 1970. *The Kaibab north deer herd*. Wildlife Bulletin no 7. Phoenix: Arizona Game and Fish Dept.

Rykiel, E. 1996. Testing ecological models: The meaning of validation. *Ecological Modelling* 90:229–44.

Saito, L., H. Segale, D. DeAngelis, and S. Jenkins. 2007. Developing an interdisciplinary curriculum framework for aquatic-ecosystem modeling. *Journal of College Science and Teaching* 37 (2): 46–52.

Schlesinger, W. 1991. *Biogeochemistry: An analysis of global change*. 2nd ed. San Diego, CA: Academic Press.

Schmickl, T., and K. Crailsheim. 2006. Bubbleworld.evo: Artificial evolution of behavioral decisions in a simulated predator-prey ecosystem. *SAB*, ed. S. Nolfi et al. 594–605. Berlin: Springer-Verlag.

Schroeder, W., and J. Strongman. 1974. Adapting urban dynamics to Lowell. *Readings in urban dynamics*. Vol. 1. Waltham, MA: Pegasus Communications.

Schweppe, F. 1973. *Uncertain dynamic systems*. Englewood Cliffs, NJ: Prentice Hall.

Senge, P. 1977. Statistical estimation of feedback models. *Simulation* (June): 177–84.

———. 1990. *The fifth discipline*. New York: Doubleday Currency.

Siegel, L. S., A. N. Alshawabkey, C. D. Palmer, and M. A. Hamilton. 2003. Modeling cesium partitioning in the rhizosphere: A focus on the role of root exudates. *Journal of Soil and Sediment Contamination* 12 (1): 47–68.

Socolow, R., R. Hotinski, J. Greenblatt, and S. Pacala. 2004. Solving the climate problem. *Environment* 46 (10): 8–19.

Soden, B. J., and I. M. Held. 2006. An assessment of climate feedbacks in coupled ocean-atmosphere models. *Journal of Climate* 19 (July): 3354–60.

Sperling, D. 1988. *New transportation fuels*. Berkeley and Los Angeles: University of California Press.

———. 1995. *Future drive*. Washington, DC: Island Press.

Stainforth, D., T. Aina, S. Christensen, M. Collins, N. Faull, D. Frame, J. Kettleborough, et al. 2005. Uncertainty in predictions of the climate response to rising levels of greenhouse gases. *Nature* 433 (Jan. 27): 403–6.

Stave, K. 2002. Using system dynamics to improve public participation in environmental decisions. *System Dynamics Review* 18 (2): 139–67.

Stephenson, K., L. Shabman, S. Langsdale, and H. Cardwell. 2007. *Computer aided dispute resolution: Proceedings from the CADRe workshop*. Albuquerque, NM: U.S. Army Corps of Engineers, Institute for Water Resources.

Sterman, J. 1992. Teaching takes off: Flight simulators for management education. *OR/MS Today* (October): 40–44.

———. 2000. *Business dynamics*. Boston: Irwin McGraw-Hill.

Swart, J. 1990. A system dynamics approach to predator prey modeling. *System Dynamics Review* 6 (1): 94–98.

Taylor, R. 1984. *Predation*. New York: Chapman and Hall.

Tidwell, V., H. Passell, S. Conrad, and R. Thomas. 2004. System dynamics modeling for community-based water planning: Application to the middle Rio Grande. *Aquatic Sciences* 66 (4): 357–72.

Turoff, M. 1970. The design of a policy Delphi. *Technological Forecasting and Social Change* 2:149–72.

[UNFCCC] United Nations Framework Convention on Climate Change. 1992. www.unfccc.int.

Vallentine, J. 1990. *Grazing management*. San Diego, CA: Academic Press.

Van den Belt, M. 2004. *Mediated modeling*. Washington, DC: Island Press.

Vennix, J. 1996. *Group model building: Facilitating team learning using system dynamics*. New York: Wiley.

———. 1999. Group model building: Tackling messy problems. *System Dynamics Review* 15 (4): 379–401.

Videira, N. 2005. Stakeholder participation in environmental decision making: The role of participatory modeling. PhD diss., New University of Lisbon, Portugal.

Voinov, A. 2008. *Systems science and modeling for ecological economics*. San Diego, CA: Academic Press.

Volterra, V. 1926. Variations and fluctuations of the number of individuals in animal species living together. Reprinted 1931 in R. N. Chapman, *Animal ecology*. New York: McGraw-Hill.

Vorster, P. 1985. *A water balance forecast model for Mono Lake, California*. U.S. Forest Service, Berkeley Service Center, Box 245, Berkeley, CA 94201.

Walters, M. 2003. *Six modern plagues and how we are causing them*. Washington, DC: Island Press.

Warren, K. 2000. *Competitive strategy dynamics*. West Sussex, UK: Wiley.

Watson, A., and J. Lovelock. 1983. Biological homeostasis of the global environment: The parable of Daisyworld. *Tellus* B35:284–89.

Watt, K. 1966. *Systems analysis in ecology*. New York: Academic Press.

———. 1968. *Ecology and resource management: A quantitative approach*. New York: McGraw-Hill.

Weart, S. R. 2003. *The discovery of global warming*. Cambridge, MA: Harvard University Press.

Webster, M., C. Forest, J. Reilly, M. Babiker, D. Kicklighter, M. Mayer, R. Prinn, M. Sarofim, A. Sokolov, P. Stone, and C. Wang. 2003. Uncertainty analysis of climate change and policy response. *Climatic Change* 61:295–320.

Wilber, C., and R. Harrison. 1978. The methodological basis of institutional economics: Pattern model, story telling and holism. *Journal of Economic Issues* 12 (1): 61–89.

Zeoli, L. 2004. An alternate explanation for Leopold's Kaibab deer herd irruption of the 1920s. M.S. thesis, Washington State University, Pullman.

Index